Cambridge Texts in Applied Mathematics

All the titles listed below can be obtained from good booksellers or from
Cambridge University Press. For a complete series listing visit
http://www.cambridge.org

Viscous Flow
H. OCKENDON AND J.R. OCKENDON

A First Course in the Numerical Analysis of Differential Equations
ARIEH ISERLES

Mathematical Models in the Applied Sciences
A.C. FOWLER

Thinking About Ordinary Differential Equations
ROBERT E. O'MALLEY

A Modern Introduction to the Mathematical Theory of Water Waves
R.S. JOHNSON

Rarefied Gas Dynamics
CARLO CERCIGNANI

Symmetry Methods for Differential Equations
PETER E. HYDON

High Speed Flow
C.J. CHAPMAN

Wave Motion
J. BILLINGHAM AND A.C. KING

An Introduction to Magnetohydrodynamics
P.A. DAVIDSON

Linear Elastic Waves
JOHN G. HARRIS

Vorticity and Incompressible Flow
A.J. MAJDA AND A.L. BERTOZZI

Infinite-dimensional Dynamical Systems
J.C. ROBINSON

Introduction to Symmetry Analysis
BRIAN J. CANTWELL

Bäcklund and Darboux Transformations
C. ROGERS AND W.K. SCHIEF

Finite Volume Methods for Hyperbolic Problems
R.J. LEVEQUE

Introduction to Hydrodynamic Stability
P.G. DRAZIN

Theory of Vortex Sound
M.S. HOWE

Scaling
G.I. BARENBLATT

Complex Variables: Introduction and Applications (Second edition)
MARK J. ABLOWITZ AND ATHANASSIOS S. FOKAS

A First Course in Combinatorial Optimization
JON LEE

Practical Applied Mathematics

Modelling, Analysis, Approximation

SAM HOWISON

Mathematical Institute, Oxford University
Director of The Oxford Centre for Industrial and
Applied Mathematics

CAMBRIDGE
UNIVERSITY PRESS

CAMBRIDGE
UNIVERSITY PRESS

University Printing House, Cambridge CB2 8BS, United Kingdom

One Liberty Plaza, 20th Floor, New York, NY 10006, USA

477 Williamstown Road, Port Melbourne, VIC 3207, Australia

314-321, 3rd Floor, Plot 3, Splendor Forum, Jasola District Centre, New Delhi - 110025, India

79 Anson Road, #06-04/06, Singapore 079906

Cambridge University Press is part of the University of Cambridge.

It furthers the University's mission by disseminating knowledge in the pursuit of education, learning and research at the highest international levels of excellence.

www.cambridge.org
Information on this title: www.cambridge.org/9780521603690

First published 2005
Reprinted with Corrections, 2005

A catalogue record for this publication is available from the British Library

ISBN 978-0-521-84274-7 Hardback
ISBN 978-0-521-60369-0 Paperback

Contents

Preface

This book was born out of my fascination with applied mathematics as a place where the physical world meets the mathematical structures and techniques that are the cornerstones of most applied mathematics courses. I am interested largely in human-sized theatres of interaction, leaving cosmology and particle physics to others. Much of my research has been motivated by interactions with industry or by contact with scientists in other disciplines. One immediate lesson from these contacts is that it is a great asset to an interactive applied mathematician to be open to ideas from any direction at all. Almost any physical situation has some mathematical interest, but the kind of mathematics may vary from case to case. We need a strong generalist streak to go with our areas of technical expertise.

Another thing we need is some expertise in numerical methods. To be honest, this is not my strong point. That is one reason why the book does not contain much about these methods. (Another is that if it had then it would have been half as long again and would have taken five more years to write.) In the modern world, with its fast computers and plethora of easy-to-use packages, any applied mathematician has to be able to switch into numerical mode as required. At the very least, you should learn to use packages such as Maple and Matlab for their data display and plotting capabilities and for the built-in software routines for solving standard problems such as ordinary differential equations. With more confidence, you can write your own programs. In many cases, a quick and dirty first try can provide valuable information, even if this is not the finished product. Explicit finite differences (remember to use upwind differencing for first derivatives) and tiny time steps will get you a long way.

Who should read this book? Many people, I hope, but there are some prerequisites. I assume that readers have a good background in calculus up to vector calculus (grad, div, curl) and the elementary mechanics of particles. I also assume that they have done an introductory (inviscid) fluid mechanics course and a first course in partial differential equations, enough to know the basics of the heat, wave and Laplace equations (where they come from, and how to solve them in simple geometries). Linear algebra, complex analysis and probability put in an occasional appearance. High-school physics is an advantage. But the most important prerequisite is an attitude: to go out and apply your mathematics, to see it in action in the world around you, and not to worry too much about the technical aspects, focusing instead on the big picture.

Another way to assess the technical level of the book is to position it relative to the competition. From that point of view it can be thought of as a precursor to the books by Tayler [58] and Fowler [19], while being more difficult than, say, Fowkes & Mahoney [18] or Fulford & Broadbridge [22]. The edited collections [9, 38] are at the same general level, but they are organised along different lines. The books [40, 56] cover similar material but with a less industrial slant.

Organisation.   The book is organised, roughly, along mathematical lines. Chapters are devoted to mathematical techniques, starting in Part I with some ideas about modelling, moving on in Part II to differential equations and distributions, and concluding with asymptotic (systematic approximation) methods in Part III. Interspersed among the chapters are case studies, descriptions of problems that illustrate the techniques; they are necessarily rather open-ended and invite you to develop your own ideas. The case studies run as strands through the book. You can ignore any of them without much impact on the rest of the book, although the more you ignore the less you will benefit from the remainder. There are long sections of exercises at the ends of the chapters; they should be regarded as an integral part of the book and at least should be read through if not attempted.

Conventions.   I use 'we', as in 'we can solve this by a Laplace transform', to signal the usual polite fiction that you, the reader, and I, the author, are engaged on a joint voyage of discovery; 'we' also signifies that I am presenting ideas within a whole tradition of thought. 'You' is mostly used to suggest that *you* should get your pen out and work though some of the 'we' stuff, a good idea in view of my fallible arithmetic, or do an exercise to fill in some details. 'I' is associated with authorial opinions and can mostly be ignored if you like.

I have tried to draw together a lot of threads in this book, and in writing it I have constantly wanted to point out connections with something else or make a peripheral remark. However, I don't want to lose track of the argument. As a compromise, I have used marginal notes and footnotes[1] with slightly different purposes in mind.

Marginal notes are usually directly relevant to the current discussion, often being used to fill in details or point out a feature of a calculation. This is a book to work through: feel free to use the empty margin spaces for calculations.

---

[1] Footnotes are more digressional and can be ignored by readers who just want to follow the main line of argument.

Acknowledgements. I have taken examples from many sources. Some examples are very familiar and I do not apologise for this: the old ones are often the best. Much the same goes for the influence of books; if you teach a course using other people's books and then write your own, some impact is inevitable. Among the books that have been especially influential are those by Tayler [58], Fowler [19], Hinch [27] and Keener [33]. Even more influential has been the contribution of colleagues and students. Many a way of looking at a problem can be traced back to a coffee-time conversation or a Study Group meeting.[2] There are far too many of these collaborators for me to attempt the invidious task of thanking them individually. Their influence is pervasive. At a more local level, I am immensely grateful to the OCIAM students who got me out of computer trouble on various occasions and found a number of errors in drafts of the book. Any remaining errors are quite likely to have been caused by cosmic ray impact on the computer memory, or perhaps by cyber-terrorists. I will be happy to hear about them.

The book began when I was asked to give some lectures at a summer school in Siena and was continued through a similar event a year later in Pisa. I am most grateful for the hospitality extended to me during these visits. I would like to thank the editors and technical staff at Cambridge University Press for their assistance in the production of the book. In particular, I am extremely grateful to Susan Parkinson for her careful, constructive and thoughtful copy-editing of the manuscript. Lastly I would like to thank my family for their forbearance, love and support while I was locked away typing. This book is dedicated to them.

Colemanballs. At the end of each section of exercises is what would normally be a wasted space. Into each of these I have put two things. One is a depiction of a wave form and is explained on p. 212. The other is a statement made by a real live applied mathematician in full flow. In the spirit of scientific accuracy, they are wholly unedited. They are mostly there for their intrinsic qualities (and it would be a miserable publisher who would deny me that extra ink), but they make a point: interdisciplinary mathematics is a collaborative affair; it involves discussions and

---

[2] Study Groups are week-long intensive meetings at which academics and industrial researchers get together to work on open problems from industry, proposed by the industrial participants. Over the week, heated discussions take place involving anybody who is interested in the problem, and a short report is produced at the end. The first UK Study Group was held in Oxford in 1968, and they have been held every year since, in Oxford and other UK universities. The idea has now spread to more than 15 countries on all the habitable continents of the world. Details of forthcoming events, and reports of problems studied at past meetings, can be found on their dedicated website www.mathematics-in-industry.org.

arguments, the less inhibited the better. We all have to go out on a limb, in the interests of pushing the science forwards. If we are wrong, we try again. And if the mind runs ahead of the voice, our colleagues won't take it too seriously (nor will they let us forget it). Here is one to be going on with, from the collection [29] of the same title:

'If I remember rightly, $\cos \pi /2 = 1$.'

Part I
Modelling techniques

1
The basics of modelling

1.1 Introduction

This short introductory chapter is about mathematical modelling. Without trying to be too prescriptive, we discuss what we mean by the term, why we might want to do it and what kind of models are commonly used. Then we look at some very standard models, which you have almost certainly met before, and we see how their derivation is a blend of what are thought of as universal physical laws, such as conservation of mass, momentum and energy, with experimental observations and, perhaps, some ad hoc assumptions in lieu of more specific evidence.

One of the themes that run through this book is the applicability of all kinds of mathematical idea to 'real-world' problems. Some of these arise in attempts to explain natural phenomena, for example in models for water waves. We will see a number of these models as we go through the book. Other applications are found in industry, which is a source of many fascinating and non-standard mathematical problems and a big 'end-user' of mathematics. You might be surprised at how little is known of the detailed mechanics of most industrial processes, although when you see the operating conditions – ferocious temperatures, inaccessible or minute machinery, corrosive chemicals – you realise how expensive and difficult it would be to carry out detailed experimental investigations. In any case, many processes work just fine, having been designed by engineers who know their job. If it ain't broke, don't fix it; so where does mathematics come in? Some important uses are in the quality control and cost control of existing processes and in the simulation and design of new processes. We may want to understand: why does a certain type of defect occur; what is the 'rate-limiting' part of a process (the slowest ship, to be speeded up); how to improve efficiency, however marginally; whether a novel idea is likely to work at all and if so, how to control it.

It is in the nature of real-world problems that they are large, messy and often rather vaguely stated. It is very rarely worth anybody's while to produce a 'complete solution' to a problem which is complicated and whose desired outcome is not necessarily well specified (to a mathematician). Mathematicians are usually most effective in analysing a relatively small 'clean' subproblem for which more broad-brush approaches run into difficulty. Very often the analysis complements a large numerical simulation which, although effective elsewhere, has trouble with this particular aspect of the problem. Its job is to provide understanding and insight in order to complement simulation, experiment and other approaches.

We begin with a chat about what models are and what they should do for us. Then we bring some simple ideas about physical conservation laws and how to use them together with the experimental evidence about how materials behave, with the aim of formulating closed systems of equations; this is illustrated with two canonical models, for heat flow and for fluid motion. There are many other models embedded elsewhere in the book, and we will deal with these as we come to them.

1.2 What do we mean by a model?

There is no point in trying to be too precise in defining the term 'mathematical model': we all understand that it is some kind of mathematical statement about a problem originally posed in non-mathematical terms. Some models are *explicative*, that is, they explain a phenomenon in terms of simpler, more basic processes. A famous example is Newton's theory of planetary motion, whereby the whole complex motion of the solar system was shown to be a consequence of 'force equals mass times acceleration' and the inverse square law of gravitation. However, not all models aspire to explain. For example, the standard Black–Scholes model for the evolution of prices in stock markets, used by investment banks the world over, says that the percentage difference between tomorrow's stock price and today's is a lognormal random variable. Although this is a great simplification, in that it says that all we need to know are the mean and variance of this distribution, it says nothing about what will cause the price change.

All useful models, whether explicative or not, are *predictive*: they allow us to make quantitative predictions (whether deterministic or probabilistic) that can be used either to test and refine the model, should that be necessary, or for use in practice. The outer planets were found using Newtonian mechanics to analyse small discrepancies between observation and theory,[1] and the Moon missions would have been impossible without this model. Every day, banks make billions of dollars worth of trades based on the Black–Scholes model; in this case, since model predictions do not always match market prices, they may use the latter to refine the basic model (here there is no simple underlying mechanism to appeal to, so adding model features in a heuristic way is a reasonable way to proceed).

Most of the models we discuss in this book are based on differential equations, ordinary or partial: in the main they are deterministic models of continuous processes. Many of them should already be familiar to you, and they are all accessible with the standard tools of real and complex analysis, partial differential equations, basic linear algebra and so on. I would like, however, to mention some kinds of models that we don't have the space to cover.

• Statistical models

Statistical models can be both explicative and predictive, in a probabilistic sense. They deal with questions of extracting information about cause and effect or making predictions in a random environment and describing that randomness. Although we touch on probabilistic models, for a full treatment see a text such as [51].

• Discrete models of various kinds

Many, many vitally important and useful models are intrinsically discrete: think, for example of the optimal scheduling of take-off slots from LHR, CDG or JFK airports, or the routeing of packets of information through the mobile phone network. Discrete mathematics is a vast area with a huge range of techniques, impinging on practically every other area of mathematics, computer science, economics and so on.

• 'Black-box' models such as neural nets or genetic algorithms, and 'lumped parameter' models

The term 'model' is often used for these techniques, in which a 'black box' is trained, using observed data, to predict the output of a system

---

[1] This is a very early example of an *inverse problem*: assuming a model, and given observations of the solution determine certain model parameters, in this case the unknown positions of Uranus and Neptune. A more topical example is the problem of constructing an image of your insides from a scan or from electrical measurements taken from electrodes on your skin. Unfortunately, such problems are beyond the scope of this book; for a discussion, see [15].

given the input. The user need never know what goes on inside the black box (it is usually some form of curve fitting and/or optimisation algorithm), so while these algorithms can have some predictive capacity they can rarely be explicative. Although often useful, the philosophy behind black-box models is more or less orthogonal to that behind the models in this book; if you are interested, see [23]. Lumped-parameter models are somewhat in the same spirit: a complex system is represented by a much simpler set of ad hoc descriptions, as for example when a complicated mechanical system is modelled by a simple spring–dashpot combination.

1.3 Principles of modelling: physical laws and constitutive relations

Many models, especially those based on mechanics or heat flow (which includes most models in this book), are underpinned by physical principles such as the conservation of mass, momentum, energy or electric charge. We may have to think about how we interpret these ideas, especially in the case of energy, which can take so many forms (kinetic, potential, heat, chemical, . . .) and be converted from one to another. Although they are in the end subject to experimental confirmation, the experimental evidence for these conservation principles is so overwhelming that, with care in interpretation, we can take them as assumptions.[2]

Work is heat and heat is work: this is loosely the first law of thermodynamics, in mnemonic form.

However, this gets us only so far. We can do very simple problems, such as the mechanics of point particles, and that's about it. Suppose, for example, that we want to derive the heat equation for heat flow in a homogeneous, isotropic, continuous solid. We can reasonably assume that at each point $\mathbf{x}$ and time t there is an energy density $E(\mathbf{x}, t)$ such that the internal (heat) energy inside any fixed volume V of the material is

$$\int_V E(\mathbf{x}, t) \, d\mathbf{x}.$$

We can also assume that there is a heat flux vector $\mathbf{q}(\mathbf{x}, t)$ such that the rate of heat flow across a plane with unit normal $\mathbf{n}$ is

$$\mathbf{q} \cdot \mathbf{n}$$

per unit area. Then we can write down the conservation of energy for V in the form

$$\frac{d}{dt} \int_V E(\mathbf{x}, t) \, d\mathbf{x} + \int_{\partial V} \mathbf{q}(\mathbf{x}, t) \cdot \mathbf{n} \, dS = 0,$$

---

[2] We are making additional assumptions that we are not dealing with quantum effects, or matter on the scale of atoms, or relativistic effects. We will deal only with models for human-scale systems.

where ∂V is the surface bounding V, on the assumption that no heat is converted into other forms of energy. Next, we use Green's theorem on the surface integral and, as V is arbitrary, the usual argument (see Section 1.4) gives us

$$\frac{\partial E}{\partial t} + \nabla \cdot \mathbf{q} = 0. \tag{1.1}$$

At this point, general assumptions fail us, and we have to bring in some experimental evidence. We need to relate both E and $\mathbf{q}$ to the temperature $T(\mathbf{x}, t)$, by what are called *constitutive relations*. For many, but not all, materials, the internal energy is directly proportional to the temperature, which is written as

$$E = \rho c T,$$

where ρ is the density and c is a constant called the *specific heat capacity*. Likewise, *Fourier's law* states that the heat flux is proportional to the temperature gradient,

$$\mathbf{q} = -k\nabla T.$$

Ask yourself why there is a minus sign. The second law of thermodynamics in loose mnemonic form: heat cannot flow from a cooler body to a hotter one.

Putting these both into (1.1), we have

$$\rho c \frac{\partial T}{\partial t} = k\nabla^2 T,$$

as expected. The appearance of material properties such as c and k is a sure sign that we have introduced a constitutive relation, and it should be stressed that these relations between E, $\mathbf{q}$ and T are material dependent and experimentally determined. There is no a priori reason for them to have the nice linear form given above, and indeed for some materials one or other may be strongly nonlinear.[3]

Another set of models where constitutive relations pay a prominent role is models for solid and fluid mechanics.

1.3.1 Example: inviscid fluid mechanics

Let us first look at the familiar Euler equations for inviscid incompressible fluid motion,

'Oiler', not 'Yewler'.

$$\rho\left(\frac{\partial \mathbf{u}}{\partial t} + \mathbf{u}\cdot\nabla\mathbf{u}\right) = -\nabla p, \qquad \nabla\cdot\mathbf{u} = 0.$$

---

[3] It is an experimental fact that temperature changes in most materials are proportional to the energy put in or taken out. However, both c and k may depend on temperature, especially if the material melts or freezes gradually, as for paraffin or some kinds of frozen fish. Such materials lead to nonlinear versions of the heat equation; fortunately, many common substances have nearly constant c and k values and so are well modelled by the linear heat equation.

Remember a *material volume* is one whose boundary moves with the fluid velocity, that is, it is made up of a fixed set of fluid particles.

Here $\mathbf{u}$ is the fluid velocity and p the pressure, both being functions of position $\mathbf{x}$ and time t, and ρ is the fluid density. The first equation is clearly 'mass × acceleration = force', bearing in mind that we have to calculate the acceleration 'following a fluid particle' (that is, we use the convective derivative), and the second expresses mass conservation (now would be a good moment for you to do the first two exercises at the end of the chapter unless this is all very familiar material).

The constitutive relation is rather less obvious in this case. When we work out the momentum balance for a small material volume V, we are encapsulating the physical law

convective rate of change of momentum in V = net force on V.

The convective rate of change of momentum in V is given by

$$\int_V \rho \left(\frac{\partial \mathbf{u}}{\partial t} + \mathbf{u} \cdot \nabla \mathbf{u} \right) \mathrm{d}V.$$

We then say that the net force on V is provided solely by the pressure and acts normally to ∂V. This is our constitutive assumption: that the internal forces in an inviscid fluid are completely described by a pressure field that acts isotropically (equally in all directions) at every point. Then, ignoring gravity, the force on V is

$$\int_{\partial V} -p\mathbf{n} \, \mathrm{d}S = -\int_V \nabla p \, \mathrm{d}V,$$

by a standard vector identity, and since V is arbitrary we do indeed retrieve the Euler equations.

1.3.2 Example: viscous fluids

Things are a little more complicated for a *viscous* fluid, namely one whose 'stickiness' generates internal forces which resist the motion. This model will be unfamiliar to you if you have never looked at viscous flow. If this is so, you can do one or more of the following.

- Just ignore it; you will then miss out on some nice models for thin fluid sheets and fibres in Chapter 20, but that's about all.
- Go with the flow: trust me that the equations are not only believable (an informal argument is given below, and in any case I am assuming you know about the inviscid part of the model) but indeed correct. As is so often necessary in real-world problems, see what the mathematics has to say and let the intuition grow.
- Go away and learn about viscous flow; try the books [45] or [2].

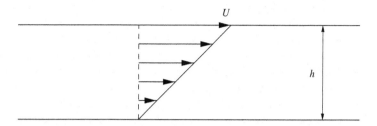

Figure 1.1 Drag on two parallel plates in shear, a configuration known as Couette flow. The upper plate is moving relative to the lower at speed U; the arrows indicate the velocity profile.

Viscosity is the property of a liquid that measures its resistance to shearing, which occurs when layers of fluid slide over one another. In the configuration of Figure 1.1, the force per unit area on either plate due to viscous drag is found for many liquids to be proportional to the shear rate U/h and is written $\mu U/h$, where the constant μ is called the *dynamic viscosity*. Such fluids are termed *Newtonian*.

Our strategy is again to consider a small element of fluid and, using the momentum balance equation, on the left-hand side work out the rate of change of momentum,

$$\int_V \rho \frac{D\mathbf{u}}{Dt} \, dV,$$

while on the right-hand side we have

$$\int_{\partial V} \mathbf{F} \, dS,$$

the net force on the boundary of the element. Then we use the divergence theorem to turn the surface integral into a volume integral and, as V is arbitrary, we are done.

Now for any continuous material, whether a Newtonian fluid or not, it can be shown (you will have to take this on trust; see [45] for a derivation) that there is a *stress tensor*, a matrix σ with entries σ_{ij}, having the property that the force per unit area exerted by the fluid in direction i on a small surface element with normal $\mathbf{n} = (n_j)$ is given by $\sigma \cdot \mathbf{n} = (\sigma_{ij} n_j)$; see Figure 1.2. It can also be shown that σ is symmetric: $\sigma_{ij} = \sigma_{ji}$. In an isotropic material (one with no built-in directionality), there are also some invariance requirements with respect to translations and rotations.

Thus far, our analysis could apply to any fluid. The force term in the equation of motion takes the form

$$\int_{\partial V} \sigma \cdot \mathbf{n} \, dS, \quad \text{with components} \quad \int_{\partial V} \sigma_{ij} n_j \, dS$$

We are using the summation convention, where by terms with repeated indices are summed over from 1 to 3; thus for example

$$\sigma_{ii} = \sigma_{11} + \sigma_{22} + \sigma_{33}.$$

It should be clear that

$$\nabla \cdot \mathbf{u} = \partial u_i / \partial x_i$$

and that

$$(\nabla \cdot \sigma)_i = \frac{\partial \sigma_{ij}}{\partial x_j}.$$

Figure 1.2 Force on a small
surface element.

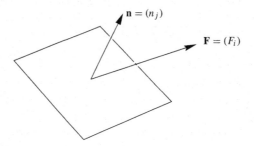

which by the divergence theorem is equal to

$$\int_{\partial V} \nabla \cdot \boldsymbol{\sigma} \, dS, \quad \text{with components} \quad \int_{\partial V} \frac{\partial \sigma_{ij}}{\partial x_j} \, dS,$$

and so we have the equation of motion

$$\frac{D(\rho \mathbf{u})}{Dt} = \nabla \cdot \boldsymbol{\sigma}. \tag{1.2}$$

We now have to say what kind of fluid we are dealing with. That is, we have to give a constitutive relation to specify $\boldsymbol{\sigma}$ in terms of the fluid velocity, pressure etc. For an inviscid fluid, the only internal forces are those due to pressure, which acts isotropically. The pressure force on our volume element is

$$\int_{\partial V} -p\mathbf{n} \, dS$$

with corresponding stress tensor

$$\sigma_{ij} = -p \, \delta_{ij},$$

Which matrix has entries δ_{ij}? Interpret $\delta_{ij} v_j = v_i$ in matrix terms.

where δ_{ij} is the Kronecker delta. This clearly leads to the Euler momentum-conservation equation

$$\rho \left(\frac{\partial \mathbf{u}}{\partial t} + \mathbf{u} \cdot \nabla \mathbf{u} \right) = -\nabla p.$$

When the fluid is viscous, we need to add on the contribution due to viscous-shear forces. From consideration of the experiment of Figure 1.1 it is very reasonable that the new term should be linear in the velocity gradients, and it can be shown, bearing in mind the invariance requirements mentioned above, that the appropriate form for σ_{ij} is

$$\sigma_{ij} = -p \, \delta_{ij} + \mu \left(\frac{\partial u_i}{\partial x_j} + \frac{\partial u_j}{\partial x_i} \right).$$

For future reference we write out the components of σ in two dimensions:

$$(\sigma_{ij}) = \begin{pmatrix} -p + 2\mu\dfrac{\partial u}{\partial x} & \mu\left(\dfrac{\partial u}{\partial y} + \dfrac{\partial v}{\partial x}\right) \\[2ex] \mu\left(\dfrac{\partial u}{\partial y} + \dfrac{\partial v}{\partial x}\right) & -p + 2\mu\dfrac{\partial v}{\partial y} \end{pmatrix}. \qquad (1.3)$$

Substituting this into the general equation of motion (1.2), and using the incompressibility condition $\nabla \cdot \mathbf{u} = \partial u_i / \partial x_i = 0$, it is a straightforward *exercise* to show that the equation of motion of a viscous fluid is

> The emphasis means that it is a good idea to do it.

$$\rho\left(\frac{\partial \mathbf{u}}{\partial t} + \mathbf{u} \cdot \nabla \mathbf{u}\right) = -\nabla p + \mu \nabla^2 \mathbf{u}, \qquad \nabla \cdot \mathbf{u} = 0. \qquad (1.4)$$

These equations are known as the *Navier–Stokes equations*. The first contains the corresponding inviscid terms, i.e. the Euler equations, but with the new term $\mu \nabla^2 \mathbf{u}$, which represents the influence of viscosity. As we shall see later, this term has profound effects.

1.4 Conservation laws

Perhaps we should elaborate on the 'usual argument' that, allegedly, leads to equation (1.1). Whenever we work in a continuous framework, and we have a quantity that is conserved, we offset changes in its *density* $P(\mathbf{x}, t)$ with equal and opposite changes in its *flux* $\mathbf{q}(\mathbf{x}, t)$. Taking a small volume V and arguing as above, we have

$$\frac{d}{dt} \int_V P(\mathbf{x}, t)\, d\mathbf{x} + \int_{\partial V} \mathbf{q} \cdot \mathbf{n}\, dS = 0,$$

the first term being the time rate of change in the quantity inside V, and the second the net flux of it into V. Applying Green's theorem to this latter integral,[4] we have

$$\int_V \frac{\partial P}{\partial t} + \nabla \cdot \mathbf{q}\, d\mathbf{x} = 0.$$

As V is arbitrary, we conclude that

$$\frac{\partial P}{\partial t} + \nabla \cdot \mathbf{q} = 0,$$

a statement that is often referred to as a *conservation law*.[5]

---

[4] Needless to say, this argument requires $\mathbf{q}$ to be sufficiently smooth, which can usually be verified *a posteriori*; in Chapter 7 we shall explore some cases where this smoothness is not present.

[5] Sometimes this term is reserved for cases in which $\mathbf{q}$ is a function of P alone.

In the heat-flow example at the start of Section 1.3, $P = \rho c T$ is the density of the internal heat energy and $\mathbf{q} = -k\nabla T$ is the heat flux. Another familiar example is the conservation of mass in a compressible fluid flow for which the density is ρ and the mass flux is $\rho \mathbf{u}$, so that

$$\frac{\partial \rho}{\partial t} + \nabla \cdot (\rho \mathbf{u}) = 0.$$

This is not as silly as it sounds: a fluid may be incompressible and have different densities in different places; in the jargon we say that it is *stratified*.

When the fluid is incompressible and of constant density, this reduces to $\nabla \cdot \mathbf{u} = 0$ as expected.

1.5 General remarks

There are, of course, many widely used models that we have not described in this short chapter. Rather than give a long catalogue of examples, we'll move on, leaving other models to be derived as we come to them. We conclude with an important general point.

As stressed above, the construction of a model for a complicated process involves a blend of physical principles and (mathematical expressions of) experimental evidence; these may be supplemented by plausible ad hoc assumptions, where direct experimental evidence is unavailable, or may be used as a 'summary' model of a complicated system from which only a small number of outputs is needed. However, the initial construction of a model is only the first step in building a useful tool. The next task is to analyse it: does it make mathematical sense? Can we find solutions, whether explicit (i.e. as a formula), approximate or numerical, and if so how? Then, crucially, what do these solutions (predictions) have to say about the original problem? This last step is often the cue for an iterative process in which discrepancies between predictions and observations prompt us to rethink the model. Perhaps, for example, certain terms or effects that we thought were small could not, in fact, safely be neglected. Perhaps some ad hoc assumption we made was not right. Perhaps, even, a fundamental mechanism in the original model does not work in the way we assumed it did (a negative result of this kind can often be surprisingly useful). We shall develop all these themes as we go on.

1.6 Exercises

1 Conservation of mass. A uniform incompressible fluid flows with velocity $\mathbf{u}$. Taking an arbitrary fixed volume V, show that the net mass flux across its boundary ∂V is

$$\int_{\partial V} \mathbf{u} \cdot \mathbf{n} \, dS.$$

Use Green's theorem to deduce that $\nabla \cdot \mathbf{u} = 0$. What would you do if the fluid were incompressible but of spatially varying density (see Section 1.4)?

2 The convective derivative. Let $F(\mathbf{x}, t)$ be any quantity that varies with position and time in a fluid with velocity $\mathbf{u}$. Let V be an arbitrary material volume moving with the fluid, so that the points on ∂V move with velocity $\mathbf{u}$. Show that

$$\frac{d}{dt} \int_V F \, dV = \int_V \frac{\partial F}{\partial t} \, dV + \int_{\partial V} F \mathbf{u} \cdot \mathbf{n} \, dS,$$

Draw a picture of $V(t)$ and $V(t + \delta t)$ to see where the second term comes from.

where the second term is there because the boundary of V moves. When the fluid is incompressible, use Green's theorem to deduce the convective-derivative formula

$$\frac{dF}{dt} = \frac{\partial F}{\partial t} + \mathbf{u} \cdot \nabla F.$$

Derive this in another way by considering the total time derivative of the quantity $F(\mathbf{x}(t), t)$, where $d\mathbf{x}/dt = \mathbf{u}$. Apply the convective derivative to the fluid velocity $\mathbf{u}$ to verify that the left-hand side of the Euler momentum equation

$$\rho \left(\frac{\partial \mathbf{u}}{\partial t} + \mathbf{u} \cdot \nabla \mathbf{u} \right) = -\nabla p$$

is the acceleration following a fluid particle.

3 Waves on a membrane. A membrane of density ρ per unit area lying close to the xy-plane is stretched to tension T. Its displacement in the normal direction is $u(x, y, t)$. Take a small element A of the membrane and derive the force balance

$$\iint_A \rho \frac{\partial^2 u}{\partial t^2} \, dA = \int_{\partial A} T \frac{\partial u}{\partial n} \, ds$$

Deduce the equation of motion

$$\frac{\partial^2 u}{\partial t^2} = c^2 \nabla^2 u,$$

where $c^2 = T/\rho$ is the wave speed.

4 Fick's law of diffusion. A certain substance diffuses in an inert medium. Its concentration, which is small, is $c(\mathbf{x}, t)$. Fick's law says that the flux of the substance is $-D\nabla c$, where D is a constant called the *(molecular) diffusivity.* Show that $c(\mathbf{x}, t)$ satisfies the diffusion equation

$$\frac{\partial c}{\partial t} = D\nabla^2 c.$$

In addition, the substance is consumed by a reaction that eats it up at a rate proportional to c. How is the diffusion equation modified?

'A sphere being squeezed on six of its sides ... '

2
Units, dimensions and dimensional analysis

2.1 Introduction

This chapter and the next cover some simple ideas to do with dimensional analysis. These ideas can be very helpful in understanding the basic physical mechanisms on which we will build mathematical models, but they are primarily the first step towards our main objective, to build up a systematic framework within which to assess such models of complex problems. Real-world situations, arising in industry or elsewhere, almost always involve many coupled physical processes. We may be able to write down models for each of them individually, and so for the whole, but faced with the resulting pages of equations, what then? Can we say anything about the 'structure' of the problem? What are the pivotal points? Are all the mechanisms we have included equally significant? If not, how can we decide which should we keep? Is it safe to put the equations on a computer? If there are many input physical parameters, which is the best way to explore the space of solutions?

We will start with some basic material on dimensions and units; in the next chapter we will move on to see how scaling reveals *dimensionless parameters* that, if small (or large) can point the way to useful approximation schemes. Along the way, we'll see gentle introductions to some models that we will use repeatedly in later chapters. Almost all these deal with reasonably familiar material and will not trouble you too much; the only possible exception is the material on electrostatics, and we don't have to do much of that.

2.2 Units and dimensions

There is just one simple idea underpinning this section. If an equation models a physical process, then all the terms in it that are separated by $+$, $-$ or $=$ must have the same physical dimensions. If they did not, we would be saying something obviously ludicrous like

$$\text{apples} + \text{lawnmowers} = \text{light bulbs} - \text{whisky}.$$

For example, is the answer real and positive etc, when it obviously should be? Is it about the right size? If we expect the temperature to increase when we increase the input heat flux does our formula do what it should? Does the acceleration point in the same direction as the force?

This is the most basic of the many consistency (error-correcting) checks that you should build into your mathematics.

To quantify this idea, we'll use a standard notation for the dimensions of a quantity: square brackets around the quantity. All the scientific units that we'll be using can written in terms of the *primary dimensions* mass [M], length [L], time [T], electric current [I] and temperature [Θ].[1] Once a specific set of units has been chosen, such as those in the SI system, the hitherto abstract primary dimensions can each be related to a specific primary unit; the SI units for our primaries are kg for kilogram, m for metre, s for second, A for ampere, K for kelvin (or we may use °C).[2]

Given the primary dimensions, we can derive all other *secondary dimensions* from them. Sometimes this is a matter of definition: for a speed u we have

$$[u] = [L][T]^{-1}.$$

In other cases we may use the dimensional structure of a physical law, as in

$$\text{force } F = \text{mass} \times \text{acceleration}, \qquad \text{so that} \quad [F] = [M][L][T]^{-2};$$

the SI unit is the newton, N. Other instances are:

$$\text{pressure } P = \text{force per unit area}, \qquad \text{so that} \quad [P] = [M][L]^{-1}[T]^{-2},$$

whose SI unit is the pascal, Pa;

$$\text{energy } E = \text{force} \times \text{distance moved}, \qquad \text{so that} \quad [E] = [M][L]^{2}[T]^{-2},$$

the SI unit being the joule, J;

$$\text{power} = \text{energy per unit time},$$

---

[1] There are two more primary dimensions, the amount of a substance (SI unit the mole) and luminous intensity (the candela), but we don't need them in this book.

[2] You might imagine that it should not be necessary to stress the importance of choosing, and sticking to, a standard set of units for the primary dimensions and of stating what units are used. Examples such as the imperial–metric cock-up (one team using imperial units, another using metric ones), that led to the failure of the Mars Climate Orbiter mission in 1999, prove this wrong.

giving the watt, $W = J s^{-1}$; and so on. The idea extends in an obvious way to physical parameters and properties of materials. For example,

density ρ = mass per unit volume,   so that $[\rho] = [M][L]^{-3}$.

2.2.1 Example: heat flow

We are going to see a lot of heat-flow problems in this book (I assume that you have already met the heat equation in an introductory course). Let's begin by working out the dimensions of thermal conductivity k. The heat flux, which means the energy flow in a material per unit area per unit time, has dimension

$$[\mathbf{q}] = [\text{energy}][L]^{-2}[T]^{-1} = [M][T]^{-3}.$$

By Fourier's law (an experimental fact), the heat flux is proportional to the temperature gradient:

$$\mathbf{q} = -k\nabla T.$$

Thus, noting that differentiation with respect to a spatial variable brings in a length scale in the denominator,

Why? What does integration do?

$$[q] = [k][\Theta][L]^{-1}.$$

Combining these two versions of $[q]$, we find that

$$[k] = [M][L][T]^{-3}[\Theta]^{-1}.$$

The usual SI units of k, $W m^{-1} K^{-1}$, are chosen to be descriptive of what this parameter measures. It is an exercise now to check that the heat equation

$$\rho c \frac{\partial T}{\partial t} = k\nabla^2 T, \tag{2.1}$$

in which c is the specific heat capacity with SI units $J kg^{-1} K^{-1}$, is dimensionally consistent.

Note also, for future reference, that the combination

$$\kappa = \frac{k}{\rho c},$$

known as the *thermal diffusivity*, has dimensions $[L]^2[T]^{-1}$. The higher is κ, the faster the material conducts heat: that is, heat put in is conducted more and absorbed less; you can see this because κ is the ratio of the heat conduction coefficient k to a parameter characterising the absorption as internal energy, ρc. By way of examples, water with its large specific heat has $\kappa = 1.4 \times 10^{-7} m^2 s^{-1}$, while for air, with much lower density, $\kappa = 2.2 \times 10^{-5} m^2 s^{-1}$. Amorphous solids such as glass

$(\kappa = 3.4 \times 10^{-7} \ \mathrm{m^2 \ s^{-1}})$ conduct less well than crystalline solids such as metals: for gold (an extreme and expensive example, much valued for its thermal properties), $\kappa = 1.27 \times 10^{-4} \ \mathrm{m^2 \ s^{-1}}$.

Given a length L, we can construct a time L^2/κ, called the thermal diffusion time, which can be interpreted as the order of magnitude of the time it takes for you to notice an abrupt temperature change a distance L away. Conversely, during a specified time t the abrupt temperature change propagates a distance of order $\sqrt{\kappa t}$.

Of course, the heat equation, being parabolic, gives an infinite speed of propagation, so in one sense the 'noticing' is instantaneous. What I mean is that this is how long it takes for the temperature at a distance L to change by an appreciable fraction of the abrupt applied temperature; see Exercise 12.

2.3 Electric fields and electrostatics

Several of the problems we look at in this book involve electromagnetic effects. We need only a small subset of the wonderful edifice of electromagnetism, and most of what we use should be a reminder of school physics, written in more mathematical terms.

Models for electricity bring with them a stack of potentially confusing units. A good place to start is Coulomb's (experimental) observation that, in a vacuum, the force between two point charges q_1, q_2 is inversely proportional to the square of the distance r between them. We need a unit for charge, and as the relevant fundamental unit is the ampere, A, which measures the flow of electric charge per unit time down a wire,[3] we find that the charge unit is 1 A s, known as the coulomb, C. So, the magnitude of the force is

$$F = \frac{q_1 q_2}{4\pi \varepsilon_0 r^2},$$

As for its direction, we know that like charges repel, as in Figure 2.1 in which q_1 and q_2 have the same sign. The constant ε_0 is known as the *(electric) permittivity of free space*, and the factor 4π is inserted to save a lot of occurrences of this factor in other formulae. Thus

Notice that 4π, being a pure number, is omitted from this dimensional balance.

$$[\varepsilon_0] = \frac{([\mathrm{I}][\mathrm{T}])^2}{[\mathrm{L}]^2 [\mathrm{M}][\mathrm{L}][\mathrm{T}]^{-2}} = [\mathrm{M}]^{-1}[\mathrm{L}]^{-3}[\mathrm{T}]^4[\mathrm{I}]^2,$$

a combination which in SI is called one farad per metre ($\mathrm{F \ m^{-1}}$) for a reason which will become clear if (when) you do Exercise 2. The numerical value of ε_0 is approximately $8.85 \times 10^{-12} \ \mathrm{F \ m^{-1}}$, from which we can deduce that one coulomb is a colossal amount of free charge. The attractive force between opposite charges of 1 C separated by 1 m is $(4\pi\varepsilon_0)^{-1}$; this is more than 10^8 N, and it would take two teams of 2000 large elephants, each pulling their bodyweight, to drag the charges apart.

---

[3] See Exercise 4 for the definition of the ampere.

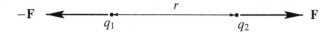

Figure 2.1 Force between two charges.

Suppose we regard charge 1 as fixed at the origin and charge 2 as a movable 'test charge' at the point $\mathbf{x}$. The force on it, now regarded as a vector, is

$$\mathbf{F} = q_2\mathbf{E},$$

where

$$\mathbf{E} = \frac{q_1\hat{\mathbf{x}}}{4\pi\varepsilon_0 r^2} = \frac{q_1\mathbf{x}}{4\pi\varepsilon_0 r^3} \qquad (2.2)$$

Here, $\hat{\mathbf{x}}$ is a unit vector along $\mathbf{x}$, and $r = |\mathbf{x}|$.

is the *electric field* due to charge 1. Since $\nabla \wedge \mathbf{E} = \mathbf{0}$ for $\mathbf{x} \neq \mathbf{0}$, and $\mathbb{R}^3 \setminus \{\mathbf{0}\}$ is simply connected, there exists an *electric potential*

If you didn't know that $\nabla \wedge (\mathbf{x}/r^3) = \mathbf{0}$, check it using the formula

$$\phi = \frac{q_1}{\varepsilon_0 r} \qquad \text{with} \qquad \mathbf{E} = -\nabla\phi$$

$$\nabla \wedge (\phi\mathbf{v}) = \nabla\phi \wedge \mathbf{v} + \phi\nabla \wedge \mathbf{v}$$

(the minus sign is conventional). Because $\nabla \cdot \mathbf{E} = 0$ away from $\mathbf{x} = \mathbf{0}$, we have

for scalar $\phi(\mathbf{x})$ and vector $\mathbf{v}(\mathbf{x})$. Check also that $\nabla \cdot \mathbf{E} = 0$, using $\nabla \cdot (\phi\mathbf{v}) = \nabla\phi \cdot \mathbf{v} + \phi\nabla \cdot \mathbf{v}$.

$$\nabla^2\phi = 0, \qquad \mathbf{x} \neq \mathbf{0}.$$

Instead of point charges, we may have a distributed charge density $\rho(\mathbf{x})$, which we can think of (in a loose way for now) as some sort of limit of a large number of point charges. Then we find that

$$\nabla^2\phi = -\frac{\rho}{\varepsilon_0}.$$

We will see a justification for this equation in Chapter 10 (see also Exercise 1).[4]

We have strayed somewhat from our theme of units and dimensions. Returning to $\mathbf{E}$, we find from (2.2) that

$$[\mathbf{E}] = [\text{M}][\text{L}][\text{T}]^{-3}[\text{I}]^{-1};$$

it is measured in volts per metre, V m^{-1}, from which the units for ϕ are volts. Perhaps more usefully, since $q_2\mathbf{E}$ is a force, the formula

$$\text{work done} = \text{force} \times \text{distance moved parallel to the force}$$

tells us that the change in electric potential is the energy per unit charge

---

[4] In fact it is rather unusual to have $\rho \neq 0$, i.e. not to have charge neutrality, in the bulk of a material. The reason is that if the material is even slightly conducting, any excess charge moves (by mutual repulsion) to form a surface layer or, if it can escape elsewhere, it does so. If the material is a good insulator then the charge cannot get into the interior anyway. In the next chapter we describe a situation where charge neutrality does not hold.

expended in moving against the electric field:

$$q \left[\phi\right]_A^B = - \int_A^B q\mathbf{E} \cdot \mathrm{d}\mathbf{x},$$

a formula that serves as a definition of ϕ.[5]

Thinking now of the rate at which work is done against the electric field (or just manipulating the definitions), we can see that

$$1 \text{ volt} \times 1 \text{ ampere} = 1 \text{ watt},$$

and hence the dimensional correctness of the formula

$$P = VI$$

for the power dissipated when a current I flows across a potential difference V. When a current is carried by free electrons through a solid, the electric field forces the free electrons through the more-or-less fixed array of solid atoms, and the work done against this resistance is lost as heat at a rate VI. In many cases the current is proportional to the voltage, giving the linear version of Ohm's law,

$$V = IR,$$

The only SI unit that is not a Roman letter?

from which the primary units of resistance R (SI unit the ohm, Ω) can easily be found. There are also many nonlinear resistors, for example diodes, in which R depends on I.

2.4 Sources and further reading

Barenblatt's book [4] has a lot of material about dimensional analysis, and is the source of the exercise below on atom bombs. For electromagnetism I suggest the book by Robinson [53] if you can get hold of it, as his physical insight was extraordinary; failing this, try [31]. McMahon's book [42] has a very interesting chapter on the dimensional analysis of animal locomotion.

2.5 Exercises

The first set of exercises is about electromagnetism; even if you have never seen this topic before, do what you can, at least to get practice in

---

[5] Likewise, gravitational potential energy is a measure of the work done per unit mass against the gravitational field. If you have ever studied the Newtonian model for gravitation, which is also governed by the inverse square law, you will see the immediate analogy between electric field and gravitational force field, charge density and matter density, and the electric and gravitational potentials. The major difference is of course that there are two varieties of charge, whereas matter apparently never repels other matter.

working out units. But I hope that you will be induced to learn more about this wonderful subject. Exercises on the rest of the chapter follow those an electromagnetism.

Electromagnetism

1 **Gauss' flux theorem.**   Consider the electric field of a point charge q at the origin (see Section 2.3). Take a volume V with boundary ∂V enclosing the sphere $|\mathbf{x}| = \epsilon$. Integrate $\nabla \cdot \mathbf{E}$ over the annular region V_ϵ between ∂V and $|\mathbf{x}| = \epsilon$ and let $\epsilon \to 0$, to show that

Note: ϵ is not to be confused with ε_0!

$$\int_{\partial V} \mathbf{E} \cdot \mathbf{n}\, dS = \frac{q}{\varepsilon_0};$$

note the absence of 4π from this formula.

Generalise to a finite number of charges. Explain informally why the result is consistent with the continuous charge density equation $\nabla^2 \phi = -\rho/\varepsilon_0$, where $\mathbf{E} = -\nabla\phi$.

2 **Capacitance.**   A capacitor is a circuit device that stores charge. The archetypal capacitor consists of two parallel conducting plates, each of area A, separated by a distance d. If one of the plates is earthed and the other raised to a voltage V, it is found that there is a proportional charge Q on it (think of the current trying to get round the circuit and piling up on one of the plates). The constant of proportionality is called the *capacitance* C, so $C = Q/V$, measured in coulombs per volt, known as farads (F).

Work out the dimensions of the farad in terms of primary quantities. Show that the formula

$$C = \frac{\varepsilon_0 A}{d}$$

is dimensionally plausible. Check (for consistency) that it does what it should as A and d vary. Explain why the units of ε_0 are F m^{-1}. Given that $\varepsilon_0 \approx 8.85 \times 10^{-12}$ F m^{-1}, how big is a 1 μF (quite a large value) capacitor if $d = 1$ mm? How big would a 1 F capacitor be? (In practice, capacitors are bulky objects which are made more compact by rolling them up and filling the space between the plates with a material of higher permittivity than ε_0.)

Based solely on this dimensional analysis, make an order of magnitude guess at the capacitance of (a) an elephant (assumed conducting); (b) a homemade parallel-plate capacitor made from two 10-metre rolls of kitchen foil 30 cm wide separated by cling-film.

If you walk across a nylon carpet you may become charged with static electricity, to a voltage of say 30 kV. (The charge appears on your shoes because of friction with the carpet. It is easily transported around you, because your body is quite a good conductor, to form a surface layer.) Estimate how much charge you accumulate. Given that air loses its insulating property and breaks down into an ionised gas at electric fields of around $3\,\mathrm{MV\,m^{-1}}$, how far is your finger from the door handle when you discharge?

It is quite easy to work out the capacitance of a sphere of radius a. The electrostatic potential ϕ satisfies $\nabla^2\phi = 0$ for $r > a$, where r is the distance from the centre of the sphere. If the sphere is raised to a voltage V relative to a potential of zero at infinity, we have $\phi = V$ on $r = a$ and $\phi \to 0$ as $r \to \infty$. We (you) can write down ϕ immediately. Now use Gauss' flux theorem, also known as the divergence theorem, on a sphere $r = a+$ to show that the total charge on the sphere is

> The notation $r = a+$ means do it for $r = a + \epsilon$ and let $\epsilon \downarrow 0$.

$$-\varepsilon_0 \iint_{r=a} \frac{\partial \phi}{\partial r}\, dS$$

and deduce that the capacitance of the sphere is $4\pi\varepsilon_0 a$.

A capacitor with capacitance C is charged up to voltage V and discharged to earth (voltage 0) through a resistor of resistance R. If the charge on the capacitor is Q and the current to earth is I, explain why

$$Q = VC, \qquad I = \frac{dQ}{dt} \qquad \text{and} \qquad V = IR.$$

Find $I(t)$ and confirm that RC has the dimension of time; interpret this time physically and explain why it increases with both R and C.

3 Slow electrons.   The charge on an electron is approximately 1.6×10^{-19} C. In copper, there are about 8.5×10^{28} free electrons per cubic metre (this calculation is based on Avogadro's number, the density and atomic weight of copper and an assumption of one free electron per atom). What is the mean speed of the electrons carrying 1 A of current down a wire of diameter 1 mm? Does the answer surprise you?

4 Forces between wires.   It is another experimental observation that the force F per unit length between long straight parallel wires in a vacuum, carrying currents I_1, I_2, is inversely proportional to the distance r between them and directly proportional to each of the currents. This is written

$$F = \frac{\mu_0 I_1 I_2}{2\pi r}; \tag{2.3}$$

> Can you think why line currents get a factor 2π but point charges get a factor 4π?

the factor 2π is included for convenience elsewhere. The constant μ_0 is known as the *permeability of free space*; what are its fundamental units? The SI units are henries per metre, $\mathrm{H\,m^{-1}}$.

Now recall that we have not yet defined the unit of current, the ampere. Because μ_0 and the currents in (2.3) occur multiplied together, there is a degree of indeterminacy in their scales: we could multiply the currents by α and divide μ_0 by α^2, for any α. We exploit this by arbitrarily (in fact it is a cunning choice from the practical point of view) setting

$$\mu_0 = 4\pi \times 10^{-7} \text{ H m}^{-1}$$

and then *defining* the ampere A as the current that makes F exactly equal to $2 \times 10^{-7} \text{ N m}^{-1}$ when the wires are infinitely long.

A current generates a magnetic field, denoted by $\mathbf{B}$. The *Lorentz force law* states that the force on a charge q moving with velocity $\mathbf{v}$ in an electric field $\mathbf{E}$ and magnetic field $\mathbf{B}$ is

Remember iron filings experiments to show the magnetic fields of bar magnets or wires? The filings line up in the direction of $\mathbf{B}$.

$$\mathbf{F} = q(\mathbf{E} + \mathbf{v} \wedge \mathbf{B}).$$

Deduce the dimensions of $\mathbf{B}$ (SI unit the tesla, T). Interpreting the currents as moving line charges, show that (2.3) is consistent with a magnetic field

$$\mathbf{B} = \frac{\mu_0 I}{2\pi r}\mathbf{e}_\theta$$

for a wire carrying current I along the z-axis of cylindrical polar coordinates (r, θ, z). How would iron filings on a plane normal to the wire line up in this case?

Show that, like the coulomb and farad, the tesla is an inconveniently large unit by working out the current required to give a field of 1 T at a distance of 1 m. How many 1 kW toasters would this current power at 250 V? (Answer: 1.25 million.) Why are electromagnets made of coils? The most powerful superconducting magnets, using coils to reinforce the field, have only recently broken the 10 T barrier.

5 The speed of light.   Show that

$$c = (\varepsilon_0 \mu_0)^{-1/2}$$

is a speed, and work out its numerical value.

6 Electromagnetic waves.   OK, the result of the previous exercise is not a coincidence. We don't have the space to derive Maxwell's famous equations for $\mathbf{E}$ and $\mathbf{B}$, but here they are: in a vacuum, $\mathbf{E}$ and $\mathbf{B}$ satisfy the four equations that follow. First,

$$\nabla \wedge \mathbf{E} = -\frac{\partial \mathbf{B}}{\partial t}.$$

This is Faraday's law of induction; it tells us that time-varying magnetic fields generate electric fields. Next,

$$\frac{1}{\mu_0} \nabla \wedge \mathbf{B} = \varepsilon_0 \frac{\partial \mathbf{E}}{\partial t} + \mathbf{j}.$$

Thus when there are currents present they appear in the form of a source term $\mathbf{j}$, the current density, on the right-hand side of this equation, which is revealed as the model for the generation of magnetic fields by currents. The term $\varepsilon_0 \partial \mathbf{E} / \partial t$ is Maxwell's inspiration, the displacement current. Finally,

$$\nabla \cdot \mathbf{B} = 0, \qquad \nabla \cdot \mathbf{E} = 0.$$

The first of these says that there are no 'magnetic monopoles' (magnetic fields are only generated by currents, and magnetic lines of force have no ends), and the second is a special case of $\nabla \cdot \mathbf{E} = \rho / \varepsilon_0$, showing the generation of electric fields by charges.

Take these equations on trust and cross-differentiate them to show that $\mathbf{E}$ and $\mathbf{B}$ satisfy wave equations:

$$\frac{\partial^2 \mathbf{B}}{\partial t^2} = c^2 \nabla^2 \mathbf{B}, \qquad \frac{\partial^2 \mathbf{E}}{\partial t^2} = c^2 \nabla^2 \mathbf{E}$$

where $c^2 = (\varepsilon_0 \mu_0)^{-1/2}$ as above. You may need the vector identity[6]

$$\nabla \wedge \nabla \wedge \mathbf{v} = \nabla (\nabla \cdot \mathbf{v}) - \nabla^2 \mathbf{v}.$$

7 Planck's constant and the fine structure constant.   This book is not the place for an account of quantum mechanics. We can note, however, that underpinning it all is Schrödinger's equation

$$\frac{\hbar}{i} \frac{\partial \psi}{\partial t} - \frac{\hbar^2}{2m} \nabla^2 \psi = V \psi$$

for the wave function ψ of a particle of mass m moving in a potential V (ψ is complex-valued and $|\psi|^2$ is the probability density of the particle's location). Find the dimensions of $\hbar$ (Planck's constant is $h = 2\pi \hbar$) and V. Show that the combination

$$\frac{e^2}{2\varepsilon_0 hc},$$

where e is the charge on an electron, is dimensionless. Such dimensionless ratios of fundamental constants are not coincidences, and this one, called the *fine structure constant*, plays an important role in quantum electrodynamics. It gets its name from its influence on the fine structure of the spectrum of light emitted by a glowing gas;

---

[6] Curl Curl is the name of an Australian beach-town, latitude $33° 46'$ S, longitude $151° 17'$ E. It is near the promontory Dee Why Head; what a pity there is no Dee Exe Head.

crudely speaking it is the ratio of the speed an electron would have if it were to orbit a hydrogen nucleus in a circle (which it does not) to the speed of light. Its numerical value is very close to 1/137, a source of some fascination to numerologists. For more, see its own website www.fine-structure-constant.org.

Other exercises

8 cgs units.   An alternative system of units to SI is the cgs system, in which the unit of mass is the gram (g) and the unit of length is the centimetre. Establish the following conversion table (which is really here for your reference) and construct the reverse table to turn SI into cgs.

	cgs	SI
velocity	1 cm s^{-1}	10^{-2} m s^{-1}
density	1 g cm^{-3}	10^{3} kg m^{-3}
dynamic viscosity	1 poise	10^{-1} kg m^{-1}s^{-1}
kinematic viscosity	1 cm^2 s^{-1}	10^{-4} m^2 s^{-1}
pressure	1 dyne cm^{-2}	10^{-1} Pa
energy	1 erg	10^{-7} J
force	1 dyne	10^{-5} N
surface tension	1 dyne cm^{-1}	10^{-3} N m^{-1}

9 Imperial to metric.   Establish the quite useful relation

$$1 \text{ mph} \approx 0.447 \text{ m s}^{-1}.$$

Using the Web or other sources for the definitions, show that

$$1 \text{ Btu} = 1 \text{ calorie},$$

a result which might be of use if you are interested in central heating.

Btu = British thermal unit, a measure of energy; kilocalories are worried about by dieters. One calorie is the amount of energy needed to heat one gram of water by 1 °C.

10 Atom bombs.   An essentially instantaneous release of an amount E of energy from a very small volume creates a rapidly expanding high-pressure fireball bounded by a very strong thin spherical shock wave across which the pressure drops abruptly to atmospheric. The pressure inside the fireball is so great that the ambient atmospheric pressure is negligible by comparison, and the only property of the air that determines the radius $r(t)$ of the fireball is its density ρ. Show dimensionally, by identifying the only possible combination of E, t and ρ, that

$$r(t) \propto E^{1/5}t^{2/5}\rho^{-1/5}.$$

This result is due to G.I. Taylor, a colossus of British applied mathematics in the last century; whatever branch of fluid mechanics you look at, you will find that 'G.I.' wrote a seminal paper on it.[7] It can be used to deduce E from observations of $r(t)$; Taylor's publication of this observation [59] apparently caused considerable embarrassment in US military scientific circles, where it was regarded as top secret.

11 **Rowing.**   A boat carries N similar people, each of whom can put power P into propelling the boat. Assuming that they each require the same volume V of boat to accommodate them, show that the wetted area of the boat is $A \propto (NV)^{2/3}$ (here, as so often, the cox is ignored). Assuming inviscid flow, why might the drag force be proportional to $\rho U^2 A$, where U is the speed of the boat and ρ the density of water? (In saying this, we are ignoring any drag due to waves created by the boat.) Deduce that the rate of energy dissipated by a boat travelling at speed U is proportional to $\rho U^3 A$ and put the pieces together to show that

$$U \propto N^{1/9} P^{1/3} \rho^{-1/3} V^{-2/9}.$$

If we suppose, very crudely, that P and V are both proportional to body mass, is size an advantage to a rower?

   This example is based on a paper by McMahon [41], described in his book [42]; the theory agrees well with observed race times.

12 **Similarity solution to the heat equation.**   Show that the problem

$$\frac{\partial T}{\partial t} = \kappa \frac{\partial^2 T}{\partial x^2}, \qquad x > 0, \quad t > 0,$$

with

$$T(x, 0) = 0, \qquad T(0, t) = T_0 > 0,$$

which corresponds to the instantaneous heating of a cold half-space from its boundary at $x = 0$, has a *similarity solution*

$$T = T_0 F \left(\frac{x}{\sqrt{\kappa t}} \right)$$

and find F in terms of the error function erf $\xi = (2/\sqrt{\pi}) \int_0^\xi e^{-s^2} ds$. Sketch F and interpret this solution in the light of the discussion at the end of subsection 2.2.1.

13 **Firewalking.**   Returning to the problem of Exercise 12, calculate the heat flux per unit area into the boundary $x = 0$ as a function of t, showing that it is proportional to $T_0(\rho ckt)^{1/2}$.

---

[7] The Taylor of Taylor's theorem was several hundred years earlier ...

Firewalkers happily walk on a bed of glowing wood embers at a temperature about 500 °C greater than that of their feet. Estimate the heat flux per unit area into their feet during a half-second step (for wood, $\rho \approx 800$, $c \approx 400$, $k \approx 0.15$, in SI units, varying with the species). If all this heat is confined to a layer of foot 1 mm deep, calculate the resulting temperature increase, assuming that the material properties of feet are similar to those of water ($c = 4200$, $\rho = 1000$, in SI units). Carry out a more accurate calculation by looking for a similarity solution (see above) to the heat conduction equation for $-\infty < x < \infty$ as well as $0 < x < \infty$, with different values of the material parameters in the two half-spaces, and with T and $k\partial T/\partial x$ continuous at $x = 0$.

The point is, of course, that the low conductivity, specific heat capacity and density of wood mean that not much heat is available to burn the feet. What answer would you get if the wood were replaced by steel ($\rho = 7860$, $c = 420$, $k = 63$), and to which material parameter(s) is the difference due?

According to my father, who recently did this, it feels like walking over crushed bark.

'The pressure in the inlet is maintained at a given temperature, which is usually around 5 atmospheres.'

3
Scaling and nondimensionalisation

3.1 Nondimensionalisation and dimensionless parameters

Like its predecessor, this chapter has one theme, which is simple but has far-reaching repercussions. The key idea is this. Any equation we write down for a physical process models balances between physical mechanisms. These may not all be equally important; experience shows that we can count ourselves unlucky if more than two are in balance at once. We can begin to assess how important they are by scaling all the variables with 'typical' values – values that have the size we expect to see, dictated by the geometry, boundary conditions and so on – so that the equation becomes *dimensionless*. Instead of a large number of physical parameters and variables, all with dimensional units, we are left with an equation written in *dimensionless variables*. All the physical parameters and typical values are collected together into a smaller number of *dimensionless parameters* (or *dimensionless groups*), which, when suitably interpreted, should tell us the relative importance of the various mechanisms.

This is much easier to see by working through an example than by waffly generalities. So here's a selection of three relatively simple physical situations where we can see the technique in action.

3.1.1 Example: advection–diffusion

We'll start with a combination of two very familiar models, those for heat conduction and fluid flow. When you stand in front of a fan to cool down, two mechanisms come into play: heat is conducted (diffuses) into the air and is then carried away by it. The process of heat transfer via a moving fluid is called *advection*, as distinct from *convection*, which is hot-air-rises heat transfer due to density changes.[1] Both advection and convection are major mechanisms for heat transfer in systems such as the earth's atmosphere, oceans and molten core; almost any industrial process (think of cooling towers as a visible example); car engines; computers; you name it. Their analysis is of enormous practical importance.

It is often easier to analyse advection because we can usually decouple the question of finding the fluid flow from the heat-flow problem. In convection, the buoyancy force that drives the flow is strongly temperature dependent – indeed, without it there would be no flow – and the problem is correspondingly more difficult. For our first example, we'll consider the two-dimensional flow of an incompressible liquid with a given (that is, we can calculate it separately) velocity **u** past a circular cylinder of radius a, with a free-stream velocity $(U, 0)$ at large distances. This is the basis for a simple model of, for example, the cooling of a hot pipe.

For the moment, it doesn't matter too much what we take for **u**. Let's just use the standard inviscid-flow model $\mathbf{u} = \nabla\phi$, where

$$\phi = U\left(r\cos\theta + \frac{a^2\cos\theta}{r}\right)$$

in plane polar coordinates. We need to generalise the heat conduction equation to include the advection. This requires us to recognise that when we write down the conservation of energy in the form

rate of change in internal energy of a fluid particle

= net heat flux into it,

we have to do this 'following a particle'. Thus, the time derivative $\partial/\partial t$ in the usual heat equation (2.1) is replaced by the material (convective) derivative $\partial/\partial t + \mathbf{u}\cdot\nabla$, giving

$$\rho c\left(\frac{\partial T}{\partial t} + \mathbf{u}\cdot\nabla T\right) = k\nabla^2 T. \tag{3.1}$$

If you don't quite believe this argument, do Exercise 1 at the end of this chapter.

Lastly we need some boundary conditions. It is simplest to have one constant temperature at infinity and another on the cylinder,[2] so that

$$T \to T_\infty \quad \text{as} \quad r \to \infty, \qquad T = T_0 \quad \text{on} \quad r = a.$$

The problem is summarised in Figure 3.1.

---

[1] The usage is changing in the loose direction; convection is often used for both processes, subdivided where necessary into forced convection for advection and natural convection for buoyancy-driven heat transport. A lot of the heat lost by a hot person in still air is by (natural) convection.

[2] The conditions on the cylinder are not especially realistic; a Newton condition of the form $-k\partial T/\partial n + h(T - T_0) = 0$ would be better. See Exercises 5 and 6.

Figure 3.1 Advection–diffusion of heat from a cylinder; T satisfies $\rho c \frac{\partial T}{\partial t} + \mathbf{u} \cdot \nabla T = k \nabla^2 T$.

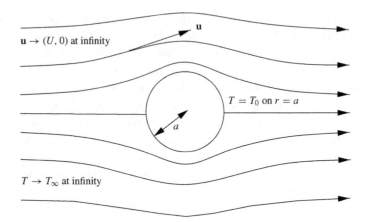

$\mathbf{u} \to (U, 0)$ at infinity

$T = T_0$ on $r = a$

$T \to T_\infty$ at infinity

In order to see the relative importance of advection (the left-hand side of (3.1)) and diffusion (the right-hand side), we scale all the variables with 'typical' values. The obvious candidate[3] for the length scale is a; then we can scale $\mathbf{u}$ with U and time with a/U, the order-of-magnitude residence time of a fluid particle near the cylinder. So, we write

$$\mathbf{x} = a\mathbf{x}', \qquad \mathbf{u} = U\mathbf{u}', \qquad t = (a/U)t'.$$

Only T has not yet been scaled. Using the two datum points T_0 and T_∞ we can write,

$$T(\mathbf{x}, t) = T_\infty + (T_0 - T_\infty) T'(\mathbf{x}', t').$$

This gives

$$\frac{\rho c U}{a}\left(\frac{\partial T'}{\partial t'} + \mathbf{u}' \cdot \nabla' T'\right) = \frac{k}{a^2}\nabla'^2 T'. \qquad (3.2)$$

Now comes a key point. All the terms in the original equation have the same physical dimensions. Our 'primed' quantities have no dimensions: they are just numbers. Thus, if we divide through (3.2) by one of the (still dimensional) quantities multiplying a 'primed' term, we will be left with a dimensionless term. All the other terms in the equation must then be dimensionless as well, and so the physical parameters (a, ρ, etc.) *must now occur in combinations that are dimensionless*.

So, divide through (3.2) by, say, k/a^2 to get

$$\frac{\rho c U a}{k}\left(\frac{\partial T'}{\partial t'} + \mathbf{u}' \cdot \nabla' T'\right) = \nabla'^2 T'.$$

[3] If the cross-section of the cylinder is another shape, we can use any measure of its 'diameter', although you can see an obvious difficulty here if it is, say, a long thin ellipse. We return to this point later.

We see that there is just one dimensionless combination in this problem,

$$\text{Pe} = \frac{\rho c U a}{k} = \frac{U a}{\kappa},$$

known as the *Peclet number*. There is no dimensionless parameter in the boundary conditions because they scale linearly, to become

$$T' \to 0 \quad \text{as} \quad r' \to \infty, \qquad T' = 1 \quad \text{on} \quad r' = 1.$$

There several things to say about this analysis. The first is the simple observation that whereas the original problem has a large 'parameter space' consisting of the seven parameters U, a, ρ, c, k, T_∞, T_0, the reduced problem contains the *single* dimensionless parameter Pe. That's quite a reduction; and if you think it is obvious, there are still plenty of mathematical subjects almost untouched by the idea (economics, for example).

Second, *all problems with the same value of* Pe *can be obtained by solving one canonical scaled problem for that value of* Pe. So, if we want to make an experimental analogue of a very large physical set-up, we can do it in a smaller setting as long as we achieve the same Peclet number.[4]

Third, and probably most important, the sizes of any dimensionless numbers in a problem tell us a great deal about the balance of the physical mechanisms involved and about the behaviour of solutions. In our example, we can write

$$\begin{aligned}
\text{Pe} &= \frac{\rho c U a}{k} \\
&= \frac{\rho c U (T - T_\infty)}{k(T - T_\infty)/a} \\
&\approx \frac{\text{advective heat flow}}{\text{conductive heat flow}}.
\end{aligned}$$

> What would have happened if we had used the Newton condition?

> It is worth noting that there may be more than one possible choice for some of the scales, and iteration may be needed to find the most appropriate one for a given problem. In our example, there are two other possible length scales, whose consequences are explored in Exercise 2. Usually, the obvious choice is the best.

If Pe is large then advection dominates over conduction, and vice versa if Pe is small. A 10-cm-radius hot head in an air stream with $\rho \approx 1.3$ kg m^{-3}, $c \approx 993$ J kg^{-1} K^{-1}, $k \approx 0.24$ W m^{-1} K^{-1}, moving at 1 m s^{-1} from a fan, has a Peclet number of about 500, undeniably large (large Peclet numbers are more common than small ones).

Any problem with a parameter equal to 500 must surely offer scope for judicious approximation: after all, $1/500$ is tiny, and we may hope to cross out the terms it multiplies without losing too much. If we do this

---

[4] We also have to ensure that the fluid velocity scales correctly. This may not be so easy given that it has its own equations of motion, which may not behave properly. In our example, it is clear that the potential-flow model does scale correctly, because it is linear.

in our equation

$$\frac{\partial T'}{\partial t'} + \mathbf{u}' \cdot \nabla' T' = \frac{1}{\mathrm{Pe}} \nabla'^2 T',$$

we see that the convective derivative of T' is approximately zero. That is, as we follow a particle its temperature does not change, and since all particles start far upstream at $x = -\infty$, T' is everywhere equal to its upstream value of 0. This is fine, until we realise that we can't satisfy $T' = 1$ on the cylinder. We shall have to wait until we have looked at asymptotic expansions before we can see how to get out of this difficulty.

A last remark: one soon gets tired of writing primes on all scaled variables. As soon as the scalings have been introduced, it's usual to write the new dimensionless equations in the original notation, a pedantically incorrect but universal practice signalled by the phrase 'dropping the primes'.

3.1.2 Example: the damped pendulum

Sometimes the correct scales for one or more variables can only be deduced from the equations, as in the following example. The basic model for a linearly damped pendulum that is displaced an angle θ from the vertical (see Figure 3.2) is

$$l \frac{\mathrm{d}^2\theta}{\mathrm{d}t^2} + k \frac{\mathrm{d}\theta}{\mathrm{d}t} + g \sin\theta = 0,$$

and let us suppose that the initial angle and angular speed are prescribed:

$$\theta = \alpha_0, \qquad \frac{\mathrm{d}\theta}{\mathrm{d}t} = \omega_0 \qquad \text{at} \quad t = 0.$$

Here k is the damping coefficient (the damping is proportional to the velocity) and g is the acceleration due to gravity; their units are

> Check that these units are consistent with the equation; remember that θ is a pure number.

$$[k] = [\mathrm{L}][\mathrm{T}]^{-1}, \qquad [g] = [\mathrm{L}][\mathrm{T}]^{-2}.$$

Combining the dimensional parameters l, k, g and ω_0, it is easy to see that there are *three* timescales built into the parameters of this problem:

$$t_1 = \sqrt{\frac{l}{g}}, \qquad t_2 = \frac{l}{k}, \qquad t_3 = \frac{1}{\omega_0}.$$

> Solve $du/dt = -ku = -u/t_2$ and see that u decreases by a fraction $1/e$ in each time interval t_2.

The first is the period of small undamped oscillations (in the linear theory). The second is the time over which the damping has an effect. The third is prescribed by us: it tells us how long it takes the pendulum to cover one radian at its initial angular speed if no other forces act.

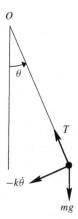

O

θ

T

$-k\dot{\theta}$

mg   **Figure 3.2** Motion of a simple pendulum.

Let's scale time with $t_1 = \sqrt{l/g}$, which is appropriate if we are expecting to see small-amplitude oscillatory behaviour. Then, writing

$$t = t_1 t',$$

we have the dimensionless model

$$\frac{d^2\theta}{dt'^2} + \frac{t_1}{t_2}\frac{d\theta}{dt'} + \sin\theta = 0,$$

with

$$\theta = \alpha_0, \qquad \frac{d\theta}{dt'} = \frac{t_1}{t_3} \qquad \text{at} \quad t' = 0.$$

The model contains two obviously dimensionless parameters,

$$\gamma = \frac{t_1}{t_2} = \sqrt{\frac{k^2}{gl}} \qquad \text{and} \qquad \beta_0 = \frac{t_1}{t_3} = \sqrt{\frac{\omega_0^2 l}{g}}.$$

The first of these, γ, is the ratio of the timescale over which the system responds to the physical mechanism of gravity (the period for small oscillations) to the timescale for damping. The second, β_0, is the ratio of the initial speed of the pendulum to the speed changes induced by gravity. As angles are automatically dimensionless there is a third dimensionless group, α_0, and so, dropping the primes, the dimensionless model is

$$\frac{d^2\theta}{dt^2} + \gamma\frac{d\theta}{dt} + \sin\theta = 0,$$

with

$$\theta = \alpha_0, \qquad \frac{d\theta}{dt} = \beta_0 \qquad \text{at} \quad t = 0.$$

We can make some immediate statements about the behaviour of the system just by looking at the sizes of our dimensionless parameters. For example, if γ is small, we expect the damping to have its effect over many cycles. If β and α_0 are both small, we hope that linearised theory will be valid; and so on. Later, in Chapter 13, we'll see how to quantify some of these ideas.

3.1.3 Example: beams and strings

We know that the motion of a string made of a material with density ρ (mass per unit volume, $[\rho] = [M][L]^{-3}$), of length L, cross-sectional area A and stretched to tension T, can be modelled by the equation[5]

$$\rho A \frac{\partial^2 y}{\partial t^2} - T \frac{\partial^2 y}{\partial x^2} = 0, \quad 0 < x < L,$$

(a) What are the dimensions of T? (b) Show that $\sqrt{T/(\rho A)}$ is a speed. (c) Show that the wave speed is exactly $c = \sqrt{T/(\rho A)}$ (with no numerical prefactor) by substituting in a solution of the form $y(x, t) = f(x - ct)$.

where $y(x, t)$ is the amplitude of small transverse displacements.[6]

In the string model, the restoring force is provided by the component of the tension normal to the string, $T \, \partial y/\partial x$. If, however, we have a stiff beam or rod then the restoring force is caused by its resistance to bending, which, as we shall see in Chapter 4, can be shown to be proportional to $-\partial^3 y/\partial x^3$. If in addition there is a force perpendicular to the wire, of magnitude F per unit length, we get the equation

What is the wave speed?

$$\rho A \frac{\partial^2 y}{\partial t^2} + E A k^2 \frac{\partial^4 y}{\partial x^4} = F.$$

Here A is the cross-sectional area of the beam, while k is a constant with the dimensions of length known as the *radius of gyration* of the cross-section of the beam. In this model, k encapsulates the effect of the shape of the beam; for a fixed cross-sectional area, k is smallest for a circular cross-section, while for a standard I-beam it is large.[7] A derivation of this model is given in Section 4.1.

The constant E is a property of the material from which the beam is made known as the *Young's modulus*. The larger it is, the more the

---

[5] You may have seen this in the form

$$\tilde{\rho} \frac{\partial^2 y}{\partial t^2} - T \frac{\partial^2 y}{\partial x^2} = 0, \quad 0 < x < L,$$

in which $\tilde{\rho}$ is the mass per unit length, or line density.

[6] Note the engineering rule of thumb

$$\text{wave speed} = \sqrt{\frac{\text{stiffness}}{\text{inertia}}},$$

which applies very generally to non-dissipative linear systems.

[7] The definition of k is: in a cross-sectional plane, take coordinates (ξ, η) with origin at the centre of mass of the cross-section. Then

$$A k^2 = \iint_{\text{cross-section}} \xi^2 + \eta^2 \; d\xi \, d\eta.$$

That is, $A k^2$ is the moment of inertia of the cross-section.

material resists being bent or sheared (bending leads to shearing). Lastly, ρ is again the underlying material density, which we use in preference to the line density because this model takes into account both the non-zero cross-sectional area of the beam and its shape.

If the force F is due to gravity and y is measured vertically upwards, we have

$$F = -\rho A g.$$

However, we might consider other forces, such as the drag from a fluid flowing past the beam. Examples of this are: wind drag on a skyscraper, flagpole, car radio aerial or hair (see the next chapter); water drag on a reed bending in a stream; the drag of gas escaping through a brush seal in a jet engine. For inviscid flows, the pressure in the liquid has a typical magnitude $\rho_l U_l^2$, where ρ_l is the liquid density and U_l a typical value of its speed (think of Bernoulli's equation for steady irrotational flow, $p + \frac{1}{2}\rho|\mathbf{u}|^2 = $ constant). Because the flow about a cylinder, even a circular one, is not symmetrical in a real (as opposed to an ideal) flow,[8] there is a net pressure force on the cylinder, and its magnitude is roughly proportional to ρU^2.

So, it is reasonable that the drag per unit length on an isolated cylinder in a flow with free-stream velocity U_l can be well approximated by

$$F = \text{geometric factor} \times \text{pressure} \times \text{perimeter}$$
$$= C_\text{d} \times \rho_l U_l^2 \times k.$$

Here the *drag coefficient* C_d depends on the fluid speed and on the shape and orientation of the cylinder (when we work out the drag force, we resolve the pressure, which acts normally to the surface, in the direction of the free stream and integrate over the perimeter of the cylinder; all this information is lumped into the drag coefficient), and we have used k as a measure of the length of the cross-sectional perimeter. Combining the expression for F with the model equation, we get

What other length might we have used? Why is k probably better?

$$\rho A \frac{\partial^2 y}{\partial t^2} + E A k^2 \frac{\partial^4 y}{\partial x^4} = C_\text{d}\rho_l U_l^2 k$$

as the equation for a beam (flagpole, reed) that is subject to a fluid drag force.

This model has a huge number of physical parameters, but we can get a lot of information from some simple scaling arguments, here presented as an exercise.

[8] D'Alembert's paradox says that there is no drag on a cylinder in irrotational inviscid (potential) flow! In real life, even a very small viscosity has a profound effect, leading to completely different flows from the ideal ones. We'll get an idea why this is so, later on.

1. Make the model dimensionless using the length of the beam L to scale x and scales to be determined, y_0 and t_0, for y and t.
2. Verify that the units of E are $[M][L]^{-1}[T]^{-2}$.
3. Roughly how big is the steady-state displacement?
4. If the drag force is switched on suddenly at $t = 0$, over what timescale does the beam initially respond?
5. What is the timescale for free oscillations?

We will return to versions of this model at several places later in the book.

3.2 The Navier–Stokes equations and Reynolds numbers

Recall from Chapter 1 that the flow of an incompressible Newtonian viscous fluid is governed by the Navier–Stokes equations

$$\rho\left(\frac{\partial \mathbf{u}}{\partial t} + \mathbf{u}\cdot\nabla\mathbf{u}\right) = -\nabla p + \mu\nabla^2\mathbf{u}, \qquad \nabla\cdot\mathbf{u} = 0, \tag{3.3}$$

where $\mathbf{u}$ is the fluid velocity and p the pressure, both of which are functions of position $\mathbf{x}$ and time t, while the physical parameters of the fluid are its density ρ and its *dynamic viscosity* μ.

Let us now look at how we should nondimensionalise the Navier–Stokes equations. We begin by noting that it is often useful to combine μ and ρ to form the *kinematic viscosity*

$$\nu = \frac{\mu}{\rho}.$$

What are the units of the dynamic and kinematic viscosities? Since, as in the discussion regarding Figure 1.1,

$$\frac{\text{force}}{\text{area}} = \frac{\mu U}{h},$$

we have

$$[\mu] = [M][L]^{-1}[T]^{-1}$$

(the SI unit is the pascal second, Pa s), and so

$$[\nu] = [L]^2[T]^{-1}.$$

Mnemonic, acres per annum; the acre is one of the many old English units of area. Hectares per megasecond?

Suppose we have flow past a body of typical size L, with a free-stream velocity $U\mathbf{e}_1$. As in the advection–diffusion problem, we scale all distances with L, time with L/U and velocities with U, writing

$$\mathbf{x} = L\mathbf{x}', \qquad t = (L/U)t', \qquad \mathbf{u} = U\mathbf{u}'.$$

Only p has not yet been scaled, and in the absence of any obvious exogenous (externally prescribed) scale we let the equations tell us what the possibilities are. For now, let's write

$$p = P_0 p'$$

and substitute all these into the momentum equation (clearly mass conservation just becomes $\nabla' \cdot \mathbf{u}' = 0$). This gives

$$\frac{\rho U^2}{L} \left(\frac{\partial \mathbf{u}'}{\partial t'} + \mathbf{u}' \cdot \nabla' \mathbf{u}' \right) = -\frac{P_0}{L} \nabla' p' + \frac{\mu U}{L^2} \nabla'^2 \mathbf{u}'. \qquad (3.4)$$

Now we can divide through by one of the coefficients to leave a dimensionless term; because all the other terms must also be dimensionless, that will tell us the pressure scale. For example, divide through by $\rho U^2 / L$; this leaves us with

$$\frac{\partial \mathbf{u}'}{\partial t'} + \mathbf{u}' \cdot \nabla' \mathbf{u}' = -\frac{P_0}{\rho U^2} \nabla' p' + \frac{\mu}{\rho U L} \nabla'^2 \mathbf{u}'.$$

It is now clear that we can choose the 'inviscid' pressure scale

$$P_0 = \rho U^2,$$

and when we do this the dimensionless equation takes the form

$$\frac{\partial \mathbf{u}'}{\partial t'} + \mathbf{u}' \cdot \nabla' \mathbf{u}' = -\nabla' p' + \frac{1}{\mathrm{Re}} \nabla'^2 \mathbf{u}', \qquad (3.5)$$

where the dimensionless combination

$$\mathrm{Re} = \frac{\rho U L}{\mu} = \frac{U L}{\nu}$$

is known as the *Reynolds number* after the British hydrodynamicist Osborne Reynolds.

So what does this tell us? The most important conclusion is that *if basic viscous effects are all we are considering,*[9] *then all flows with the same Reynolds number are scaled versions of each other.* This is the idea behind the wind tunnel: we don't need to build full-scale prototype aeroplanes or cars to test for lift and drag; we can use a scale model as long as we get the Reynolds number right. Furthermore, by forming dimensionless groups, we reduce the size of our parameter space as far as possible. In our example above, instead of the four physical parameters ρ, μ, L and U we have the single combination Re. So, for a given shape of body, in principle all we need to do is sweep through

---

[9] For example, we are ignoring temperature changes due to viscous dissipation, which may themselves affect the viscosity or density of the fluid.

the Reynolds numbers from 0 to ∞ to find all the possible flows past a body of that shape.

Thinking now of the dimensionless parameters as encapsulating the competing (or balancing) physical mechanisms that led to our original equation, we can write the Reynolds number as

$$\text{Re} = \frac{\rho U L}{\mu} = \frac{\rho U^2}{\mu U / L}.$$

The numerator of the last fraction is clearly a measure of the pressure force per unit area due to fluid inertia on a surface while, as we saw above, the denominator is a measure of viscous shear forces. So the Reynolds number tells us the ratio of inertial forces to viscous ones. When it is large the inertial forces dominate, while for small Re it is viscosity that wins.

In the former case, it is tempting to cross out the term multiplied by $1/\text{Re}$ in the dimensionless equation (3.5); this leaves us with the Euler equations

$$\frac{\partial \mathbf{u}}{\partial t} + \mathbf{u} \cdot \nabla \mathbf{u} = -\nabla p, \qquad \nabla \cdot \mathbf{u} = 0$$

(we have dropped the primes). There is a large class of exact solutions of the Euler equations when the flow is irrotational, so that $\nabla \wedge \mathbf{u} = \mathbf{0}$. In this case, there is a velocity potential ϕ which satisfies Laplace's equation

$$\nabla^2 \phi = 0$$

in the fluid. However, we must be very careful in making this approximation. One obvious reason is that for most viscous fluids we should apply the *no-slip condition* on rigid boundaries; this says that the fluid velocity at the boundary must equal the velocity of the boundary itself, i.e. the fluid particles at the boundary stick to it. Most solutions of the Euler equations do not satisfy this condition, and the reconciliation of the two ideas led to boundary layer theory and the theory of matched asymptotic expansions. The latter is a triumph of twentieth-century applied mathematics, and we will look at it briefly in Chapter 16. A second reason for proceeding with caution is the everyday observation that very fast (very large Reynolds-number) flows are turbulent and so intrinsically unsteady. For these reasons one may worry that the inviscid model exists in theory but is never seen in practice, but that would be unduly pessimistic. Boundary layer theory is useful, and in many interesting flows either the Reynolds number is large but not enormous or the flow takes place on a timescale short enough that turbulence does not have time to become a nuisance.

Returning to the theme of nondimensionalisation, what if Re is small, as for slow viscous flow? Is is safe to say that since $\nabla'^2 \mathbf{u}'$ is divided by

See Exercise 11, where failure to satisfy no-slip is proved for potential flows.

Re, we can simply set $\nabla'^2 \mathbf{u}'$ equal to zero? No, it is not. If we did this, we would be saying that pressure forces are not important, and it is common experience that they are. In such a situation, we should check whether there is an alternative scaling of the equations. It is not hard to see that there is indeed a second possible pressure scale,

$$\tilde{P}_0 = \frac{\mu U}{L},$$

and it is an exercise to show that scaling p in this way leads to an alternative version of (3.5),[10]

$$\mathrm{Re}\left(\frac{\partial \mathbf{u}'}{\partial t'} + \mathbf{u}' \cdot \nabla' \mathbf{u}'\right) = -\nabla' p' + \nabla'^2 \mathbf{u}'.$$

If Re is small, maybe we can neglect the convective derivative (inertial) terms on the left to get the *Stokes flow* model for slow flow:

$$\mathbf{0} = -\nabla p + \nabla^2 \mathbf{u}, \qquad \nabla \cdot \mathbf{u} = 0.$$

As we continue, we shall see how we might justify dropping terms in this way (and why it might go wrong).

This is obvious from the definition of (the dimensionless) Re; why?

3.2.1 Water in the bathtub

Question: Is it true that water flows out of the bathtub with an anticlockwise swirl in the northern hemisphere and a clockwise swirl south of the equator? Answer: Only under very carefully controlled circumstances. Here's why.

Remember the Coriolis theorem about transferring the equations of motion to a rotating coordinate system: if $\mathbf{v}$ is a vector, and we want its time derivative as measured in a frame rotating with angular velocity $\mathbf{\Omega}$, then

$$\left.\frac{d\mathbf{v}}{dt}\right|_{\text{fixed}} = \left.\frac{d\mathbf{v}}{dt}\right|_{\text{rotating}} + \mathbf{\Omega} \wedge \mathbf{v}.$$

There is nothing difficult about this: it is just the chain rule in disguise.

The Navier–Stokes equations in a rotating frame are then

$$\rho\left(\frac{\partial \mathbf{u}}{\partial t} + 2\mathbf{\Omega} \wedge \mathbf{u} + \mathbf{u} \cdot \nabla \mathbf{u}\right) = -\nabla p + \mu \nabla^2 \mathbf{u} + \rho\big(\mathbf{g} - \mathbf{\Omega} \wedge (\mathbf{\Omega} \wedge \mathbf{r})\big),$$

where in this case $\mathbf{\Omega}$ is the angular velocity of the earth, equal to 2π per 24 hours, about 7.3×10^{-5} radians per second, in the direction of the earth's axis of rotation. Now consider water moving at 1 m s^{-1} in a bath of size about 1 m. If we scale the variables with representative values based on these figures (all of which are 1 in SI units), the ratio of the

The origin is at the centre of the earth, and the $\mathbf{\Omega} \wedge \mathbf{\Omega} \wedge \mathbf{r}$ term is incorporated into the gravitational body force to give the 'apparent gravity' term.

---

[10] Pedantically speaking, note that the p' in this equation is not the same as the p' in the other dimensionless version of Navier–Stokes (3.5), as it has been scaled differently . . .

This is an exercise that you should carry out ...

Coriolis term $2|\mathbf{\Omega} \wedge \mathbf{u}|$ to the other acceleration terms is about twice the value of $|\mathbf{\Omega}|$ in SI units, i.e. about 10^{-4}. Thus the rotation effect is tiny. In practice, other effects such as residual swirl from the way the water was put into the bath, or asymmetry in the plughole or the way the plug is pulled out, completely swamp the Coriolis effect unless the experiment is carried out under very carefully controlled conditions. However, if we look at rotating air masses on the scale of a hurricane or typhoon, the much greater length scale means that the Coriolis effect is enormously important. As air masses leave the equator and travel north or south, they carry their angular momentum (whose direction is along the earth's axis of rotation) with them. As they move north or south round the curve of the earth this angular momentum is transformed into rotatory motion in the tangent (locally horizontal) plane and this can be followed by intensification into a localised storm.

3.3 Buckingham's Pi-theorem

Let us take a short detour to state the only quasi-rigorous result in the area of dimensional analysis: the Buckingham Pi-theorem.

Suppose that we have n independent physical variables and parameters $Q_1, \ldots, Q_n$ (in the discussion above, these are $\mathbf{x}$, t, μ etc.) and that the solution of a mathematical model gives us one of these in terms of the others:

$$Q_n = f(Q_1, \ldots, Q_{n-1}).$$

Suppose also that there are r independent basic physical dimensions ([M], [L], [T] etc.)

Then there are $k = n - r$ dimensionless combinations[11] $\Pi_i(Q_j)$ and a function g such that

$$\Pi_k = g(\Pi_1, \ldots, \Pi_{k-1}).$$

The proof is a counting exercise.

Example: the drag on a cylinder

Suppose a cylinder of length L and radius a is held in a viscous fluid moving with far-field velocity U normal to the axis of the cylinder. How does the drag force depend on the parameters of the problem? What happens as $L \to \infty$?

There are six independent physical quantities in this problem:

- L and a, which are properties of the cylinder and which both have dimension [L];

---

[11] From which the name of the theorem comes.

- μ and ρ, which are properties of the fluid and have dimensions $[M][L]^{-1}[T]^{-1}$, $[M][L]^{-3}$ respectively;
- the force F on the cylinder ($[M][L][T]^{-2}$) and the free-stream velocity U ($[L][T]^{-1}$); F is an output (or dependent) variable, and we aim to express it in terms of the inputs.

In this case, $r = 3$ (for $[M]$, $[L]$ and $[T]$), and so there must be $k = n - r = 3$ dimensionless quantities. One natural choice is the aspect ratio L/a, which we call Π_1, and another is the Reynolds number $\mathrm{Re} = Ua(\mu/\rho)^{-1} = Ua/\nu = \Pi_2$. For the third, a little experimentation shows that something of the form

In choosing a as the length to appear in Re, we are looking ahead to when we let $L \to \infty$.

$$\frac{F}{\rho U^2 [L]^2}$$

will do, and we need to choose which lengths to use to replace $[L]^2$. Here it will help if we think what physical balance is expressed by this parameter. The numerator, F, is a force, while the denominator is the inviscid pressure scale ρU^2 multiplied by an area. Hence it makes sense to use aL, which is a measure of the surface area of the cylinder, and our third dimensionless parameter is thus $\Pi_3 = F/(\rho U^2 aL)$.

Remember that pressure = force per unit area.

Putting this all together, we have $\Pi_3 = g(\Pi_1, \Pi_2)$; that is, on dimensional grounds we have shown that the drag force is related to the other parameters by an equation of the form

$$F = \rho U^2 aL \times g(\mathrm{Re}, L/a)$$

for some function g.

If we assume further that our cylinder is very long, so that we have translational invariance along it, then instead of F and L as independent physical quantities we have only the force per unit length F' (dimensions $[M][T]^{-2}$). Then we get

$$F' = \rho U^2 a C_d(\mathrm{Re})$$

for some function $C_d(\mathrm{Re})$, which is just the drag coefficient mentioned earlier in the chapter.

There is clearly some indeterminacy in the choice of the parameters and the representation of the drag coefficient. For example, we could have written the Reynolds number as UL/ν, or we could have introduced more convoluted parameters such as $UL^2/(\nu a)$, which equals L/a times our definition above, but this would not have had such clear physical implications. It helps to make choices of parameters that correspond as closely as possible with the physical situation, although we can't always hope to get it right first time. Moreover, there are often genuine alternatives. In our example, we chose ρU^2 as our measure of the fluid pressures; this says that we expect inertia to be significant and is the clearest way of writing the drag when the flow has a large Reynolds number. However, as we saw earlier, we could have chosen $\mu U/a$ for the pressure scale, and this would have led to

$$F' = \mu U \widetilde{C}_{\mathrm{d}}(\mathrm{Re}),$$

which might be more convenient if we were looking at slow flow. Of course, the drag coefficient is uniquely determined[12] as a function of the Reynolds number, so this is merely a relabelling exercise: it is easy to see that $\widetilde{C}_{\mathrm{d}}(\mathrm{Re}) = \mathrm{Re}\, C_{\mathrm{d}}(\mathrm{Re})$.

We haven't done anything very technical in this chapter. This whole business of scaling is a combination of experience and plain common sense. The main point is that sensible scalings should reveal the primary balances between the physical mechanisms in equations, leaving the remaining terms as smaller corrections, at least at first sight (it often happens that what we thought was a small correction later on rises up and hits us between the eyes, but that's all part of the experience). If we have the wrong scalings, it usually becomes apparent fairly soon. In later chapters we give an introduction to asymptotic analysis, a framework that allows us to make the idea of approximate solutions more systematic.

3.4 Sources and further reading

Acheson's book *From Calculus to Chaos* [1] has a lot more on the pendulum problem. The flagpole problem was lifted from the book of Fowkes & Mahoney [18], where many more details will be found. It is here partly as an exercise in scaling, but also as an introduction to the beam equation.

3.5 Exercises

1 **Advection–diffusion.** If $T(\mathbf{x}, t)$ is the temperature in an incompressible fluid moving with velocity $\mathbf{u}$, explain why the heat flux is

$$\rho c T \mathbf{u} - k \nabla T.$$

Take an arbitrary small volume V *fixed in space*, write energy conservation in the form

$$\frac{\mathrm{d}}{\mathrm{d}t} \int_V \rho c T \,\mathrm{d}V + \int_{\partial V} (\rho c T \mathbf{u} - k \nabla T) \cdot \mathbf{n} \,\mathrm{d}S = 0,$$

---

[12] At large Reynolds number the flow is turbulent and so unsteady; the drag coefficient must then be interpreted as a time average.

and then use the divergence theorem and $\nabla \cdot \mathbf{u} = 0$ to derive (3.1).
(Note that in the derivation of (3.1) on p. 29 we used incompressibility to say that the density in the material volume remains constant.
If the fluid is compressible then we have to worry about what we
mean by the specific heat, because the density changes. That is, we
have to think carefully about the thermodynamics of the problem.
Fortunately, for most liquids the density change with temperature
or pressure is small enough to be neglected in the convective
derivative, although not necessarily in the buoyancy body force. In
gas dynamics, two specific heats are considered, one for changes at
constant pressure and one for changes at constant volume.)

2 Peclet numbers. Consider the advection–diffusion problem of
Figure 3.1 on p. 30. Show that other possible length scales are

$$\frac{\kappa}{U}, \qquad \frac{k(T_\infty - T_0)}{\rho U^3}.$$

If we use the first, what happens to the boundary $r = a$, and why
might this be inconvenient? Explain why the denominator of the
second is a kinetic energy flux and hence why it is an inappropriate
length scale for this problem.

3 The Boussinesq transformation. Consider the steady-state
dimensionless advection–diffusion problem

$$\mathrm{Pe}\left(u\frac{\partial T}{\partial x} + v\frac{\partial T}{\partial y}\right) = \nabla^2 T,$$

in which the velocity is given by the potential flow past a two-
dimensional body (not necessarily a circular cylinder) with potential
ϕ and stream function ψ:

$$u = \frac{\partial \phi}{\partial x} = \frac{\partial \psi}{\partial y}, \qquad v = \frac{\partial \phi}{\partial y} = -\frac{\partial \psi}{\partial x}, \qquad \nabla^2 \phi = \nabla^2 \psi = 0.$$

Switch from x, y to ϕ, ψ as independent variables, so that

$$\frac{\partial}{\partial x} = \frac{\partial \phi}{\partial x}\frac{\partial}{\partial \phi} + \frac{\partial \psi}{\partial x}\frac{\partial}{\partial \psi} = u\frac{\partial}{\partial \phi} - v\frac{\partial}{\partial \psi}$$

etc.,[13] to show that the problem becomes

$$\mathrm{Pe}\,\frac{\partial T}{\partial \phi} = \frac{\partial^2 T}{\partial \phi^2} + \frac{\partial^2 T}{\partial \psi^2}$$

Note that the left-hand side is the directional derivative of T along streamlines, which are orthogonal to the equipotentials (why?).

---

[13] A short cut: Because $\phi + i\psi$ is an analytic (holomorphic) function $w(z)$ of $z = x + iy$,
the Cauchy–Riemann equations allow us to simplify the Laplacian operator to

$$\frac{\partial^2}{\partial x^2} + \frac{\partial^2}{\partial y^2} = \left|\frac{dw}{dz}\right|^2 \left(\frac{\partial^2}{\partial \phi^2} + \frac{\partial^2}{\partial \psi^2}\right).$$

in the $\phi\psi$-plane. If the flow is symmetric, what are the boundary conditions in the new variables? (This problem can be solved by the Wiener–Hopf technique, but it is a complicated business.)

4 The Kirchhoff transformation. Suppose that the thermal conductivity of a material depends on the temperature. Show that the steady-state heat equation

$$\nabla \cdot (k(T)\,\nabla T) = 0$$

can be transformed into Laplace's equation for the new variable $u = \int^T k(s)\,ds$.

5 Newton's law of cooling and Biot numbers. The process of cooling a hot object is a complicated one. In addition to conduction to the surroundings, it may involve both forced and natural convection if the body is immersed in a liquid or gas; there may be boiling or thermal radiation. A very widely used model lumps all these effects into a single linear law, known as *Newton's law of cooling*, according to which the heat flux per unit area from the body is given by

$$-k\left.\frac{\partial T}{\partial n}\right|_{\text{boundary}} = h(T - T_\infty),$$

where h is the *heat transfer coefficient* and T_∞ is a measure of the ambient temperature. What are the units of h? Explain in general terms why this law is reasonable (including the minus sign).

There are many empirical laws giving h in specific circumstances. For the specific example of black-body radiative transfer, it can be derived from more basic thermodynamics. Recall that the *Stefan–Boltzmann law* says that the heat flux is

$$KT^4 - KT_\infty^4,$$

where T is the *absolute* temperature and K is a constant (what are its units?). Show that the Newton law is a good approximation if T is not too far from T_∞ and find h in this case.

If the body has typical temperature T_0 and size L, write the Newton law in dimensionless form as

$$-\frac{\partial T'}{\partial n'} = \text{Bi}\, T'$$

where $\text{Bi} = hL/k$ is known as the *Biot number*. How is T' related to T?

6 Coffee time. Alphonse takes milk in his coffee, and he has to carry the cup a long way from the machine to his desk. He wants the coffee to be as hot as possible when he gets there. Make a simple model to decide whether it is better to add the (cold) milk to the coffee at the machine or at his desk.

Still on the subject of coffee, Bérénice takes sugar in hers. At time $t = 0$ she puts a lump in. If $V(t)$ is the volume and $A(t)$ the surface area of the undissolved lump, and the coffee is well mixed, explain why a crude model for the evolution of the lump is

$$\frac{dV}{dt} \propto -A,$$

and why, on dimensional grounds,

$$A \propto V^{2/3}.$$

Solve the model and show that V reaches zero in a finite time.

Now solve the differential equation

$$\frac{dV}{dt} = V^{2/3}, \quad t > 0; \qquad V(0) = 0.$$

I hope you found *all* the solutions:

$$V(t) = \begin{cases} 0, & 0 < t < t_0, \\ \left(\frac{1}{3}(t - t_0)\right)^3, & t \geq t_0, \end{cases}$$

for any $t_0 \geq 0$; the nonuniqueness arises because the right-hand side $V^{2/3}$ is not Lipschitz in V. (The solution $V \equiv 0$ of the differential equation is tangent to all the solutions $V = \left(\frac{1}{3}(t - t_0)\right)^3$.)

Hercule asks the question: If I observe the state of the sugar in Bérénice's coffee, can I deduce when she put it in? Show that he can do this if the lump is only partly dissolved but he can't if it is wholly gone. Hence give a physical interpretation of the nonuniqueness mentioned above. (Ivar Stakgold told me this example. I am happy to return the favour by recommending his book [57] on Green's functions etc. see also the end of Chapter 9.)

7 **Boiling an egg.** A spherical homogeneous (i.e. purely mathematical) egg of radius a is placed in cold water at temperature T_0, the egg being initially at this temperature too. Over a time t_0 the water temperature T_w is increased linearly to T_1, where it remains. The temperature T in the egg is modelled by the heat conduction equation

$$\rho c \frac{\partial T}{\partial t} = k \nabla^2 T,$$

where ρ is the density, c the specific heat capacity and k the thermal conductivity of the egg, all assumed constant, with the Newton

This is an opportunity to review the Picard theorem on the existence and uniqueness of first-order differential equations.

boundary condition

$$-k \left. \frac{\partial T}{\partial r} \right|_{r=a} = h(T - T_w).$$

Make the model dimensionless and identify the dimensionless parameters. What possible regimes might there be and how can you identify them by looking at the sizes of your dimensionless parameters?

Is there any difference in your analysis if the egg is boiled by the traditional method of putting it into boiling water, assuming that the water temperature remains constant, and leaving it there while you sing your national anthem or a song of the right duration (in England, the hymn 'Onward Christian Soldiers' is traditional for this purpose)?

8 Flagpole in an earthquake. Suppose that a flagpole is in still air but that its base $y = 0$ is oscillated horizontally by an earthquake, so that the condition $y(0, t) = 0$ is replaced by

$$y(0, t) = a \cos \omega t, \tag{3.6}$$

the other boundary conditions remaining as

$$y_x(0, t) = 0, \qquad y_{xx}(L, t) = y_{xxx}(L, t) = 0, \tag{3.7}$$

where the subscripts denote partial differentiation. Nondimensionalize the unsteady unforced flagpole (beam) equation

$$\rho A y_{tt} + E A k^2 y_{xxxx} = 0 \tag{3.8}$$

using the timescale $1/\omega$ implicit in the boundary condition (3.6). What is the appropriate scale for y?

What is the radius of gyration of a circular cylinder of radius a?

A circular pole is 10 m high and has a radius of 10 cm. It is made of steel, for which $E_s = 2.0 \times 10^7$ kg m^{-1} s^{-2} and $\rho_s = 7.8 \times 10^3$ kg m^{-3}, and the oscillations are at a frequency of 1 Hz. What is ω? It is desired to simulate the behaviour of this pole using a wooden model of radius 1 cm and with the same value of ω. Given that $E_w \approx E_s/20$, $\rho_w \approx \rho_s/13$, how long should the model be?

9 Flagpole under gravity. Show from a vertical force balance that a vertical flagpole is subject to a compressive force $C(x)$ that satisfies

$$\frac{dC}{dx} = -A \rho g$$

(g is the acceleration due to gravity), with $C(L) = 0$. Hence find C. What is its value at $x = 0$, and why?

It can be shown (see Section 4.1) that the effect of gravity is to modify the flagpole equation to

$$\rho A y_{tt} + (C y_x)_x + E A k^2 y_{xxxx} = 0$$

(the new term is just like the tension in the equation for waves on a string, but it is on the other side of the equation because C is a compression, i.e. a negative tension). What is the dimensionless parameter that measures the relative importance of gravity for the pole of Exercise 8? How big is it in that situation?

10 Normal modes of strings and flagpoles. A string with mass density ρ per unit length is stretched between $x = 0$ and $x = L$ to tension T. The end $x = L$ is held fixed, while the end $x = 0$ is oscillated transversely at frequency ω so that its displacement there is $y(0, t) = a \cos \omega t$. Find the time-periodic solution $y(x, t) = f(x) \cos \omega t$; does it exist for all ω and, if not, what happens at the exceptional value(s)?

Show that the dimensionless unsteady flagpole equation of Exercise 8 has solutions of the form $\exp^{i \omega t} \times \cos / \sin \alpha x \times \cosh / \sinh \alpha x$ and find α. (Strictly optional, because it is rather hard work: find the time-periodic solution; you may want to use a symbolic manipulator such as Maple.)

11 Potential flow has slip. Suppose that the potential flow in an inviscid liquid satisfies the no-slip condition $\mathbf{u} = \nabla \phi = \mathbf{0}$ at a fixed boundary. Show that the tangential derivatives of ϕ vanish at the surface, so that ϕ is a constant (say zero) there. Show also that the normal derivative of ϕ vanishes at the surface and deduce from the Cauchy–Kowalevskii theorem (see [44]) that $\phi \equiv 0$, so the flow is static. (In two dimensions, you might prefer to show that $\partial \phi / \partial x - i \partial \phi / \partial y$ is analytic, i.e. holomorphic, vanishes on the boundary curve and hence vanishes everywhere.)

12 A layer of viscous fluid flowing on a surface. A uniform layer of viscous fluid, of thickness h, flows down a plane inclined at an angle θ to the horizontal, so that the acceleration due to gravity down the plane is $g \sin \theta$. Show on dimensional grounds that the flux (per unit length 'into the page') is proportional to $h^3 g \sin \theta / \nu$, where ν is the kinematic viscosity.

Explain (in terms of a force balance) why appropriate conditions at the free surface $y = h$ are $\sigma_{ij} n_j = 0$, where σ_{ij} is the stress tensor.

Take coordinates x downhill along the plane and y normal to it. Show that there is a solution $\mathbf{u} = (u(y), 0, 0)$ and that the free-surface conditions reduce to $p = 0$, $\partial u / \partial y = 0$. Find $u(y)$ and verify the dimensional analysis for $Q = \int_0^h u(y) \, dy$. Show also that this solution corresponds to one-half of the Poiseuille flow of Exercise 14 below (see Figure 20.2 on p. 266).

13 Dimensional analysis of Poiseuille flow.   In a *Poiseuille flow* down a pipe, a Newtonian viscous fluid is forced down a circular tube of cross-sectional area A (or radius a) and length L by a pressure drop ΔP. Confirm that there are the following six independent physical quantities in this problem, and state their dimensions: L and a (or $\sqrt{A}$), which are properties of the pipe; μ and ρ, which are properties of the fluid; the input or output variables (specify one and find the other) ΔP and either a volume flux Q or an average velocity U. How is U related to Q and A?

Use Buckingham's Pi-theorem to show that there are three independent dimensionless quantities and find them in their most useful forms. If this problem has a steady solution show that these three quantities are related by an equation of the form

$$\Delta P = \rho U^2 F(\mathrm{Re}, L/a),$$

Re being the Reynolds number Ua/ν.

If we assume further that the pipe is very long, so that we have translational invariance along it, then instead of ΔP and L as independent physical quantities we only have the pressure gradient P'. Show that

$$P' = \frac{\rho U^2}{a}\,\phi(\mathrm{Re})$$

for some function ϕ.[14]

Using the information in the footnote, find relations between the volume flux and the pressure drop (a) for slow flow, (b) for fast flow with $\mathrm{Re} < 2000$. How does the flux depend on the radius in each case?

Why does water come out of the tap (or a garden hose) in a thin but very fast jet when you put your finger over the end, but not when you take it away? If your water closet is refilling and you turn on a cold tap connected to the same water supply, why does the cistern stop making that 'sshh' noise?

Obviously you can take products etc., but try to single out the best combinations.

---

[14] As $\mathrm{Re} \to 0$, we have the analytical result (see Exercise 14) that $\phi \sim (8\mathrm{Re})^{-1}$. Even though the flow from which this is derived is an exact solution of the Navier–Stokes equations for all Re, it is unstable. The effective drag for large Reynolds numbers is derived from measurements of time averages of much more complicated unsteady flows. This leads to empirical approximations such as, for $\mathrm{Re} < 2000$, $\phi \approx 32/\mathrm{Re}$ and, for $\mathrm{Re} > 3000$, that ϕ is approximately half the root of

$$\frac{1}{\sqrt{\Phi}} = 2\log_{10}(\mathrm{Re}\sqrt{\Phi}) - 0.8.$$

14 Poiseuille flow: exact solution. Consider the two-dimensional version of the flow of the previous exercise, in which a viscous liquid flows in the x-direction between two parallel plates at $y = \pm a$ under a pressure gradient P'. Assuming that

$$\frac{\partial}{\partial x}(\text{everything except } p) = 0, \qquad \frac{\partial p}{\partial x} = P',$$

show that there is an exact steady solution of the Navier–Stokes equations in which the velocity is $(u(y), 0)$, where

$$\mu \frac{\partial^2 u}{\partial y^2} = P'.$$

Applying the no-slip condition at $y = \pm a$, find $u(y)$ and hence the flux per unit length in the x-direction. Repeat the calculation for a circular pipe, using cylindrical polar coordinates with the z-axis along the pipe. Note that when the flow is radially symmetric with velocity $(0, 0, w(r))$ the Navier–Stokes equations reduce to

$$\frac{\partial^2 w}{\partial r^2} + \frac{1}{r}\frac{\partial w}{\partial r} = \frac{P'}{\mu}.$$

'It's constant for all time.'
'What, the same constant?'

4

Case studies: hair modelling
and cable laying

In the next three chapters, we look at three 'real-world' problems, which
all arose in industry. One reason for presenting these case studies is
simply to give some examples of modelling in action (the only way to
get good at it is to do it). Another is to illustrate the techniques of the
previous chapter in a less academic setting. Finally we shall use these
case studies, and others presented later in the book, to illustrate the
techniques we develop later, although we do not have room to give full
details of all that has been done on these problems, much of which is,
ultimately, numerical. References to the literature are given at the end of
the chapter.

You can skip these chapters and still read much of this book. Although you won't have wasted your money entirely, you will miss out
on some nice applications of the methodology we describe later.

Both the models in this chapter are based on the Euler–Bernoulli
beam equation for the bending of a slender elastic beam. This is such
a common model (we have seen it already in the context of flagpoles)
that it merits its own section, following which I have included a short
section on the topical problem of the modelling of hair. Then we turn to
the problem of building a model for cable laying.

4.1 The Euler–Bernoulli model for a beam

We wrote down the Euler–Bernoulli model for the displacement of a
slender nearly straight beam in subsection 3.1.3. In fact there is no requirement for the beam to be straight, but it must be slender for a crucial
assumption in the following model to hold. Let us therefore consider a

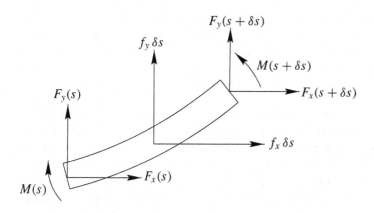

Figure 4.1 Forces and moments on an element of a beam.

beam, or slender elastic rod, lying along the curve $\mathbf{r} = (x(s), y(s))$ in the plane (the equations are much more complicated in three dimensions). Here s is arclength, and if we let $\theta(s)$ be the angle between the curve and the x-axis we have

$$\frac{dx}{ds} = \cos\theta, \qquad \frac{dy}{ds} = \sin\theta, \qquad \frac{d\theta}{ds} = \kappa,$$

where κ is the curvature.

Now look at Figure 4.1, which shows a small element of the beam, of length δs. The forces acting on the boundary of the element are simply the internal elastic forces acting on its ends (there are no such forces on its curved surface), and there is also a body force with components f_x, f_y per unit length. We write the elastic forces at the ends as $\left(F_x(s), F_y(s)\right)$ and $\left(F_x(s + \delta s), F_y(s + \delta s)\right)$ respectively. In equilibrium the difference between these must cancel the body force $(f_x, f_y)\delta s$, and taking $\delta s \to 0$ we find the force balance equations

> If the beam is nearly straight and lies along the x-axis, you can think of F_x as a tension or compression.

$$\frac{dF_x}{ds} + f_x = 0, \qquad \frac{dF_y}{ds} + f_y = 0.$$

Now, unlike a string, a beam resists being bent by generating an internal bending moment $M(s)$ to balance the moment of the internal forces. This leads to a net moment on the element of $M(s + \delta s) - M(s)$. Balancing this by the result of resolving the internal forces normally to the beam and taking moments about the left-hand end of our element, we find the equation

$$\frac{dM}{ds} - F_x \sin\theta + F_y \cos\theta = 0.$$

Let us pause for a moment and count equations and unknowns. The unknowns are θ (from which we can find x and y by integration), F_x, F_y and M, and we have three equations. However, we haven't yet said anything about the material from which the beam is made: we need a constitutive relation to tell us something about how the forces and displacements are related. For a beam that started off straight and is bent into a curve, a good model is

$$M = b\frac{d\theta}{ds},$$

that is, the bending moment is proportional to the curvature (stop and think why this is reasonable). A systematic derivation of this condition starting from the equations of linear elasticity is surprisingly difficult, but it can be done by the methods of Chapter 20. At any rate, it can be shown that the constant of proportionality b, known as the bending stiffness, is equal to EAk^2, where, as before, E is the Young's modulus, A the cross-sectional area of the beam, and k the radius of gyration of that cross-section. The missing equation is thus

$$\frac{dM}{ds} = EAk^2\frac{d\theta}{ds}.$$

It is easy to eliminate M, and we find the system

$$\frac{dF_x}{ds} + f_x = 0, \qquad \frac{dF_y}{ds} + f_y = 0,$$

$$EAk^2\frac{d^2\theta}{ds^2} - F_x\sin\theta + F_y\cos\theta = 0 \qquad (4.1)$$

for F_x, F_y and θ. It is then straightforward to show that when the beam is straight and nearly along the x-axis, so that $\theta \approx dy/dx$, we recover the system

$$EAk^2\frac{d^4y}{dx^4} - \frac{d}{dx}\left(F_x\frac{d\theta}{dx}\right) - f_y = 0, \qquad \frac{dF_x}{dx} + f_x = 0,$$

which is a generalisation of the flagpole equation to include body forces in both directions.

4.2 Hair modelling

One of the fastest growing customers for mathematical modelling is the entertainment industry. The main drivers are the demand for realistic real-time simulation in computer games and the trend towards photo-realistic animated characters. Long hair and clothes are notoriously difficult to model; for example in the 2001 film *Final Fantasy*, about 20% of the production time was devoted to the 60 000 strands of lead character Aki's hair.[1] In this short section we look at a very simple model for hair, in

---

[1] Water, with its longer mathematical pedigree, has been more successfully treated, a famous example being the ocean in *Titanic*, much of which was computer generated. It

which each strand is treated individually and does not interact with its neighbours. This is only one of several possible models for hair, and at the time of writing this is a wide-open research area.

The idea is to treat a hair as an elastic rod of cross-sectional area A and density ρ, under gravity. Thus we simply use the model of the previous section, with gravity providing the body force,

$$\frac{\mathrm{d}F_x}{\mathrm{d}s} = 0, \qquad \frac{\mathrm{d}F_y}{\mathrm{d}s} - \rho g A = 0, \qquad b\frac{\mathrm{d}^2\theta}{\mathrm{d}s^2} - F_x \sin\theta + F_y \cos\theta = 0,$$

and with the constitutive relation $M = b\,\mathrm{d}\theta/\mathrm{d}s$. Now the hair has a free end, at which $F_x = F_y = 0$, so measuring s from there we can easily find F_x and F_y, leaving the equation

$$b\frac{\mathrm{d}^2\theta}{\mathrm{d}s^2} + \rho g A s \cos\theta = 0$$

for θ. Appropriate boundary conditions are quite easy in this case, as we can expect to prescribe θ where the hair enters the head (say normally), and we'll set $\mathrm{d}\theta/\mathrm{d}s = 0$ at $s = 0$ because that end of the hair is free. This is a two-point boundary value problem that is relatively straightforward to solve numerically using any of a variety of packages, although because this nonlinear system may have bifurcations the software must be able to handle these. Solutions of this equation do indeed do more or less what real hair does, although the neglect of hair–hair interactions is a serious defect of the model. See the end-of-chapter exercises for further properties of this problem.

4.3 Undersea cable laying

Cables and pipelines have been laid under the sea since the early electric telegraphs; nowadays they often hold optical fibres. Several factors compete in the design of cables; for example, strength and durability dictate thick cables, while expense and speed of laying dictate thin ones. The process of laying is a dangerous time in the life of a cable, and very precise control of the operation is necessary to avoid damage while maximising the laying speed. In this case study, which recurs in Chapter 16, we look at a model of the process of laying a cable from a ship, as shown in Figure 4.2. We consider a steady-state model, as a first step towards developing a dynamic model to enable real-time control of the operation.

---

is said that a mathematician pointed out that the algorithm for waves did not conserve mass, and received the Hollywood mogul's reply, 'I don't give a flying fish (actually, he used another word) if it loses mass, so long as it looks good'. Ho hum.

Figure 4.2 Cable laying from a ship.

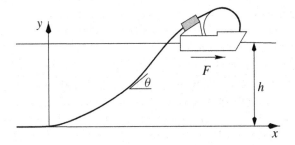

As the ship moves forward the cable is unreeled from a large drum and passes through a 'tensioner', shown as a box above the stern of the ship. This has the effect of prescribing the angle at which the cable leaves the ship. The ship exerts a force F on the cable, which also experiences buoyancy forces as it sinks to the sea bed. Our objective is to set up a model that allows us to calculate the shape of the cable (where does it feel the greatest stresses?) and the thrust needed from the ship.

4.4 Modelling and analysis

Taking an origin at the point where the cable touches the sea bed, a distance h below the surface, we denote its position by $(x(s), y(s))$ for $0 < s < L$, where the wetted length L is as yet unknown. (For simplicity we are going to ignore the small length of cable in the air astern of the ship.) The angle between the cable and the horizontal is θ, as before, and the unit tangent and normal to the curve are $\mathbf{t} = (\cos\theta, \sin\theta)$ and $\mathbf{n} = (-\sin\theta, \cos\theta)$ respectively.

We are going to solve the beam system (4.1), and the main difficulty is in writing down the external forces f_x and f_y. There are three external forces on the cable: one is its weight, a second is buoyancy, and one third is drag as it moves through the water. The weight of the cable is easy, just contributing a term $-\rho_c g A$ to the equation for F_y, where ρ_c is the density of the cable. We will focus on the buoyancy (the drag is dealt with in an exercise). Having completed the model, we then need to decide what boundary conditions to apply at $s = 0$ and $s = L$. Almost inevitably, the solution of the resulting two-point boundary value problem for a system of ordinary differential equations will be carried out numerically, again using a two-point boundary value problem solver, although in Chapter 16 we also look at an approximation for cables with low bending stiffness.

The buoyancy force (B_x, B_y) per unit length on the cable is entirely due to hydrostatic pressure, and rather surprisingly it is not just the usual 'weight of water displaced', but instead has two constituents. They both

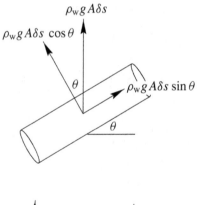

$\rho_{\text{w}}gA\delta s$

$\rho_{\text{w}}gA\delta s \, \cos\theta$

$\rho_{\text{w}}gA\delta s \, \sin\theta$

θ

θ

Figure 4.3 Hydrostatic force on a cylinder.

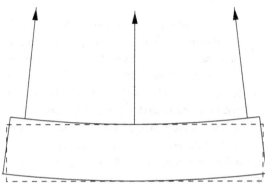

Figure 4.4 Normals on a slightly bent cylinder.

stem from the ordinary Archimedes force, but we have to be careful in evaluating it. Consider a cylindrical element as in Figure 4.3. If the ends of the cylinder were exposed to the water, the buoyancy force would be equal to the weight of the water displaced, namely $\rho_{\text{w}}gA$ per unit length, acting vertically upwards (see Exercise 4 to prove this). Remembering that the pressure acts normally to a surface, and resolving along and normal to the cylinder, what remains after we have subtracted the contribution from the ends is a force per unit length along the normal of equal to $\rho_{\text{w}}gA \cos\theta$.

However, our cylinder is not quite straight. As can be seen in Figure 4.4, the surface area on the 'outside' of a bend is bigger than that on the 'inside', and so if a pressure p, here the hydrostatic pressure $\rho_{\text{w}}g(h - y)$, acts on a curved cylinder like this (but not on its endpoints), there is a net force in the normal direction. It is fairly clear that this extra force is proportional to the curvature – the area surplus or deficit is proportional to the rate of change of θ with s – and it can be shown (see Exercise 5) that the magnitude of this contribution to the buoyancy force is $pA\kappa$ per unit length, so that the total buoyancy force is $\rho_{\text{w}}gA \cos\theta + pA\kappa$ along **n**.

Old chestnut, claims new victims every year: you are in a boat on a lake, and you throw a brick over the side. Does the water level rise, fall or stay the same?

Substituting B_x and B_y into the beam model, we have finally

$$\frac{\mathrm{d}F_x}{\mathrm{d}s} + B_x = 0, \qquad \frac{\mathrm{d}F_y}{\mathrm{d}s} + B_y - \rho_{\mathrm{c}}gA = 0,$$

$$E A k^2 \frac{\mathrm{d}^2\theta}{\mathrm{d}s^2} - F_x \sin\theta + F_y \cos\theta = 0, \qquad (4.2)$$

where

$$(B_x, B_y) = \left(\rho_{\mathrm{w}} g A \cos\theta + p A \frac{\mathrm{d}\theta}{\mathrm{d}s} \right) (-\sin\theta, \cos\theta). \qquad (4.3)$$

4.4.1 Boundary conditions

We need four boundary conditions for this system in order to fix F_x, F_y and θ (there are two first-order equations and one second-order equation). Having done this, we will integrate $\mathrm{d}x/\mathrm{d}s = \cos\theta$, $\mathrm{d}y/\mathrm{d}s = \sin\theta$ from $s = 0$ to $s = L$, with the initial conditions $x(0) = y(0) = 0$. Then the condition $y(L) = h$ will tell us L and $x(L)$ will tell us the horizontal distance between touchdown of the cable and the ship. Of course, these integrations must be carried out numerically, and on the face of it the equations are even more nonlinear than the hair model; but read on.

Let us first think about the conditions at $s = 0$. We expect the cable to leave the sea bed smoothly, so we impose

$$\theta = 0, \qquad \frac{\mathrm{d}\theta}{\mathrm{d}s} = 0 \qquad (4.4)$$

at $s = 0$. The second of these conditions says that the bending moment is continuous at this point. As we shall see in subsection 9.5.2, only a point force could cause a discontinuity in M.

Nothing else obvious can be applied at $s = 0$, so let us look at $s = L$. Here we know the angle at which the cable leaves the ship, and we know the horizontal force F_x:

In the language of beam equations, we are imposing 'clamped' boundary conditions.

$$\theta = \theta^*, \qquad F_x = F \qquad (4.5)$$

at $s = L$. These complete the boundary conditions.

4.4.2 Effective forces and nondimensionalisation

Before we scale equations (4.2)–(4.5), we note a potentially serious difficulty and a neat extrication from it. Because p is hydrostatic, we have $p = \rho_{\mathrm{w}}g(h - y)$. But this means that the p in (4.3) depends on y, and our scheme of solving for y only *after* finding θ looks doomed: it seems that the system is fully coupled. However, there is a *deus ex machina*. We first note that $\mathrm{d}p/\mathrm{d}s = -\rho_{\mathrm{w}}g\,\mathrm{d}y/\mathrm{d}s = -\rho_{\mathrm{w}}g\sin\theta$. Then, we define *effective horizontal and vertical forces* by

$$F_x^e = F_x + pA\cos\theta, \qquad F_y^e = F_y + pA\sin\theta,$$

so that

$$\frac{\mathrm{d}F_x^e}{\mathrm{d}s} = \frac{\mathrm{d}F_x}{\mathrm{d}s} - pA\sin\theta\,\frac{\mathrm{d}\theta}{\mathrm{d}s} + A\cos\theta\,\frac{\mathrm{d}\theta}{\mathrm{d}s},$$

and similarly for $\mathrm{d}F_y^e/\mathrm{d}s$. When we substitute in (4.3) all the terms involving p vanish and, as $F_x^e = F_x$ at $y = 0$ where $p = 0$, there is no problem in applying the boundary condition (4.5). The variables F_x and F_y are only steps on the way to θ, which is the quantity that we really need; so there is no loss in our not calculating them.

Carrying out this simplification, we arrive at the system

$$\frac{\mathrm{d}F_x^e}{\mathrm{d}s} = 0, \qquad \frac{\mathrm{d}F_y}{\mathrm{d}s} = \rho_c g A,$$

$$EAk^2\frac{\mathrm{d}^2\theta}{\mathrm{d}s^2} - F_x^e\sin\theta + F_y^e\cos\theta = 0,$$

with the boundary conditions

$$\theta = 0, \qquad \frac{\mathrm{d}\theta}{\mathrm{d}s} = 0$$

at $s = 0$, and

$$\theta = \theta^*, \qquad F_x^e = F$$

at $s = L$.

We should clearly scale x, y and s with h, and we choose to scale F_x^e, F_y^e with $\rho_c g A L$. Immediately dropping the primes, we have the dimensionless model

$$\frac{\mathrm{d}F_x^e}{\mathrm{d}s} = 0, \qquad \frac{\mathrm{d}F_y}{\mathrm{d}s} = 1, \qquad \epsilon\frac{\mathrm{d}^2\theta}{\mathrm{d}s^2} - F_x^e\sin\theta + F_y^e\cos\theta = 0,$$

$$(4.6)$$

Consistency: Check that the scale for F_x^e, F_y^e is indeed a force. It is probably slightly preferable to use this scale rather than F because F may be an unknown.

with the boundary conditions

$$\theta = 0, \qquad \frac{\mathrm{d}\theta}{\mathrm{d}s} = 0 \qquad (4.7)$$

at $s = 0$, and

$$\theta = \theta^*, \qquad F_x^e = F^* \qquad (4.8)$$

at $s = \lambda$. Here the three dimensionless parameters are

$$\epsilon = \frac{Ek^2}{\rho_c g h^3}, \qquad F^* = \frac{F}{\rho_c g A}, \qquad \lambda = L/h;$$

note that λ is unknown.

We can do a little better still: we can find F_x^e and F_y^e explicitly. Substituting into the equation for θ we have

$$\epsilon \frac{d^2\theta}{ds^2} - F^* \sin\theta + (F_0 + s)\cos\theta = 0, \qquad (4.9)$$

in which $F_0 = F_y^e(0)$ is an unknown constant. Although this is a second-order equation, there are *three* boundary conditions for this equation, namely the relevant parts of (4.7) and (4.8), and so we have an extra equation to tell us the unknown constant F_0.

We return to this problem in Chapter 16, where we show how to construct an approximate solution when ϵ, as its name suggests, is small; this is the case when the cable is heavy or the water is deep (it is clear that ϵ measures the relative importance of bending stiffness and cable weight). This kind of boundary value problem, where a small parameter multiplies the highest derivative, is often known as 'stiff' in a numerical context, and there are many specialised 'stiff solvers' to handle these problems.

Even though the 'beam' is anything but stiff!

4.5 Sources and further reading

The cable-laying problem was proposed by the British company BICC; it is a simplified version of more complicated three-dimensional 'up-winding' problems to do with the winding of wire onto a reel (the twist, or *torsion*, of the wire plays an important role in these situations).

4.6 Exercises

1 The Euler strut (i).   A thin rod of length L and bending stiffness b is clamped at each end and is compressed by forces F, as shown in Figure 4.5. Adapt the analysis of Section 4.1 to derive the dimensionless boundary value problem

$$\frac{d^2\theta}{ds^2} + \alpha^2 \sin\theta = 0, \qquad \theta(0) = \theta(1) = 0,$$

for the angle between the rod and the x-axis, where $\alpha^2 = FL^2/b$. Show that $\theta = 0$ is always a solution; what does it represent?

Now suppose that θ is small. Assuming that $\sin\theta \approx \theta$ (we will do this in more detail in Exercise 7 of Chapter 13), write down an approximate linear two-point boundary value problem for θ, and show

F

F

Figure 4.5 The Euler strut.

that its only solution is $\theta \equiv 0$ unless $\alpha = n\pi$ for integer n. Deduce that, as F is increased from zero, it is first possible to have a non-trivial solution ($\theta \neq 0$) when $FL^2/b = \pi^2$. Sketch the resulting solution. What happens when $FL^2/b = 4\pi^2$?

The appearance of a non-trivial solution as a parameter varies is known as a *bifurcation*. This one is easy to illustrate in practice with, say, a plastic ruler. On a larger scale, when putting up a modern tent with a carbon fibre pole you have to bend the pole to fit it into its sockets. As you do so by bringing the ends together, starting with a straight pole, you initially go through the first buckling mode $\alpha = \pi$. You may also see the second mode if the pole is long enough. (Do not try to put your tent up in a thunderstorm. If you were struck by lightning, why might it be more likely to hit the end of the pole than the middle, even if the latter is higher?)

Buckling can also occur when the pole is held vertically, so that gravity supplies the compression, as the next example shows.

2 Groan. Take the hair model

$$b\frac{d^2\theta}{ds^2} + \rho gs \cos\theta = 0$$

with the boundary conditions $\theta = \theta_0$ (given) at $s = 0$ and $d\theta/ds = 0$ at $s = L$; explain what these conditions model. Look for a solution for a nearly vertical hair. That is, write $\theta = \pm\pi/2 + \phi$ and derive two versions (related by $\xi \leftrightarrow -\xi$) of the *Airy equation* That's 'orrible.

$$\frac{d^2\phi}{d\xi^2} \pm \xi\phi = 0, \qquad 0 < \xi < \xi_0 = L\sqrt{\frac{\rho g}{b}},$$

where ξ is a suitably scaled version of s. Which of $\pm$ is for upward-pointing hair and which for downward-pointing?

Taking the minus sign for the conventional definition, Airy's equation has the standard linearly independent solutions $\mathrm{Ai}(\xi)$, $\mathrm{Bi}(\xi)$, which for $\xi < 0$ both oscillate with an amplitude that decays as $|\xi|^{-1/4}$ (in Chapter 23 we see why), while for $\xi > 0$ one solution grows rapidly and one decays, as seen in Figure 4.6. What sort of solutions do you Compare $y'' + \lambda y = 0$ for $\lambda > 0$, expect to see for (a) upward-pointing (b) downward-pointing hair? $\lambda < 0$.

Show that, as ξ_0 varies, an upward-pointing hair can buckle via a bifurcation away from the vertical solution, and find the shortest length at which it does so in terms of Ai and Bi. Using the fact that, for $x > 0$, $\mathrm{Ai}(x)$ is decreasing and $\mathrm{Bi}(x)$ is increasing, show that downward-pointing hairs cannot buckle away from the vertical solution.

3 Waving hair and unsteady beams. Consider an unsteady version of the Euler–Bernoulli beam model, in which the beam is parametrised

Figure 4.6 The Airy functions. $\mathrm{Ai}(0) = 3^{-2/3}/\Gamma(\frac{2}{3})$, $\mathrm{Bi}(0) = 3^{-1/6}/\Gamma(\frac{2}{3})$.

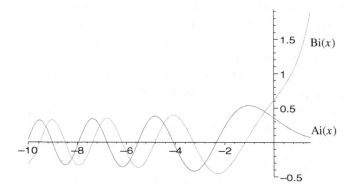

by $(x(s, t), y(s, t))$. Justify the model

$$\frac{\partial F_x}{\partial s} + f_x = \rho A \frac{\partial^2 x}{\partial t^2}, \qquad \frac{\partial F_y}{\partial s} + f_y = \rho A \frac{\partial^2 y}{\partial t^2},$$

$$\frac{\partial M}{\partial s} - F_x \sin\theta + F_y \cos\theta = 0,$$

provided that the rate of change of angular momentum of the element can be ignored. Show that for a straight beam under constant tension the equation of motion for small displacements is

$$\rho A \frac{\partial^2 y}{\partial t^2} + b \frac{\partial^4 y}{\partial x^4} - T \frac{\partial^2 y}{\partial x^2} = 0,$$

where T is the tension (that is, F_x) and gravity has been ignored.

4 **Eureka!**   Dot both sides with a constant vector and use the divergence theorem to show that the identity

$$\int_{\partial V} \Phi \mathbf{n}\, \mathrm{d}S = \int_V \nabla \Phi\, \mathrm{d}V$$

hold for sufficiently regular scalar functions Φ and volumes V. Put

$$\Phi = \rho_w g(\text{constant} - y)$$

to derive Archimedes' principle: the buoyant force on a body immersed in water of density ρ_w is equal to the weight of the water displaced. Well worth jumping out of the bath for.

Inspect the first term on the right-hand side of formula (4.3), which gives the buoyant force on an element of a submerged cable. What happens if $\theta = \pi/2$? Do you believe this? Would a cylindrical stick held vertically in water rise when let go? Resolve the apparent paradox.

Calculate the total moment of the pressure forces on the body by integrating $\mathbf{r} \wedge (-p\mathbf{n})$ over its surface, and deduce that the couple they exert is the same as if the buoyancy force acted through the centre of mass of the body (assuming it to have uniform density). You will need to prove the vector identity

$$\iint_{\partial V} \phi \mathbf{x} \wedge \mathbf{n} \, dS = \iiint_V \mathbf{x} \wedge \nabla \phi \, dV$$

for sufficiently smooth vectors $\mathbf{x}$ and scalars ϕ.

5 **Hydrostatic force on a bent cylinder.** Consider the slightly bent cylinder of Figure 4.4, and suppose that the value of the arclength, measured along the centreline, is s_0 at the left-hand end and $s_0 + \delta s$ at the right-hand end. Suppose that the centreline has position $\mathbf{r}_0(s) = (x(s), y(s), 0)$ and that the cylinder is circular in each plane normal to this line, with radius ϵ. Suppose furthermore that a constant pressure p acts normally to the curved surface of the cylinder (but not the ends).

The correction due to hydrostatic variation in pressure over the element vanishes when $\delta s \to 0$.

Show that the vector tangent to the centreline is

$$\mathbf{t}(s) = (\cos \theta(s), \sin \theta(s), 0)$$

and that the normal (in the xy-plane) is

$$\mathbf{n} = (-\sin \theta, \cos \theta, 0).$$

Show that $\mathbf{t}$, $\mathbf{n}$ and $\mathbf{b} = \mathbf{t} \wedge \mathbf{n} = (0, 0, 1)$ are orthonormal. Show that a point on the surface can be written in the form

This will be familiar if you have met the Serret–Frenet formulae from differential geometry.

$$\mathbf{r} = \mathbf{r}_0(s) + \epsilon \cos \phi \, \mathbf{n}(s) + \epsilon \sin \phi \, \mathbf{b}(s),$$

for $s_0 < s < s_0 + \delta s, 0 \le \phi < 2\pi$. Explain why

$$\frac{\partial \mathbf{r}}{\partial \phi} \wedge \frac{\partial \mathbf{r}}{\partial s}$$

is normal to the surface and why the unit normal to the surface and the surface area element are

$$\mathbf{N} = \mathbf{n} \cos \phi + \mathbf{b} \sin \phi,$$

$$dS = \left| \frac{\partial \mathbf{r}}{\partial \phi} \wedge \frac{\partial \mathbf{r}}{\partial s} \right| d\phi \, ds = \epsilon(1 - \epsilon \kappa \cos \phi) \, d\phi \, ds,$$

where $\kappa = d\theta/ds$ is the curvature of the centreline. Deduce that

$$\int_S p\mathbf{N} \, dS = \pi \epsilon^2 \mathbf{n} \, \delta s,$$

and hence confirm formula (4.3).

6 Water drag on a cylinder.   If a cylinder of radius a, with axis in the xy-plane and lying along a line making an angle θ with the x-axis, is placed in a nearly inviscid fluid moving with far-field velocity $(U, 0, 0)$, explain why it is reasonable that the drag per unit length on it is $\rho_w U^2 a C_d (\cos\theta, \sin\theta, 0)$ (see Chapter 3). Incorporate this force into the cable-laying model when the ship moves forward with speed U, and identify the new dimensionless parameter that tells you the relative importance of drag and buoyancy.

'If 10 corresponds to 23, then 23 is half of 45.'

5
Case study: the thermistor (1)

5.1 Heat and current flow in thermistors

A thermistor is a temperature-dependent resistor. A typical thermistor is a penny-shaped piece of a special ceramic material, about 1 mm thick and with a radius of 5 mm, having metal contacts on the flat faces. The kind in which we are interested becomes more resistive as it gets hotter, so it can be used as a fuse: if the current through the thermistor surges for any reason, the resulting ohmic ($I^2 R$) heating increases the resistance and so cuts the current. The beauty of this is that when the current surge goes away the thermistor just cools down, and normal operation can resume without anybody having to replace a fuse. Televisions have dozens of thermistors in them as a protection against overheating, and so do hairdryers, which is why they switch themselves off for a while if they get too hot.

There are various reasons for wanting to analyse the heat and current flow in thermistors. One is the obvious question of design: how do the characteristics, such as the switch-off time in response to a current surge, depend on the physical parameters? Another is quality control: some thermistors can crack because the rapid thermal expansion caused by large temperature gradients stresses the material too much. The full analysis of cracking requires a model for thermoelasticity, the discussion of which is beyond the scope of this book; however, even an order of magnitude estimate of the temperature gradients could be used as an input to an 'engineering' rule of thumb for the likelihood of cracking.

5.1.1 A black-box model

We can start by treating a thermistor as a single unit, all of which is at the same temperature $T(t)$, which depends on time alone. That is, we entirely neglect the details of any heat flow within the component. We can then assign a temperature-dependent resistance $R(T)$ to the thermistor; for the thermistors we consider, $R(T)$ increases with T.

Suppose that we apply a constant voltage V_0 across the device. By Ohm's law the current $I(t)$ is $V_0/R(t)$, and the heat generated by this current is $V_0^2/R(t)$. Let us also suppose that heat loss to the environment can be represented by an increasing function $H(T)$ that vanishes when T is equal to the ambient temperature T_a (for Newton cooling, $H(T)$ is proportional to $T - T_a$). Then the overall energy balance for a thermistor of mass m and specific heat capacity c is

$$mc\frac{\mathrm{d}T}{\mathrm{d}t} = \frac{V_0^2}{R(T)} - H(T),$$

an ordinary differential equation that can easily be solved, either numerically or by separating the variables if the forms of $R(T)$ and $H(T)$ allow us to evaluate the resulting integrals. Moreover, a simple argument in which the two functions on the right-hand side are plotted on the same graph shows that there is always a steady state and that it is unique.

This is all right as far as it goes, but it tells us nothing about the temperature distribution within the thermistor. Moreover, it misses the important fact that for some resistivity–temperature laws there may be no solution to a model in which the temperature varies spatially, even though the 'black box' model above always has a unique solution.

5.1.2 A simple model for the heat flow

We now discuss a simple model for a thermistor on its own, with a voltage V_0 applied across it at $t = 0$; then we will extend this to a thermistor in a simple circuit. We need first to think about how electric current flows through a solid. That is, we need a generalisation of Ohm's law $I = V/R$ for a resistor. This is straightforward. We assume that there is a local version of Ohm's law relating the current density $\mathbf{j}$ (units A m^{-2}) to the electric field E (V m^{-1}) linearly:

$$\mathbf{j} = \sigma(T)\mathbf{E},$$

You may be more familiar with the *resistivity* $\rho(T) = 1/\sigma(T)$. What are its units?

where $\sigma(T)$ is a material property called the *conductivity*, whose dependence on temperature T, which is intrinsic to the proper working of the device, is shown explicitly. Now remember that there is an electric potential ϕ, such that $\mathbf{E} = -\nabla\phi$, and that current is conserved: $\nabla \cdot \mathbf{j} = 0$.

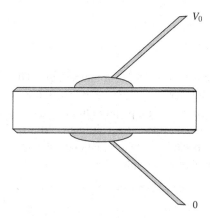

Figure 5.1 A thermistor: the shaded regions are good (metallic) conductors.

Putting these together, we have

$$\mathbf{j} = -\sigma(T)\nabla\phi, \qquad \nabla \cdot (\sigma(T)\nabla\phi)) = 0.$$

Also, we can easily write down boundary conditions for the potential. If, as in Figure 5.1, the top and bottom of the thermistor are coated with an excellent conductor then the potential is very nearly constant on each, while there is no current through the sides. Thus, for $t > 0$, we have

$$\phi = V_0 \quad \text{on} \quad z = H, \qquad \phi = 0 \quad \text{on} \quad z = 0$$

and

$$\frac{\partial \phi}{\partial n} = 0 \quad \text{on} \quad r = a,$$

where r, θ, z are cylindrical polar coordinates with origin at the centre of the bottom face; H is the thickness and a is the radius of the cylinder.

Now we need to write down a model for heat generation and conduction in the thermistor. That means we have to find a local version of the law for the power generated in a resistor, $VI = I^2 R = V^2/R$. In a bulk flow, the local rate of heat production (volumetric heating) is

$$\mathbf{j} \cdot \mathbf{E} = \sigma |\nabla\phi|^2.$$

See the discussion of energy and work on p. 19. You should check that the units are correct at W m^{-3}.

This appears as a source term in the heat equation for the temperature $T(\mathbf{x}, t)$,

$$\rho c \frac{\partial T}{\partial t} = k\nabla^2 T + \sigma |\nabla\phi|^2.$$

Consistency: the heating term is positive so it acts to make T increase in time.

Boundary conditions for the heat equation are often problematical. The isothermal or perfectly insulated conditions beloved of exam question setters are rarely strictly applicable in practice. It is safer to write down the more general Newton cooling law (see Exercise 5 at the end of Chapter 3),

$$-k\frac{\partial T}{\partial n} = h(T - T_a),$$

for the sides of the thermistor, where T_a is the ambient temperature and h the heat transfer coefficient. Taking some liberties, we may then hope to model the cooling effect of the conducting top and bottom surfaces, together with the connecting wire and its solder, by a similar condition but with a larger value of h. Finally, because the heat equation is forward parabolic we need an initial condition, for example

$$T(\mathbf{x}, 0) = T_a(\mathbf{x}).$$

5.2 Nondimensionalisation

This problem is not too hard to nondimensionalise. In the first instance, let us scale r and z with the thickness H and time with the heat conduction scale H^2/κ, where $\kappa = \rho c/k$ is the thermal diffusivity and ρ, c and k are the density, specific heat capacity and thermal conductivity respectively. Let us now think about the temperature scale. The conductivity must change noticeably as the temperature varies, or the device would be pointless, and we should be able to identify a temperature change ΔT over which it does so. Let us therefore use this as the temperature scale, writing $T - T_a = \Delta T \, u(\mathbf{x}', t')$. Lastly we use the external voltage V_0 as the scale for ϕ and the 'cold' value of the conductivity, σ_0 as the scale for $\sigma(T)$. Notice that we have to scale known functions of T as well as T itself.

Scaling and immediately dropping the primes, we have the dimensionless equations

$$\nabla \cdot (\sigma \, \nabla \phi) = 0, \qquad \frac{\partial u}{\partial t} - \nabla^2 u = \gamma \sigma(u)|\nabla \phi|^2$$

for $0 < z < 1, 0 \leq r < \alpha = a/H$. The boundary conditions are

$$\phi = 0, 1 \quad \text{on} \quad z = 0, 1 \quad \text{respectively}, \qquad \frac{\partial \phi}{\partial r} = 0 \quad \text{on} \quad r = \alpha$$

and

$$\frac{\partial u}{\partial n} + \beta(\mathbf{x})u = 0 \quad \text{on the boundary},$$

where the $\mathbf{x}$-dependence of β models the difference between the top or bottom and the side, β taking different values in the two cases.

There are now just three dimensionless parameters:

$$\alpha = \frac{a}{H}, \qquad \beta = \frac{hH}{k} \qquad \text{and} \qquad \gamma = \frac{\sigma_0 V_0^2}{k \Delta T}.$$

Of these, α measures the aspect ratio, β the heat transfer and γ the competition between heat generation and conduction. When we put in typical physical values, namely

$$\rho = 5.6 \times 10^3 \text{ kg m}^{-3}, \qquad c = 540 \text{ J kg}^{-1} \text{ K}^{-1},$$
$$k = 2 \text{ W m}^{-1} \text{ K}^{-1}, \qquad \sigma_0 = 2 \, \Omega^{-1} \text{ m}^{-1}, \qquad \Delta T = 100 \text{ K},$$
$$V_0 = 250 \text{ V}, \qquad r = 5 \times 10^{-3} \text{ m}, \qquad H = 10^{-3} \text{ m},$$
$$h = 10 \text{ (sides)}, 10^2 \text{ (top) W m}^{-2} \text{ K}^{-1},$$

we find that

$$\alpha = 5, \qquad \beta = 10^{-2} \text{ (sides)}, 10^{-1} \text{ (top)}, \qquad \gamma = 625.$$

Already we have learned a lot. We know that there are just three dimensionless parameters, and that two are large and one is small. The large value of the aspect ratio α suggests that a one-dimensional model should perform well, and this notion is reinforced by the fact that β is especially small at the sides of the device: most of the heat generated will be lost through the top and bottom. The fact that γ is very large suggests that we may have chosen the wrong timescale, at least for the initial heating-up stage. However, the device does work, so the decrease in conductivity as the temperature increases must eventually switch off the heating term in the temperature equation, large though it appears to be. If we rescale time by writing $t = \gamma^{-1}\tau$, we find that

$$\frac{\partial u}{\partial \tau} = \sigma(u)|\nabla \phi|^2 + \frac{1}{\gamma}\nabla^2 u,$$

and with luck we can neglect the last term to simplify the problem considerably. However, it is not likely that we can explain spatial variations in the temperature without the last term, so there must be more to it than this. The full story is outlined in Chapter 23.

5.3 A thermistor in a circuit

In practice, our thermistor is likely to be part of a circuit, as shown in Figure 5.2, where the rest of the circuit is represented by a resistor of resistance R_0. This introduces some minor complications, as we no longer know the voltage drop across the thermistor but instead just have a relationship between this voltage and the current through the device. The model inside the thermistor is much as before, and there is no need to repeat the equations for T and ϕ. At the top and bottom of the thermistor, though, we have

$$\phi = 0, \quad \text{on} \quad z = 0, \qquad \phi = V(t), \quad \text{on} \quad z = H,$$

Figure 5.2 A thermistor in a circuit. The switch is closed at $t = 0$.

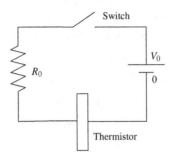

Figure 5.2 A thermistor in a circuit. The switch is closed at $t = 0$.

where $V(t)$ is not yet known. We can use Ohm's law for the resistor, however, to say that the voltage drop across it is $I(t)R_0$, where $I(t)$ is the current in the circuit, and then we have

$$V_0 = I(t)R_0 + V(t)$$

by Kirchhoff's second law, which says that the voltages round a closed circuit sum to zero.[1] We also have an expression for $I(t)$, as it is equal to the current flowing through the thermistor, namely

Exercise: Show from the equations that this is the same as the current density integrated over the top face.

$$\iint_{z=H} \sigma(T)\,\nabla\phi \cdot \mathbf{n}\,dS,$$

which is just the current density integrated over the bottom face. Thus, in this case, the boundary condition for ϕ on $z = H$ is

$$\phi = V_0 - 2\pi R_0 \int_0^a \sigma(T)\frac{\partial\phi}{\partial z}\bigg|_{z=H} r\,dr.$$

The effect of the external resistance in the dimensionless model is to bring in another parameter, from the boundary condition for ϕ. It is left as an exercise to show that, with the same scales as above, the dimensionless model is

$$\nabla \cdot (\sigma \nabla \phi) = 0, \qquad \frac{\partial u}{\partial t} - \nabla^2 u = \gamma\sigma(u)|\nabla\phi|^2$$

for $0 < z < 1, 0 \le r < \alpha = a/H$. The boundary conditions are

$$\phi = 0 \quad \text{on} \quad z = 0, \qquad \frac{\partial\phi}{\partial r} = 0 \quad \text{on} \quad r = \alpha,$$

and

$$\frac{\partial u}{\partial n} + \beta(\mathbf{x})u = 0$$

Can you see why α^2 has been separated out, and why we've put the factor 2 in?

as before, with the new condition

$$\phi = 1 - \frac{2}{\delta\alpha^2} \int_0^\alpha \sigma(u)\frac{\partial\phi}{\partial z}\bigg|_{z=1} r\,dr,$$

where the factor 2 has been included for later convenience, and the new

[1] See Exercise 2.

dimensionless parameter is

$$\delta = \frac{2}{\pi H R_0 \sigma_0 \alpha^2}.$$

A typical value for this parameter, given $R_0 = 400\ \Omega$, is 10^{-1}, which is quite small; note that formally setting $\delta = \infty$ we retrieve the case where there is no external resistance.

5.3.1 The one-dimensional model

As we saw earlier, the large value of α and the small value of β on the sides of the thermistor suggest that a one-dimensional model should be a good approximation (we will have to wait until later in the book to see how to justify this). In such a model ϕ and u are independent of r, and so we have the simpler problem

$$\frac{\partial}{\partial z}\left(\sigma \frac{\partial \phi}{\partial z}\right) = 0, \qquad \frac{\partial u}{\partial t} - \frac{\partial^2 u}{\partial z^2} = \gamma \sigma(u) \left|\frac{\partial \phi}{\partial z}\right|^2,$$

for $0 < z < 1$, while the boundary conditions are

$$\phi = 0 \quad \text{on} \quad z = 0, \qquad \phi = 1 - \frac{1}{\delta} \sigma(u) \frac{\partial \phi}{\partial z}\bigg|_{z=1}$$

and

$$\frac{\partial u}{\partial n} + \beta u = 0 \quad \text{on} \quad z = 0, 1.$$

Notice that we can integrate the equation for $\phi(z, t)$ once: can you see the physical interpretation of the result?

Some rather mathematical properties of this model are developed in the exercises at the end of the chapter, and we return to this case study in Chapter 17. In the meantime, we move on to another case study, also with an electrical flavour.

5.4 Sources and further reading

The thermistor problem was brought to the Oxford Study Groups with Industry by the British company STC, and it has provoked a large mathematical literature for which [20] is a starting point.

5.5 Exercises

1 Zero-dimensional thermistor in a circuit. Write down a black-box model for a thermistor in a circuit under the assumption that the temperature in the thermistor is a function of time only. Does the resulting ordinary differential equation always have a unique steady-state solution?

2 Thermistor in a circuit; validity of Kirchhoff's law. Strictly speaking, adding together voltages in our circuit is not valid because the changing current generates a magnetic field that in turn generates a 'back emf', a voltage that opposes the current change. Show that the back emf is small in our thermistor case study, as follows (you may want to refer back to the exercises at the end of Chapter 2). If the circuit has typical length L, show that it is reasonable that a typical magnetic field strength is $B_0 = \mu_0 I_0/(2\pi L)$ and that the current has size $I_0 = V_0/R_0$. If $\mathbf{E}$ is the electric field, the typical back-emf magnitude is L times the scale for $\mathbf{E}$. Show from Maxwell's equation

$$\nabla \wedge \mathbf{E} = \frac{\partial \mathbf{B}}{\partial t}$$

that the back-emf scale works out as $\mu_0 I_0 L/(2\pi t_0)$, where t_0 is the timescale for changes in $I(t)$. If, say, $L = 10$ cm, $V_0 = 250$ V and $R_0 = 500\ \Omega$, this is $10^{-8}/t_0$; verify that t_0 is a lot bigger than the timescale of $100/10^{-8}$ seconds necessary for the back emf to have magnitude 100 V. We can thus neglect $\partial \mathbf{B}/\partial t$, so that $\nabla \wedge \mathbf{E} \approx \mathbf{0}$; show that the change in voltage round the circuit due to $\partial \mathbf{B}/\partial t$ is $\oint \mathbf{E} \cdot d\mathbf{s} \approx 0$.

3 One-dimensional thermistors. Consider the steady-state version of the one-dimensional thermistor problem, with the (not very realistic) boundary conditions that $T = T_\text{a}$ on $z = 0, H$ and with no external resistance. Show that the dimensionless model is

$$\frac{\partial}{\partial z}\left(\sigma \frac{\partial \phi}{\partial z}\right) = 0, \qquad \frac{\partial^2 u}{\partial z^2} = -\gamma \sigma(u)\left(\frac{\partial \phi}{\partial z}\right)^2$$

for $0 < z < 1$, with

$$\phi = 0, 1, \quad u = 0 \quad \text{on} \quad z = 0, 1 \quad \text{respectively.}$$

Explain why $\phi = \frac{1}{2}$, $\partial u/\partial z = 0$ on $z = \frac{1}{2}$. Integrate the equation for ϕ once to show that

$$\sigma(u)\frac{\partial \phi}{\partial z} = I$$

where I is a constant (what is its physical interpretation?). Substitute for $\sigma \, \partial \phi/\partial z$ in the equation for u to show that

$$\frac{\partial u}{\partial z} = -\gamma \left(\phi - \tfrac{1}{2}\right) I,$$

and then substitute for $(\partial \phi/\partial z)^2$ in the same equation to show that

$$\tfrac{1}{2}\gamma \left(\phi - \tfrac{1}{2}\right)^2 = \int_0^{u_m} \frac{ds}{\sigma(s)},$$

where $u_m = u(\frac{1}{2})$ is the largest value of u, attained at $z = \frac{1}{2}$ (why?). Deduce that there can be a steady solution only if $\sigma(u)$ is such that

$$\int_0^\infty \frac{ds}{\sigma(s)} < \frac{1}{8}\gamma,$$

restate this inequality in dimensional terms and interpret it physically. Give an example of a function $\sigma(u)$ for which it does not hold. What do you think happens to the solution of an initial-value problem if the inequality does not hold? Is the existence of a steady state more or less likely with Newton cooling conditions for u?

4 **Thermistors and conformal mapping.** Consider the steady-state thermistor equations

$$\nabla \cdot (\sigma(u)\nabla\phi) = 0, \qquad \nabla^2 u = -\sigma(u)|\nabla\phi|^2,$$

in two space dimensions but not necessarily in a rectangle. Suppose that the boundary of the thermistor has two conducting segments, on which $u = 0$ and $\phi = 0, 1$ respectively, separated by two insulating segments on which $\partial u/\partial n = 0$, $\partial\phi/\partial n = 0$.

Show that this system remains invariant under conformal maps $\xi + i\eta = f(x + iy)$ for analytic (holomorphic) f. Given that it is possible to map the thermistor region onto a certain rectangle with an obvious correspondence of boundary parts, show that the restriction on existence derived in the previous question holds irrespective of the geometry. (Note: The Riemann mapping theorem guarantees that the thermistor can be mapped onto any rectangle we choose and that we can map *three* specified boundary points onto three specified points on the boundary of the rectangle. For example, we can map three of the 'changeover' points, where the boundary conditions switch, onto three corners of the rectangle. However, we can only map the fourth changeover point onto the fourth corner if the rectangle has a specific aspect ratio. For more on conformal mapping see [6, 13] or, for Matlab users, the Schwarz–Christoffel Toolbox of that package.)

> Use the chain rule and the Cauchy–Riemann equations.

'We need the decrease of a constant sphere ... [Muttering] ... no, when the sphere is constant we need it to be decreasing ... slowly, I mean.'

6
Case study: electrostatic painting

6.1 Electrostatic painting

Many paints are based on organic solvents which, after application, evaporate and contribute to air pollution and global warming, and so they are coming under increasing regulatory pressure. A more environmentally friendly alternative for painting a metal object (or workpiece) is to cover it with a layer of very small resin paint particles that, when the workpiece is put into an oven, melt and flow into into a smooth coating. A similar process is used in 'flocking': here an object is coated in glue and then covered with short lengths of charged fibre. The charge makes the fibres stand on end, which is crucial to the final grass-like effect. To cover the workpiece, the particles are ejected from a gun that gives them an electric charge; a potential difference is maintained between the gun and the workpiece, so that the particles feel an electrostatic force that moves them towards the workpiece (see Figure 6.1). However they are also blown about, both by imposed air currents and also by air currents generated by their own drag.

We would like to know something about the controlling parameters of this process. In particular, we would like to get most of the particles to hit the workpiece and not to be carried away by the air flow (which must of course go round it). Do the particles influence the air flow, or are they passive? How thick is the final layer of resin? An attractive feature of this method of painting is that the electric field is strong near outward-pointing corners, and so particles are especially attracted to these areas, which are hard to cover well using traditional methods. A less attractive

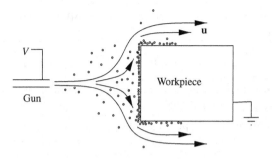

Figure 6.1 Electrostatic painting of an earthed metal workpiece.

feature is that it is very hard to see what is happening in the cloud of paint particles. A mathematical model may help to answer some of these questions.

6.2 Field equations

In this section, we begin to build a model of the painting process, by writing down 'field equations' to describe the gas (air) and particle velocities. In this problem, it helps to start with some data, as that points to a reasonable model for the gas–particle interaction. The workpiece has a typical size $L \approx 1$ m, and the observed air velocities have size $U_g \approx 1$ m s^{-1}. The air density and dynamic viscosity are $\rho_g \approx 1.3$, $\mu_g \approx 1.8 \times 10^{-5}$ in SI units (the kinematic viscosity ν_g is thus about 1.4×10^{-5} m^2 s^{-1}). The particles are tiny: they have radius $a \approx 10^{-5}$ m and their mass $m_p \approx 10^{-12}$ kg. There is an enormous number of them, a representative number density being $n_0 \approx 10^9$ per cubic metre of air. Lastly, turning to the electrical aspects, each particle carries a charge q_p of about 10^{-15} C, and the applied voltage is $V_0 \approx 10^5$ V.

Now because the number density of particles is so large, the average particle separation, 10^{-3} m, is very small compared with the workpiece dimension L but very large compared with their mean radius a. It is reasonable to consider them as isolated particles when we work out the force on them, but when we work on larger length scales we hope to get away with averaging their effects. We therefore consider the evolution of their local number density, which we think of as a continuous function $n(\mathbf{x}, t)$ representing the number of particles per unit volume measured over a small volume whose diameter is much bigger than the mean separation but much smaller than L. We proceed similarly that there is a local average particle velocity $\mathbf{v}_p(\mathbf{x}, t)$ and in calculating the force exerted by the particles on the air: this is plausible if we can convince ourselves that neighbouring particles all feel very similar influences from the fluid and that all have a similar effect on it.

We now start to write down some equations. The first expresses the conservation of particles:

$$\frac{\partial n}{\partial t} + \nabla \cdot (n\mathbf{v_p}) = 0. \tag{6.1}$$

Next, we think about the particle equation of motion. Unlike flagpoles or cables, the particles feel a fairly slow flow past them. The 'local' Reynolds number for flow at 1 m s^{-1} past a 10-micron-radius spherical particle is

$$\mathrm{Re_p} = \frac{U_g a}{\nu_g}$$

$$\approx 0.7,$$

and this is a considerable overestimate since we should really use the relative (slip) velocity, which is likely to be smaller than 1 m s^{-1}. Now the force on a spherical particle in slow flow can be shown to be

See any good book on viscous flow.

$$-6\pi a \mu_g \left(\mathbf{v_p} - \mathbf{v_g}\right)$$

where $\mathbf{v_g}$ is the 'local' gas velocity, that is, the gas velocity many particle radii away from the particle of interest (but not so far as to be near neighbouring particles). Our particles are not spherical, but we'll still assume that they feel a force proportional to the slip velocity; we'll write it as

$$-K \left(\mathbf{v_p} - \mathbf{v_g}\right),$$

where, working on the basis of near-spherical particles, we expect that $K \approx 10^{-9}$ kg s^{-1}. Lastly we need to include the the gravitational force $m_p\mathbf{g}$ and the electrostatic force $q_p\mathbf{E}$, where $\mathbf{E}$ is the electric field. Then, assuming that all neighbouring particles feel the same slip velocity and have the same particle velocity, we can write down an equation of motion

Note that D/Dt here is the convective (total) derivative.

for the particles:

$$m_p \frac{D\mathbf{v_p}}{Dt} = -K \left(\mathbf{v_p} - \mathbf{v_g}\right) + m_p\mathbf{g} + q_p\mathbf{E}. \tag{6.2}$$

Correspondingly, we have momentum and mass[1] conservation equations for the gas (let's keep things simple by writing down an inviscid model, leaving gravity out as it merely generates hydrostatic pressure):

$$\rho_g \frac{D\mathbf{v_g}}{Dt} = -\nabla p + nK \left(\mathbf{v_p} - \mathbf{v_g}\right), \qquad \nabla \cdot \mathbf{v_g} = 0. \tag{6.3}$$

---

[1] Another simplification I've slipped in is that, because the volume fraction of particles is so small, I've taken the gas volume fraction to equal 1. Technically, we should write down a two-phase model; a later exercise, on p. 202, justifies our approximation.

Perhaps the only unfamiliar term here is the body force (the force per unit volume) on the gas due to the particles. Whereas the particle equation of motion is for individual particles, and thence for the averaged-out particles because of our assumption that all nearby particles behave similarly, the force by the particles on the gas is just the force $K(\mathbf{v}_p - \mathbf{v}_g)$ on one representative particle multiplied by the number density n.

It only remains to write down Poisson's equation

$$\nabla \cdot (\varepsilon_g \mathbf{E}) = nq_p \tag{6.4}$$

for the averaged electric field (here the permittivity of air $\varepsilon_g \approx 10^{-11}$ in SI units; it is very close to ε_0), and we have collected all the field equations of the model.[2]

6.3 Boundary conditions

For the sake of completeness, we should briefly discuss the boundary conditions for our model, although we aren't going to use them much. The main issue we should address is how to deal with the thin layer of particles on the workpiece. There is little doubt that it is very thin – we don't want a centimetre-thick coating of paint! – and as far as the fluid is concerned we can assume that the workpiece forms a rigid boundary in order to complement whatever inflow conditions we impose at the gun. The particles satisfy the first-order equation (6.1), whose characteristics are the particle paths (see Chapter 7). This dictates that we impose an initial condition at the gun, and that is all we need.

Lastly, consider the electric potential. We can impose $\phi = V_0$ on the gun with a clear conscience, and likewise $\phi = 0$ at the workpiece, but we may worry that charge building up in the paint layer on the workpiece will alter the 'effective boundary condition' felt by ϕ. This depends to a large extent on the details of what happens in this layer. For example, if the charge can 'leak' off the particles to the workpiece (i.e. if electrons can move onto the particles if their charge is positive) then the layer should be relatively passive and we can ignore it. However, if the charges remain in situ, we can still ignore the effect of this layer as long as the total charge it contains is small enough (see Exercise 1). Later in the book, we shall see how we can make this kind of ad hoc approximation more systematic.

---

[2] Caveat: we haven't dealt properly with the thin layer of paint on the workpiece, as seen in Figure 6.1. Clearly the particles cannot move freely in this layer and we need to treat it separately; see the next section and the exercises.

6.4 Nondimensionalisation

We have made great progress in producing a useful model, but the result is undeniably complicated. Do we really need all the terms in these equations? Obviously we're going to have to solve them numerically, and for this reason if no other we should do what we can to check that the model is robust and suitable for a numerical attack.

Let's scale all the variables with typical values as before. Scale $\mathbf{x}$ with L, t with L/U_g, $\mathbf{v_g}$ and $\mathbf{v_p}$ with U_g, p with $\rho_g U_g^2$ and n with n_0. We have two choices for the scale for $\mathbf{E}$: one is the applied voltage V_0 while the other, $q_p n_0 L/\epsilon_g$, is derived from the Poisson equation (6.4) once all other scalings in it are fixed. It so happens that both are about the same size, 10^5 V m^{-1}, so let's save ink and use V_0.

Start with the particle equation of motion (6.2). Scaling, and immediately dropping the primes, we get

$$\frac{m_p U_g^2}{L}\frac{d\mathbf{v_p}}{dt} = -KU_g\left(\mathbf{v_p} - \mathbf{v_g}\right) - m_p g\mathbf{k} + \frac{q_p V_0}{L}\mathbf{E}$$

and, dividing by KU_g,

$$\frac{m_p U_g}{KL}\frac{d\mathbf{v_p}}{dt} = -\left(\mathbf{v_p} - \mathbf{v_g}\right) + \frac{q_p V_0}{KU_g L}\mathbf{E} - \frac{m_p g}{KU_g}\mathbf{k}. \qquad (6.5)$$

We see that the dimensionless quantity

$$\frac{m_p U_g}{KL} \approx 10^{-3}$$

is very small. With luck, we can neglect the term it multiplies, the particle acceleration: apart from an initial transient as the particles get up to speed near the gun, inertial forces on them are dominated by viscous drag. The second dimensionless parameter,

$$\frac{m_p g}{KU_g} \approx 2 \times 10^{-2},$$

is also small. It decides the competition between gravity and viscous forces in favour of the latter: these particles fall slowly compared to the rate at which they are pushed about by the air. The third dimensionless parameter,

$$\frac{q_p V_0}{KU_g L} \approx 10^{-1}, \qquad (6.6)$$

compares the electrostatic forces with the drag forces. It too is small, though larger than the other two; we shall see some consequences of this in Chapter 13.

This simple analysis has told us quite a lot. The scaled equation says that the particles follow the gas quite closely, with a small influence

Check this scaling.

It's obvious that the conservation of particle mass equation (6.1) is unchanged.

from the electrostatic force and very minor contributions from gravity and inertia.[3] We can get a good approximation to the particle motion if we write

$$\mathbf{v}_p = \mathbf{v}_g + \frac{q_p V_0}{K U_g L} \tilde{\mathbf{v}}_p, \tag{6.7}$$

where, from (6.5),

$$\tilde{\mathbf{v}}_p = \mathbf{E} + \text{smaller terms.}$$

Remember that this is a dimensionless equation: we are not directly equating a velocity to an electric field.

A preliminary conclusion from this analysis is that the device is not working terribly well: the particles are being swept along too much by the air.

Now let's look at the gas momentum equation (6.3). After scaling, this becomes

$$\frac{d\mathbf{v}_g}{dt} = -\nabla p + \frac{n_0 K L}{\rho_g U_g} n \left(\mathbf{v}_p - \mathbf{v}_g \right).$$

On the face of it, the dimensionless quantity

$$\frac{n_0 K L}{\rho_g U_g} \approx 1$$

is not small, indicating that the particles exert a body force on the air that is not small. However, remember that we decided above that the slip velocity $\mathbf{v}_p - \mathbf{v}_g$ (which this dimensionless parameter multiplies) *is* small. So, the particles do after all have a small effect on the gas, confirming that it would be a good idea to try to slow the air down to improve performance. We return to this problem in Chapter 13.

6.5 Sources and further reading

Electrostatic painting was also a Study Group problem, from Courtaulds plc, and is documented further in [3].

6.6 Exercises

1 **Paint layer.** Suppose that a thin layer of paint particles, deposited electrostatically as in the text, is growing on $y = 0$ and that its thickness is $y = h(x, t)$. Show that the normal velocity of the layer

---

[3] If the parameter in (6.6) had been *large*, not small, that would have told us that we had chosen a wrong scaling somewhere. There is no reason for $\mathbf{E}$ to be large – it has a perfectly good equation of its own in which there are no large parameters – and so we would have an equation with one large term in it and nothing to balance it. When the parameter (6.6) is small, we have the a priori plausible approximation (6.7).

The subscripts x and t indicate partial derivatives here.

boundary is $(1 + h_x^2)^{1/2} h_t$, and relate this to $\mathbf{v}_p \cdot \mathbf{n}$ at the interface. How thick will the layer grow in 10 seconds? Justify the approximate boundary condition

$$\frac{\partial h}{\partial t} = \mathbf{v}_p \cdot \mathbf{n}$$

on the workpiece (we discuss this 'linearisation' of boundary conditions in Chapter 13).

Derive an order of magnitude estimate for the potential drop across the layer, assuming that the potential approximately satisfies

$$\frac{\partial^2 \phi}{\partial y^2} = -\frac{\rho}{\varepsilon_0},$$

where ρ is the density of charge in the layer. Assuming a reasonably close packing for the particles, express this order of magnitude in terms of the average particle radius a and charge q_p, and hence assess how thick the layer needs to be before the potential drop across it rivals the applied voltage.

'$\alpha = \alpha$ if $\alpha < 1$, $\alpha = 1$ if $\alpha > 1$.'

Part II
Analytical techniques

7

Partial differential equations

This chapter is a short overview of partial differential equations, with applications in mind. The emphasis is largely on first-order quasilinear equations, for which many standard textbooks don't provide much in the way of real-life examples; we'll see applications to e-mail and, in a case study, to traffic. We will also take a brief look at the fully nonlinear case, with an eye to using it in Chapter 23 to see, among other things, why we say that light travels in straight lines. Last, we will have a brief run through the standard theory of second-order linear equations in two variables, for which the canonical examples of the wave equation, the heat equation and Laplace's equation, and their physical interpretations, are so well known that we don't need to give a full treatment of them here.

7.1 First-order quasilinear partial differential equations: theory

We begin with a review of the elementary theory for the partial differential equation[1]

Remember that c has to be on the right-hand side of the equation, or you will get an extraneous minus sign.

$$a(x_1, x_2, u)\frac{\partial u}{\partial x_1} + b(x_1, x_2, u)\frac{\partial u}{\partial x_2} = c(x_1, x_2, u),$$

where a, b and c are given smooth functions, with initial values given in parametric form on a curve Γ; that is, we can write $u = u_0(s)$ on $x_1 = x_{10}(s), x_2 = x_{20}(s)$. This should be familiar material. As illustrated

---

[1] Here x_1 and x_2 are generic independent variables: sometimes we will use x and y, sometimes x and t.

Figure 7.1 Solution of a first-order quasilinear equation by characteristics. The subscripts x_1 and x_2 indicate partial derivatives.

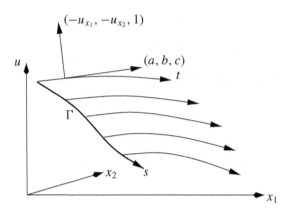

in Figure 7.1, the above equation written in the form

$$(a, b, c) \cdot \left(-\frac{\partial u}{\partial x_1}, -\frac{\partial u}{\partial x_2}, 1 \right) = 0$$

shows that (a, b, c) is orthogonal to $(-\partial u/\partial x_1, -\partial u/\partial x_2, 1)$, which itself is normal to the solution surface $u - u(x_1, x_2) = 0$.

It follows that if we solve the *characteristic equations*

$$\frac{dx_1}{dt} = a(x_1, x_2, u), \qquad \frac{dx_2}{dt} = b(x_1, x_2, u),$$

$$\frac{du}{dt} = c(x_1, x_2, u),$$

where t is a parameter along the characteristics, with the initial values

$$x_1(0) = x_{10}(s), \qquad x_2(0) = x_{20}(s), \qquad u(0) = u_0(s),$$

then the curves so generated, known as *characteristics*, are tangent to the solution surface at each point. Gluing the characteristics together gives the solution surface in parametric form,

$$(x_1, x_2, u) = \big(x_1(s, t), x_2(s, t), u(s, t) \big),$$

and in principle we are done, at least near Γ. We see immediately the central role of characteristics (or their projections onto the $x_1 x_2$-plane, known as *characteristic projections* or occasionally as *characteristic traces*). They are curves along which information propagates: indeed, the left-hand side of the partial differential equation is just the directional derivative of u along the characteristic projections, so in this direction the partial differential equation reduces to an ordinary one. The characteristics can be found one at a time, each one independently of all the others and determined only by the initial data at its starting point.

There are several caveats to make about this procedure. The first is that, as is clear from the geometrical point of view, the initial curve must not be tangent to a characteristic, for if it is, near the point of tangency we expect more than one value of u at each point (x_1, x_2). This is easily seen by noting that if there is one characteristic projection through such a point, in general there will be another also, carrying a different value of u. This insight is confirmed by a calculation in which we try to find all the partial derivatives of u at a point on Γ with the aim of constructing its Taylor series at that point. We know by differentiating $u_0(s)$ along Γ that

$$\frac{dx_{10}}{ds}\frac{\partial u}{\partial x_1} + \frac{dx_{20}}{ds}\frac{\partial u}{\partial x_2} = \frac{du_0}{ds}$$

and that

$$a(x_{10}, x_{20}, u_0)\frac{\partial u}{\partial x_1} + b(x_{10}, x_{20}, u_0)\frac{\partial u}{\partial x_2} = c(x_{10}, x_{20}, u_0).$$

Regarding these as equations for $\partial u/\partial x_1$ and $\partial u/\partial x_2$, there is a unique solution provided that the determinant of coefficients

$$\begin{vmatrix} a & b \\ \dfrac{dx_{10}}{ds} & \dfrac{dx_{20}}{ds} \end{vmatrix} \neq 0;$$

if it does vanish, so that there is no unique Taylor series, we get precisely the first two characteristic equations. Thus, if we require a unique solution u, Γ cannot be a characteristic.

Exercise: if it does vanish, show that the consistency condition for there to be a solution, albeit nonunique, is the third characteristic equation.

The second caveat concerns the region of existence of the solution. It is an obvious remark that if Γ has ends then we can only hope to find the solution in a region between the characteristics through those endpoints. Furthermore, by standard Picard theory, the characteristic equations have a unique solution for at least small t, i.e. in a small strip near Γ. How far we can go beyond this strip depends on two things. First, the local solution must not blow up, which it might well do for nonlinear equations. Second, and of more practical importance, we must (in principle) be able to find s and t uniquely from the solution of the first two characteristic equations in order to calculate u uniquely. That means that the Jacobian

$$\left| \frac{\partial(x_1, x_2)}{\partial(s, t)} \right|$$

must not vanish or become infinite. This in turn means that the characteristic projections are not allowed to cross, for if they did then the different values of u propagated along the different characteristic projections would lead to many values of u at a single point. So even though we may have a perfectly good parametric solution of the form above, it does not correspond to a single-valued solution $u(x_1, x_2)$. In general,

Figure 7.2 Solution of a first-order quasilinear equation with a gradient discontinuity.

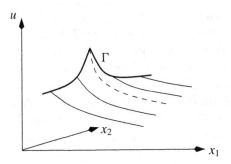

one can think of this situation as corresponding to some kind of 'folding over' of the parametrised surface in $x_1 x_2 u$-space, as in Figure 7.4. In Section 7.3, we return to the question of what to do if this happens.

The third caveat is the degree of smoothness of the solution. At first sight, we expect the solution to have single-valued first partial derivatives, so that the partial differential equation makes sense. However, it is possible to extend our idea of a solution by joining together smooth solution surfaces at curves at which the first partial derivatives have jumps. Such a solution surface might look like a sheet slung over a rope and pulled out on either side, or like a curved roof with a ridge, as in Figure 7.2. Such a discontinuity can only occur across a characteristic. We can see this by noting that when we showed that we can find the first partial derivatives of u uniquely unless they are given on a characteristic, we were in effect showing that jumps in these derivatives can only occur across a characteristic. That is, smooth solution surfaces intersect in characteristics. Alternatively, if $(x_1(t), x_2(t), u(t))$ is a curve in the solution surface along which u is continuous but there are jumps in its derivatives from left to right, then we can differentiate the statement

$$[u] = 0$$

along the curve, where $[\cdot]$ means the jump across the curve, to get

$$\left[\frac{\mathrm{d}u}{\mathrm{d}t} \right] = \left[\frac{\partial u}{\partial x_1} \right] \frac{\mathrm{d}x_1}{\mathrm{d}t} + \left[\frac{\partial u}{\partial x_2} \right] \frac{\mathrm{d}x_2}{\mathrm{d}t} = 0.$$

The partial differential equation also holds on either side of the curve, and taking its difference across the curve gives

Remember that $[u] = 0$, so the coefficients a, b, c are continuous.

$$a \left[\frac{\partial u}{\partial x_1} \right] + b \left[\frac{\partial u}{\partial x_2} \right] = 0.$$

These two homogeneous equations for $[\partial u / \partial x_1]$ and $[\partial u / \partial x_2]$ have a nonzero solution only if the determinant of coefficients

$$\begin{vmatrix} \dfrac{\mathrm{d}x_1}{\mathrm{d}t} & \dfrac{\mathrm{d}x_2}{\mathrm{d}t} \\ a & b \end{vmatrix}$$

vanishes, and this is precisely the condition for $(x_1(t), x_2(t))$ to be a characteristic projection.

We see that characteristics are curves along which gradient discontinuities propagate from the initial curve. However, this does not let us deal with the multivalued 'blow-up' described above, which can happen even with perfectly smooth initial data. In Section 7.3, we shall see how to do that by introducing solutions that are themselves discontinuous, so-called *shocks*.

7.2 Example: Poisson processes

First-order equations are often found in connection with various generating functions in probability. Many of these have their roots in the *Poisson process*, which is often used as a model for the number of occurrences in a given time of independent random events such as e-mails arriving in your inbox or calls coming to a telephone exchange. (I hesitate to propose the model for the arrival of London buses as there are good reasons for them to arrive in pairs or even threes.)

Suppose that we say that in a short time δt there is a probability $\lambda \, \delta t$ that an e-mail arrives, a probability $1 - \lambda \, \delta t$ that none arrive, and a negligible probability that two or more arrive at once. The constant λ, known as the *intensity*, measures your popularity (spam or otherwise). Starting with an empty inbox, and staying online, define the Poisson counter $N(t)$ as the number of e-mails you have received by time t, with $N(0) = 0$. Thus

$$N(t + \delta t) = \begin{cases} N(t) & \text{with probability } 1 - \lambda \, \delta t, \\ N(t) + 1 & \text{with probability } \lambda \, \delta t. \end{cases}$$

What is the probability distribution of $N(t)$? Define

$$p_n(t) = P(N(t) = n).$$

There are two, and only two, ways that $N(t + \delta t)$ can equal n: either $N(t) = n$ and no new message arrives, or $N(t) = n - 1$ and one message arrives. These events are disjoint and so we have

$$\begin{aligned} P\left(N(t + \delta t) = n\right) &= p_n(t + \delta t) \\ &= P\left(N(t) = n\right) P(\text{no new message}) \\ &\quad + P\left(N(t) = n - 1\right) P(\text{one new message}) \\ &= p_n(t)(1 - \lambda \, \delta t) + p_{n-1}(t)\lambda \, \delta t. \end{aligned}$$

Expanding $p_n(t + \delta t)$ in a Taylor series

$$p_n(t + \delta t) = p_n(t) + \delta t \frac{\mathrm{d} p_n(t)}{\mathrm{d} t} + \cdots$$

and taking $\delta t \to 0$, we get the system of differential equations

$$\frac{\mathrm{d}p_n}{\mathrm{d}t} = -\lambda\,(p_n - p_{n-1}), \qquad n = 1, 2, 3, \dots,$$

while separately

$$\frac{\mathrm{d}p_0}{\mathrm{d}t} = -\lambda p_0.$$

The initial conditions are $p_0(0) = 1$, $p_n(0) = 0$ for $n > 0$. Although in principle one can solve the equations sequentially, it's easier to define

$$G_N(x, t) = \sum_{n=0}^{\infty} x^n\, p_n(t);$$

then, multiplying the nth differential equation by x^n and summing over n, we obtain

$$\frac{\partial G_N}{\partial t} = -\lambda(1 - x)G_N,$$

whence

$$G_N(x, t) = e^{-\lambda t} e^{\lambda t x} = \sum_{n=0}^{\infty} e^{-\lambda t} \frac{(\lambda t)^n}{n!} x^n.$$

That is, the probabilities are those of the Poisson distribution with mean λt.

This generating function only satisfies an ordinary differential equation, but now suppose that a virus is doing the rounds, spread by e-mail. Let $V(t)$ be the number of computers infected, and suppose that the probability of a new infection over the next δt is $\lambda V(t)\,\delta t$, i.e. it is proportional to the number of infected computers. With $p_n(t) = P(V(t) = n)$ as before (but now n measures the number of infected computers, rather than e-mails, now for $n = 1, 2, \dots$ and with $p_1(0) = 1$ to model one source of infection), we find

$$\frac{\mathrm{d}p_n}{\mathrm{d}t} = -\lambda\,(np_n - (n-1)p_{n-1}), \qquad n = 1, 2, 3, \dots$$

(here $p_0 = 0$). Then, following the calculation above, the generating function $G_V(x, t)$ satisfies

$$\frac{\partial G_V}{\partial t} + \lambda x(1 - x)\frac{\partial G_V}{\partial x} = 0$$

Note the consistency check $G(1, t) = 1$ for both these examples: the probabilities sum to 1.

with $G_V(x, 0) = x$. The solution is easily found to be

$$G_V(x, t) = \frac{x e^{-\lambda t}}{1 - x\left(1 - e^{-\lambda t}\right)}$$

$$= \sum_{n=1}^{\infty} x^n e^{-\lambda t} \left(1 - e^{-\lambda t}\right)^{n-1}.$$

The mean of this distribution,

$$\sum_{n=1}^{\infty} np_n(t) = \frac{\partial G_V}{\partial x}(1, t) = e^{\lambda t},$$

grows exponentially in t, as we would expect.

7.3 Shocks

We started our analysis of quasilinear equations by considering smooth solution surfaces with a unique normal at each point. Then we realised that we could extend our class of solutions by allowing gradient discontinuities, as long as these occur across (propagate along) characteristics. However, blow-up still occurs when characteristic projections cross, because then we might get several values of u at the same place.

To see this in action consider the equation

$$\frac{Du}{Dt} = \frac{\partial u}{\partial t} + u \frac{\partial u}{\partial x} = 0.$$

This equation is known as a *kinematic wave equation* and, if we think of $u(x, t)$ as the speed of a particle moving along the x-axis, it says that the speed of any given particle remains constant because its derivative following that particle is zero. The equation is easy to solve by characteristics, with initial data $u(x, 0) = u_0(x)$, say, corresponding to a snapshot at $t = 0$ of the speeds all along the x-axis. The characteristic equations are

$$\frac{dt}{d\tau} = 1, \qquad \frac{dx}{d\tau} = u, \qquad \frac{du}{d\tau} = 0,$$

so u remains constant along a characteristic whose projection has slope $dx/dt = u$. This simply repeats that the particles move along characteristics with constant speed u. So, to construct the solution we simply draw all the characteristic projections through the initial line $t = 0$ and read off the value of u at any point x and later time t. This procedure works fine if $u_0(x)$ is increasing, since then the characteristics spread out as in Figure 7.3.

But if $u_0(x)$ is decreasing, we inevitably have a collision of characteristic projections – and particles – after a finite time, as in Figure 7.4. It has an obvious physical interpretation that fast particles have caught up with slow ones and are trying to occupy the same space.

On the one hand we could take the defeatist view that the solution ceases to exist at the moment when the characteristic projections first cross and that this is the end of the matter. On the other hand, we could try to extend our notion of what constitutes a solution to allow not only gradient discontinuities but also *discontinuities in u itself.*

Figure 7.3 Solutions of the kinematic wave equation with increasing initial data $\frac{1}{2}(1 + \tanh x)$. The steep rise in $u_0(x)$ moves to the right and spreads out. The mesh lines that start out parallel to the t-axis correspond to characteristics.

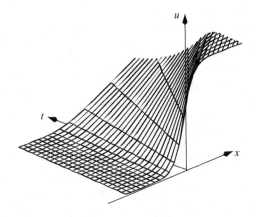

Figure 7.4 Solutions of the kinematic wave equation with decreasing initial data $\frac{1}{2}(1 - \tanh x)$. The steep fall in $u_0(x)$ gets steeper until, when the characteristic projections cross, it is vertical. (Note that the viewpoint is different from that in Figure 7.3.)

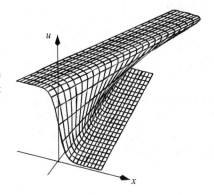

By 'physical example' we mean a physical situation that can be modelled by equations, here the shallow water equations and the equations of gas dynamics, that have shock solutions.

That is, there may be a curve (or curves) $x = S(t)$ across which u has a jump. These jumps are called *shocks*. Famous physical examples include tidal bores[2] and the shock waves created when Concorde (RIP) flew supersonically.

As one might expect, a drastic step like this is fraught with dangers. It does indeed turn out that we have generalised a bit too far, because we can have nonuniqueness of solutions with shocks. The ideas needed to deal with this are rather delicate and we refer to [44] for a fuller discussion. However, when the partial differential equation is a conservation law we can give a heuristic derivation of a necessary condition at a shock, which in practice is sometimes sufficient as well.

---

[2] A tidal bore is an abrupt change in water level that propagates *upstream* from the river mouth, often appearing as a continually breaking wave. Prerequisites for them to form are a large tidal range and a slowly convergent estuary. Examples include the Severn bore and the Trent Aegir in the UK, and the Hooghly bore in India.

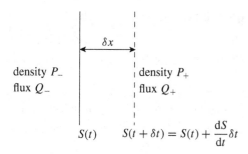

Figure 7.5 Derivation of the
Rankine–Hugoniot condition
for a shock $x = S(t)$.

7.3.1 The Rankine–Hugoniot conditions

Suppose that a quasilinear partial differential equation is in the conservation form (see Chapter 1)

$$\frac{\partial P}{\partial t} + \frac{\partial Q}{\partial x} = 0,$$

where $P(x, t, u)$ is the density and $Q(x, t, u)$ the flux of a conserved quantity; usually they depend only on u. Now suppose that there is a curve $x = S(t)$ across which $u(x, t)$ has a jump discontinuity, i.e. a shock occurs at $x = S(t)$. This means that P and Q also have jump discontinuities across this curve. However, we assume that overall the quantity, whatever it is ('stuff', say), *is* conserved. Let us do a simple-minded 'box' argument to see what this implies for the discontinuities in P and Q. Figure 7.5 sets the scene.

Compare this with the (in some ways more subtle) argument of Section 1.4.

In a small time δt, the shock moves by a small distance

$$\delta x \approx \frac{dS}{dt} \delta t.$$

The net flux into the region crossed by the shock is $Q_+ - Q_-$ in an obvious notation, $+$ referring to $x > S(t)$. Hence the amount of stuff flowing into this small box is $(Q_+ - Q_-)\delta t = [Q]_-^+ \, \delta t$. The amount of stuff in the box before the shock arrives is $P_+ \, \delta x$, and afterwards it is $P_- \, \delta x$. The difference must be accounted for by the net flux; thus,

$$[P]_-^+ \delta x = [Q]_-^+ \delta t.$$

Eliminating $\delta x / \delta t \approx dS/dt$, we arrive at

$$\frac{dS}{dt} = \frac{[Q]_-^+}{[P]_-^+}.$$

This relation between the two jump discontinuities is known as the *Rankine–Hugoniot condition*; it is a necessary, but not sufficient, condition for a unique solution with a shock, as shown in Exercise 6. Further discussion of this situation moves to Exercises 5–7 and is dealt with in more detail in [44].

7.4 Fully nonlinear equations: Charpit's method

Rather remarkably, it is possible to extend the method of attack described above to solve fully nonlinear first-order equations of the form

$$F(x, y, u, p, q) = 0,$$

where, to fit in with common notation, we are using x, y as independent variables and

$$p = \frac{\partial u}{\partial x}, \qquad q = \frac{\partial u}{\partial y}.$$

To reduce the density of the formulae, again we occasionally use the subscript notation for partial derivatives.

What we do looks wildly optimistic. We differentiate the equation $F = 0$ with respect to x, to get

$$F_x + pF_u + F_p p_x + F_q q_x = 0.$$

The mixed partial derivatives are equal.

Then we use the fact that $q_x = p_y$ to get

$$F_p \frac{\partial p}{\partial x} + F_q \frac{\partial p}{\partial y} = -F_x - pF_u,$$

which we regard as a *quasilinear* equation for p, with characteristics

$$\frac{dx}{dt} = F_p, \qquad \frac{dy}{dt} = F_q, \qquad \frac{dp}{dt} = -F_x - pF_u.$$

Of course, the coefficients in this equation may depend on q and u, which we do not yet know. However, if we repeat the whole procedure, but now differentiate with respect to y, we have the corresponding equation

$$F_p \frac{\partial q}{\partial x} + F_q \frac{\partial q}{\partial y} = -F_y - qF_u,$$

with characteristics

$$\frac{dx}{dt} = F_p, \qquad \frac{dy}{dt} = F_q, \qquad \frac{dq}{dt} = -F_y - qF_u$$

(note that the equations for x and y are the same as those arising from the equation for p). Lastly, differentiation along the curves thus found gives

$$\frac{du}{dt} = p\frac{dx}{dt} + q\frac{dy}{dt} = pF_p + qF_q.$$

Because two equations were duplicated, we are left with a 5×5 system of ordinary differential equations, summarised as

$$\frac{dx}{dt} = F_p, \qquad \frac{dy}{dt} = F_q,$$

$$\frac{dp}{dt} = -F_x - pF_u, \qquad \frac{dq}{dt} = -F_y - qF_u,$$

$$\frac{du}{dt} = pF_p + qF_q.$$

These are known as *Charpit's equations* (the name of Lagrange is also sometimes found in this context) and their solution curves are called *rays*. They are the analogue of the three characteristic equations for a quasi-linear equation, and the rays are the generalisation of the characteristics occuring in that case. They are solved with initial data

$$x = x_0(s), \qquad y = y_0(s), \qquad u = u_0(s)$$

just as in the quasilinear case, from which the initial values p_0 and q_0 of p and q can be found from the two equations

$$\frac{du_0}{ds} = p_0 \frac{dx_0}{ds} + q_0 \frac{dy_0}{ds}, \qquad F(x_0, y_0, u_0, p_0, q_0) = 0;$$

note that this prescription may, unlike in the quasilinear case, lead to more than one ray through each point on the initial curve, because of the nonlinearity of F.

Where is the catch? It is certainly true that Charpit's equations, with suitable initial data, have a unique solution, by the usual Picard argument. The missing step is rather subtle. If we find the solution to Charpit's equations in the parametric form $x = x(s, t)$, $y = y(s, t)$, $u = u(s, t)$, $p = p(s, t)$ and $q = q(s, t)$, how do we know that $p = \partial u/\partial x$ and $q = \partial u/\partial y$, when s and t are eliminated? After all, these derivatives undoubtedly involve differentiation in the s-direction, which is notably absent from our argument above. The manoeuvres needed to establish this are rather intricate, and are described in any good book on partial differential equations (such as [44] or [35], where the geometric interpretation of the Charpit approach is also described). We content ourselves with accepting that the result holds, and looking at some examples.

Example.   Solve the equation

$$\frac{\partial u}{\partial x} \frac{\partial u}{\partial y} = u \qquad \text{with} \quad u = 1 \quad \text{on} \quad xy = 1.$$

Here $F(x, y, u, p, q) = pq - u$, so Charpit's equations are

$$\frac{dx}{dt} = q, \qquad \frac{dy}{dt} = p, \qquad \frac{du}{dt} = 2pq = 2u,$$

$$\frac{dp}{dt} = p, \qquad \frac{dq}{dt} = q.$$

The initial curve is

$$x_0(s) = s, \qquad y_0(s) = 1/s, \qquad u_0(s) = 1,$$

from which we find

$$p_0(s) - \frac{1}{s^2} q_0(s) = 0.$$

Figure 7.6 A particle in limiting equilibrium on a sandpile.

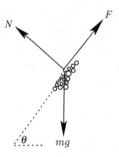

This relation, with $p_0(s)q_0(s) = 1$, gives

$$p_0(s) = \pm\frac{1}{s}, \qquad q_0(s) = \pm s$$

(note that there are two solutions). The solutions are, in order of calculation,

$$p = \pm\frac{1}{s}e^t, \quad q = \pm se^t, \quad x = s \pm s(e^t - 1),$$

$$y = \frac{1}{s} \pm \frac{1}{s}(e^t - 1), \quad u = e^{2t}.$$

The plus signs give the solution $u = xy$, and the minus signs lead after a short calculation to $u = \left(2 - \sqrt{xy}\right)^2$. (If alert, you may have spotted the possibility of looking for $u(x, y)$ as a function of xy alone.)

Example: sand piled on a table.   Here is a nice realisation of a non-linear first-order equation. Suppose we take a table, and pour dry sand onto it until we can pour no more. A very simple modelling approach[3] says that any particle on the surface of the pile is in limiting equilibrium. This means that the frictional force on such a particle, F, is as large as it can be (otherwise we could pile more sand on and make the slope steeper), and furthermore F is proportional to the normal reaction N (see Figure 7.6). That is, $F = \mu N$ where μ is called the coefficient of friction. Resolving horizontally, we have $N \sin\theta = F \cos\theta$, where θ is as shown, and hence $\tan\theta = \mu$. If the height of the sandpile is given by $z = h(x, y)$ we know that $\cos\theta = (0, 0, 1) \cdot \mathbf{n}$, where

$$\mathbf{n} = \frac{\nabla(z - h(x, y))}{|\nabla(z - h(x, y))|}$$

is the unit normal to the surface. From this, it is easy to show that

$$\left(\frac{\partial h}{\partial x}\right)^2 + \left(\frac{\partial h}{\partial y}\right)^2 = \mu^2, \tag{7.1}$$

[3] Real sandpiles are more complicated; they are continually subject to sandslides of all sizes, and have been much studied by physicists as an example of what is called a self-organised critical system.

a famous equation usually known as the *eikonal equation*. We shall see a lot more of this equation in Chapter 23. For now, we just note that it is very easy to solve by Charpit's method, as the rays are straight, and that when $h(x, y)$ is required to be a single-valued function the solutions typically form 'ridgelines' like the roof of a house, where the slope of the pile changes discontinuously.

Example: spray forming.   Many industrial processes involve building up or carving away a solid object, layer by layer, by spraying or otherwise adding material at a known rate, or by removing it. One example is the production of large aluminium billets (bars) from a controlled spray of tiny droplets of liquid metal; on a much smaller scale, components in semiconductor chips are built up by vapour deposition, or can be etched away. The new three-dimensional printers are another example. In all these processes, the key to a high-quality finished product is precise control of the deposition. And to do that, we need to know how to formulate (and solve) the direct problem: given a deposition rate, how does the solid grow?

This is easy to state in words: the object grows in such a way that the normal velocity of its boundary is equal to the deposition rate. Suppose that the boundary of the solid is given by the equation $f(\mathbf{x}, t) = 0$ and the deposition rate is a known flux $\mathbf{q}(\mathbf{x}, t)$ (that is, at each point there is a mass flux $|\mathbf{q}|$ in the direction $\mathbf{q}$). The unit normal to the boundary is

$$\mathbf{n} = \frac{\nabla f}{|\nabla f|},$$

and the mass flux normal to the interface, the deposition rate, is then $\mathbf{q} \cdot \mathbf{n}$. By a simple conservation of mass argument (compare with the Rankine–Hugoniot conditions), this is equal to ρV_n, where ρ is the density of the solid and V_n is the normal velocity of the interface. Our only remaining task is to write V_n in terms of the derivatives of $f(\mathbf{x}, t)$. To do this, consider a point $\mathbf{x}(t)$ which remains in the boundary throughout. Its velocity is $\mathbf{v} = d\mathbf{x}/dt$, which in general has components both normal and tangential to the moving boundary. However, by the chain rule,

$$\frac{\mathrm{d}}{\mathrm{d}t} f(\mathbf{x}(t), t) = \frac{\partial f}{\partial t} + \mathbf{v} \cdot \nabla f = 0. \qquad (7.2)$$

Now $\mathbf{v} \cdot \nabla f / |\nabla f|$ is simply the component of the velocity $\mathbf{v}$ along the unit normal $\mathbf{n}$, namely V_n. Hence, dividing both sides of (7.2) by $|\nabla f|$, we find that

$$V_n = -\frac{1}{|\nabla f|} \frac{\partial f}{\partial t}$$

Figure 7.7 Growth by deposition.

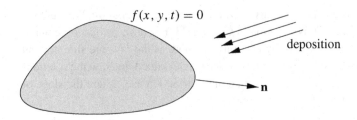

$$f(x, y, t) = 0$$

deposition

n

or, putting it all together,

$$\frac{\partial f}{\partial t} + |\nabla f| \mathbf{q} \cdot \mathbf{n} = 0.$$

This equation can, in principle, be treated by the Charpit method, although the details tend to be complicated.

7.5 Second-order linear equations in two variables

All mathematics courses with any applied component cover the basics of second-order linear equations in two variables, because they occur so very frequently in practice. This section contains a brief executive summary for the sake of completeness. Any standard text has more details.

The kind of equation we consider is one of the form

$$a(x, y)\frac{\partial^2 u}{\partial x^2} + 2b(x, y)\frac{\partial^2 u}{\partial x \partial y} + c(x, y)\frac{\partial^2 u}{\partial y^2}$$

$$= \text{terms linear in } u, \frac{\partial u}{\partial x} \text{ and } \frac{\partial u}{\partial y},$$

where a, b and c are given functions. One starting point is to ask whether there are any curves across which the second derivatives of u can have jump discontinuities, thereby generalising the 'third caveat' argument starting on p. 84. Repeating that argument, the continuity of $\partial u/\partial x$ and $\partial u/\partial y$ along such a curve $(x(t), y(t))$, together with the fact that the original partial differential equation holds on either side of it, leads to the equations

$$\dot{x}\left[\frac{\partial^2 u}{\partial x^2}\right] + \dot{y}\left[\frac{\partial^2 u}{\partial x \partial y}\right] \qquad = 0,$$

$$\dot{x}\left[\frac{\partial^2 u}{\partial x \partial y}\right] + \dot{y}\left[\frac{\partial^2 u}{\partial y^2}\right] = 0,$$

$$a\left[\frac{\partial^2 u}{\partial x^2}\right] + b\left[\frac{\partial^2 u}{\partial x \partial y}\right] + c\left[\frac{\partial^2 u}{\partial y^2}\right] = 0,$$

where the dot means d/dt. These homogeneous equations have a non-trivial solution only if

$$\begin{vmatrix} \dot{x} & \dot{y} & 0 \\ 0 & \dot{x} & \dot{y} \\ a & b & c \end{vmatrix} = 0,$$

that is, if

$$a\dot{y}^2 - 2b\dot{y}\dot{x} + c\dot{x}^2 = 0.$$

Note the minus sign and the $\dot{y}$ in front of the coefficient of the second x-derivative, and vice versa for the second y-derivative.

This is the characteristic equation for our second-order equation. We distinguish three cases, depending on whether it has two real roots for $dy/dx = \dot{y}/\dot{x}$, no roots or a repeated root.

Two real roots: hyperbolic equations

Equations with two real families of characteristics are called *hyperbolic*; such equations are the closest relatives to the quasilinear equations we studied earlier. There are two distinct families of characteristics and it is possible to reduce the equation to the *canonical form*

$$\frac{\partial^2 u}{\partial \xi \partial \eta} = \text{terms involving lower-order derivatives}$$

by changing to variables ξ, η that are constant on the characteristics. A typical well-posed problem for this equation requires both u and its normal derivative to be given on an initial curve that is nowhere tangent to a characteristic (and so, in particular, cannot be closed). Discontinuities can propagate along characteristics and the solution cannot be expected to be any smoother than the initial data: it continues to jangle forever (possibly with some damping).

In addition to use of characteristic variables, solution methods include separation of variables (especially if the coefficients are independent of time and the equation contains $\partial^2 u/\partial t^2$, when solutions proportional to $e^{-i\omega t}$ are sought as part of a normal-modes calculation), transform methods and the use of the Riemann function.

The quintessential hyperbolic equation is the wave equation

$$\frac{\partial^2 u}{\partial t^2} - c^2 \frac{\partial^2 u}{\partial x^2} = 0,$$

for which we can take $\xi = x - ct$, $\eta = x + ct$; the canonical form is

$$\frac{\partial^2 u}{\partial \xi \partial \eta} = 0,$$

revealing immediately the general solution as $f(x - ct) + g(x + ct)$ for arbitrary functions f and g.

No real roots: elliptic equations

Equations with no real characteristics are called *elliptic*. It is still possible to reduce them to a canonical form, in this case

$$\frac{\partial^2 u}{\partial \xi^2} + \frac{\partial^2 u}{\partial \eta^2} = \text{terms involving lower-order derivatives},$$

where ξ and η are such that the roots of the characteristic equation are $\xi \pm i\eta$. Now the typical problem asks for a *single* piece of information about u, for example u itself or some combination of u and its normal derivative, to be given on a closed curve containing the region of interest. If the domain is infinite, a careful description of the behaviour of u at infinity must also be given. As we see in Exercise 12, there are various possible degrees of existence and/or uniqueness of the solution, but one thing that all the solutions share is that they are smooth functions of x and y inside the domain: discontinuities in the boundary data cannot propagate into the interior.[4]

Solution methods include separation of variables and transform methods as above, Green's functions (the equivalent of Riemann functions), and the use of complex variables when the equation is Laplace's equation, the *eminence grise* of elliptic equations.

Double roots: parabolic equations

Between elliptic and hyperbolic equations lies a debatable ground where the characteristics are real and coincident. These are called *parabolic* equations. Some parabolic equations are just ordinary differential equations in disguise[5] but genuinely parabolic equations include the heat equation

$$\frac{\partial u}{\partial t} = \frac{\partial^2 u}{\partial x^2}$$

and many variants on it. As we know, these equations are ubiquitous in real-world models, because the world is full of processes that run on heat. The fact that linear diffusion also leads to the heat (or, indeed, diffusion) equation adds extensively to the list of applications.

Complex characteristics cannot coincide: why?

Cooking, to name but one.

---

[4] They can, however, be generated by discontinuities in the coefficients of the equation; this is hardly surprising.

[5] Try, for example,

$$\frac{\partial^2 u}{\partial x^2} - 2\frac{\partial^2 u}{\partial x \partial y} + \frac{\partial^2 u}{\partial y^2} = \left(\frac{\partial}{\partial x} - \frac{\partial}{\partial y} \right)^2 u = 0;$$

the change of variables $\xi = x - y$, $\eta = x + y$ reduces it to $\partial^2 u / \partial \xi^2 = 0$. It is argued in [44] that equations like this should be regarded as almost hyperbolic.

The double characteristic of the heat equation is $t = $ constant, corresponding to the double root of the characteristic equation. This shows straightaway that the solution may have discontinuities in t (i.e. across characteristics) but should be expected to be smooth in x. That is reasonable: the equation itself tells us that, as heat flows from hot to cold, regions of the graph of u having a large positive curvature are pulled up, and vice versa for negative curvature, so that the solution becomes smoother (draw a picture to see this). However, it also says that if we are finding the temperature in a bar we can feel the effect of changing the end value of the temperature immediately, which may be a little too quick for some.[6] At any event, a typical well-posed problem has u given at $t = 0$ (even though this is a characteristic) and also on spatial boundaries, say at $x = 0$ and $x = 1$; other problems frequently involve radiation boundary conditions, modelling heat transfer to an exterior medium.

You might think of this as being one initial condition corresponding to the single time derivative, plus two spatial conditions corresponding to the elliptic part $\partial^2 u/\partial x^2$ – remember that in more dimensions the heat equation is

$$\frac{\partial u}{\partial t} = \nabla^2 u.$$

7.6 Further reading

See [37] for an accessible introduction to Poisson processes. The spray-forming of an aluminium billet is described in [21].

7.7 Exercises

1 Solution blow-up. Consider the equation

$$\frac{\partial u}{\partial t} + u\frac{\partial u}{\partial x} = 0;$$

show that the general solution is given implicitly by

$$u(x, t) = f(x - tu(x, t))$$

for arbitrary smooth f. Consider the initial value problem in which $u = u_0(x)$ on $t = 0$. Show that the parametric form of the solution is

$$t = \tau, \qquad x = \tau u_0(s) + s, \qquad u = f(s).$$

Deduce that u is constant on the characteristic projections $dx/dt = u$, which are thus straight lines.

Draw the characteristic projections and sketch the solution surface for the two kinds of initial data,

$$u_{0\pm}(x) = \pm \tan^{-1}\left(\frac{x}{\epsilon}\right).$$

---

[6] It is possible to amend the equation so that it is 'very slightly' hyperbolic, with a large but finite speed of propagation; see the paper [34] for a readable account. Note also that, as we see in Chapter 9, the effect of any change in a boundary condition decays very rapidly as we move away from that boundary.

Which solution remains single-valued?

Now suppose that the partial differential equation is

$$\frac{\partial u}{\partial t} + u\frac{\partial u}{\partial x} = -cu,$$

where c is a positive constant, and that initially

$$u(x, 0) = A \tanh x.$$

Show that if $A > -c$ then $|\partial(x, t)/\partial(s, \tau)| \neq 0$ for all s, τ. Deduce that the solution exists for all t for this range of A-values. (In this example, the tendency of the characteristic projections to cross is counteracted by the exponential decay of u along them: how is this manifested geometrically?)

2 General solution of a quasilinear equation.   Given the quasilinear equation

$$a(x_1, x_2, u)\frac{\partial u}{\partial x_1} + b(x_1, x_2, u)\frac{\partial u}{\partial x_2} = c(x_1, x_2, u),$$

suppose that we can find two first integrals of the characteristic equations in the form $f(x_1, x_2, u) = C_1$, $g(x_1, x_2, u) = C_2$, where C_1 and C_2 are arbitrary constants of integration. Show that the general solution of the original partial differential equation is given by $F(C_1, C_2) = 0$, where F is an arbitrary function of two variables.

Euler's theorem on homogeneous functions says that a function $u(x_1, x_2)$ that is homogeneous of degree n (that is, for all real t $u(tx_1, tx_2) = t^n u(x_1, x_2)$) satisfies the equation

$$x_1\frac{\partial u}{\partial x_1} + x_2\frac{\partial u}{\partial x_2} = nu.$$

Show that the general solution of this equation is

$$u(x_1, x_2) = x_1^n G(x_2/x_1),$$

where G is an arbitrary function.

3 Waiting times for a Poisson process.   Suppose that we start a Poisson process at time 0. What is the distribution of T, the time until the first event (clearly this is the same as the distribution of the interval between any two events)? We find it as follows. Let $F_T(t) = P(T < t)$. Explain why

$$F_T(t + \delta t) = (1 - \lambda \delta t)F_T(t)$$

and deduce that T has the negative exponential distribution with density $f_T(t) = \mathrm{d}F_T(t)/\mathrm{d}t = \lambda e^{-\lambda t}, t > 0$. Figure 7.8 shows a histogram of the inter-arrival times of trades in the S&P 500 futures contract in Chicago (an open outcry market) with normal and log scales for

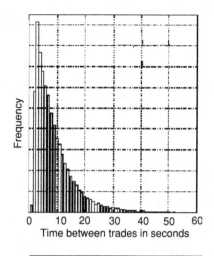

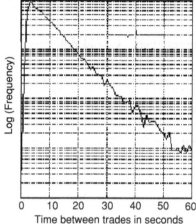

Figure 7.8 Standard and log-linear plots of the inter-arrival-time frequencies for a stock market.

the frequency; the latter is a good approximation to a straight line except for very short times between trades (thanks to Rashid Zuberi for these plots).

4 Viral antidote. Suppose that N computers are infected with a virus and, at time $t = 0$, I send them all an antidote which will cure the problem as soon as they open their inbox. Also assume that if n users are still infected at time t then in the short time interval $(t, t + \delta t)$ one and only one user will log on, with probability $\mu n p_n(t)\, \delta t$. Why is this a reasonable model?

If $p_n(t)$ is the probability that there are still n infected computers at time t, use the decomposition into disjoint events

$$P(n \text{ infected at } t + \delta t) = P(n + 1 \text{ at } t \text{ and one user logs on})$$
$$+ P(n \text{ at } t \text{ and no users log on})$$

to show that

$$p_n(t + \delta t) = \mu(n + 1)p_{n+1}(t)\delta t + (1 - \mu n\, \delta t)p_n(t).$$

Letting $\delta t \to 0$, show that the generating function

$$G_A(x, t) = \sum_{n=0}^{\infty} p_n(t)x^n$$

satisfies

$$\frac{\partial G_A}{\partial t} + \mu(x - 1)\frac{\partial G_A}{\partial x} = 0.$$

Show that if there are N victims initially, the solution is

$$G_A(x, t) = \left(1 + (x - 1)e^{-\mu t}\right)^N.$$

What is the mean of this distribution?

Modify the argument to allow for infection and as well as cure.

5 A shock in the kinematic wave equation.   Consider the equation

$$\frac{\partial u}{\partial t} + \frac{\partial(\frac{1}{2}u^2)}{\partial x} = 0,$$

which is equivalent to the kinematic wave equation

$$\frac{\partial u}{\partial t} + u\frac{\partial u}{\partial x} = 0.$$

Show that the shock speed is

$$\frac{dS}{dt} = \frac{1}{2}(u_+ + u_-).$$

Show that if

$$u(x, 0) = \begin{cases} 1, & x < 0, \\ 0, & x > 0, \end{cases}$$

there is a solution that has a shock on $x = \frac{1}{2}t$.

6 Rankine–Hugoniot is not enough.   Consider the kinematic wave equation of the previous question and show that it can be written in conservation form $\partial P/\partial t + \partial Q/\partial x$ in infinitely many ways (start by taking $P = u$, then try $P = u^2$, finding the corresponding Q in each case). Conclude that the Rankine–Hugoniot relation is not sufficient to determine the shock structure uniquely for the initial value problem in which $u(x, 0)$ is given on $t = 0$; the solutions are nonunique.

Even if we have good physical reasons for choosing a particular P and Q, Rankine–Hugoniot may still be insufficient. Consider the kinematic wave equation above with

$$u(x, 0) = \begin{cases} 0, & x < 0, \\ 1, & x > 0. \end{cases}$$

Taking $P(u) = u$, show that there is a solution with $S(t) = \frac{1}{2}t$ and

$$u(x, t) = \begin{cases} 0, & x < S(t), \\ 1, & x > S(t). \end{cases}$$

Show that there is another solution

$$u(x, 0) = \begin{cases} 0, & x < v_1 t, \\ \alpha, & v_1 t < x < v_2 t \\ 1, & x > v_2 t, \end{cases}$$

where $0 < v_1 < v_2 < \frac{1}{2}$, and find the necessary relations between α, v_1 and v_2. How could you find still further solutions?

Show that there are also solutions of the form $u(x, t) = f(x/t)$ and hence find a solution that is continuous everywhere but has gradient discontinuities across the characteristic projections $x = 0$ and $x = t$. (Note that on these characteristics $dx/dt = 0, 1$; these are the limiting values of $u(x, 0)$ as $x \to 0$ from below and above.) This solution is known as an *expansion fan*.

Now do the next exercise.

7 **Causality.** The situation of the previous exercise is throughly un-satisfactory. One way to resolve it is through the idea of *causality*, which is itself a combination of two ideas: 'information flows along characteristics in a forward direction (as measured by time, if that is one of the variables)' and 'the number of equations must equal the number of unknowns'. We think of information as propagating along characteristics, starting from the initial data; in particular, the value of u evolves along a characteristic. Now think about what happens at a shock. To determine the shock path, we need to know the value of u on either side in order to implement the Rankine–Hugoniot condition. This means that, at each point on the shock, we need two *incoming characteristics* that emanate from the initial data, each of which car-ries its own value of u. Such a solution is called *causal*. If any of the characteristics at the shock are *outgoing*, that is they do not intersect the initial data, the solution is not causal and is rejected on physical grounds. Often (but not always), the causal solution is unique.

Draw the characteristic diagram for Exercise 5 and show that the shock in it is causal.

Draw a characteristic diagram for the single-shock solution of Exercise 6 with the x-axis horizontal, making sure that you have a characteristic through each point. Draw arrows on the characteristics to indicate the direction of time flow (upwards). Deduce that this solution is not causal. (For this problem, the only causal solution is the expansion fan.)

8 Practice at Charpit.   Solve $xp + yq = pq$ with $u(x, 0) = \frac{1}{2}x$.

9 Shortest distance to a curve.   Let C be a curve, and let $D(x, y)$ be the shortest distance from the point (x, y) to C. Take coordinates n, s with origin at (x, y); measure n along the normal to C from (x, y) and s parallel to the tangent at the foot of this normal. Show that $\partial D/\partial s = 0$ at the point (x, y) and that $|\nabla D|^2 = (\partial D/\partial n)^2$. Deduce from the invariance of the gradient under rotation of the axes that

$$\left(\frac{\partial D}{\partial x}\right)^2 + \left(\frac{\partial D}{\partial y}\right)^2 = 1, \qquad D = 0 \quad \text{on} \quad C.$$

10 Sandpiles.   Calculate the shape of a limiting sandpile on a square table, (a) by common sense, (b) by solving Charpit's equations for (7.1) on p. 92. Where are the ridgelines? Repeat for a circular table and then an elliptical one (the second of these is harder).

11 Curvature flows.   A plane curve moves so that the magnitude of its normal velocity V_n is equal to its curvature κ, the sign being chosen so that a circle shrinks. Noting that $\kappa = d\theta/ds$, where s is arclength and θ is the angle between the tangent to the curve and the x-axis, show that the area inside a simple closed curve decreases at a rate 2π and deduce that the curve eventually vanishes.

If a curve moving in this way is described by $y = f(x, t)$, show that f satisfies

$$f_t = \frac{f_{xx}}{1 + f_x^2},$$

where $f_t = \partial f/\partial t$ etc. Find the travelling-wave shapes when $f(x, t) = F(x - Ut)$ with U constant.

12 Boundary-value problems for elliptic equations.   Give a physical interpretation in terms of heat flow for Poisson's equation

$$\nabla^2 u(\mathbf{x}) = F(\mathbf{x})$$

in a region V bounded by a smooth curve ∂V. When u is prescribed to be equal to a given function $f(\mathbf{x})$ on ∂V (the *Dirichlet problem*), show that the solution (if it exists) is unique, by considering

$$\int_V \nabla \cdot (v\nabla v)\, dV,$$

where $v(\mathbf{x})$ is the difference between two solutions with the same boundary values.

Now suppose that $\partial u/\partial n$ is prescribed as equal to $g(\mathbf{x})$ on ∂V (the *Neumann* problem). Show that there is a solution only if

$$\int_V F(\mathbf{x})\, d\mathbf{x} = \int_{\partial V} g(\mathbf{x})\, dS,$$

and interpret this physically. Given this condition, show that the solution is unique only up to addition of a constant (this is an example of the Fredholm Alternative).

Finally, show that for the *Robin* problem, in which

$$\frac{\partial u}{\partial n} + h(\mathbf{x})u = q(\mathbf{x})$$

is given on ∂V, uniqueness is to be expected only if $h(\mathbf{x}) > 0$. Again, interpret this physically.

'The stagnation point might go halfway to infinity and then stop.'

8

Case study: traffic modelling

8.1 Simple models for traffic flow

Mathematicians and physicists have long been interested in the problem of traffic, and the area is one of active research. A variety of models have been suggested with a view to understanding, for example, how and why traffic jams form, how to maximise carrying capacity of roads, or how best to use signals, speed limits and other controls to reduce journey times (the feedback effect whereby quicker journeys encourage more people to take to the roads is strangely absent from these analyses). Some models are based on discrete simulations of the movement of individual cars; as you may imagine, such models can be very large and complicated, and indeed they fall into the trendy area of 'complex systems'. There is, however, a strand of traffic research that treats the cars as a continuum with a local number density and velocity that are more or less smooth functions of space and time, much as in the treatment of charged particles in the case study in Chapter 6. Models of this kind are unlikely ever to forecast the fine details of gridlock in New York City or indeed Oxford; but they do offer insights into the way in which traffic can behave, and to some extent they can be calibrated to, or at least compared with, observations. On the scale from parsimony (as few parameters and mechanisms as possible) to complexity, these models are very much at the parsimonious end; the cost, a lack of realism, is balanced by a gain in understanding. They fit in well with my recommended philosophy of always trying to do the easiest problem first.

Let us, then, start with a toy model for cars travelling in one direction down a single-lane road (no overtaking) that is long and straight.

Suppose that x measures distance along the road and that we work on a large enough lengthscale, or we look from far enough away, that the cars can be treated as a continuum with number density $\rho(x, t)$ (cars per kilometre) and speed $u(x, t)$. Supposing further that no cars join or leave the road, we immediately write down 'conservation of cars' in the form

$$\frac{\partial \rho}{\partial t} + \frac{\partial(\rho u)}{\partial x} = 0,$$

as the flux of cars is equal to ρu.

Given the continuum assumption, this equation is uncontroversial; but it is only one equation for two unknowns. We need some kind of 'constitutive relation' to close the system.

Blinkered drivers

As the basis for a very simple model we might say that, as they enter the road, drivers choose the constant speed they want to drive at, and then they drive at that speed no matter what happens. Of course, this is ludicrously unrealistic, but let's see what features it predicts. If the speed u of an individual car is constant, then the derivative of u following that car is zero:

$$\frac{Du}{Dt} = \frac{\partial u}{\partial t} + u \frac{\partial u}{\partial x} = 0.$$

This kinematic wave equation is easy to solve by characteristics with initial data $u(x, 0) = u_0(x)$, say, corresponding to a snapshot at $t = 0$ of the speeds all along the road. The characteristic equations are

$$\frac{dt}{d\tau} = 1, \qquad \frac{dx}{d\tau} = u, \qquad \frac{du}{d\tau} = 0,$$

so that u remains constant along a characteristic whose projection has slope $dx/dt = u$. This simply says that the cars move along characteristics with constant speed u. So, to construct the solution, we simply draw all the characteristic projections through the initial line $t = 0$ and read off the value of u at any point x and later time t. This procedure works fine if $u_0(x)$ is increasing, since then the characteristics spread out as in Figure 7.3. But if $u_0(x)$ is decreasing, we inevitably have a collision of characteristic projections – and cars – after a finite time, as in Figure 7.4. This is an example of the solution blow-up we discussed in Chapter 7, and here it has an obvious physical interpretation that fast cars have caught up with slow ones and are trying to occupy the same bit of road. That is, the model predicts that cars with different speeds will end up in the same place. Clearly, this model is inadequate as a description of how real traffic behaves. Its predictions are realistic within its severe limitations, but they are so far off the mark that we need to do something more sophisticated.

Local speed–density laws

In our quest for greater realism, we should try to describe how drivers respond to the traffic around them. A simple way to do this is to propose a (constitutive) relation between the speed of cars at a point x and their density there. That is, we assume that

$$u = U(\rho)$$

for a suitable function U. This function should be determined experimentally from observations of local speed and density, or at least written down in a parametric form and the parameters calibrated (fitted) to observations of global features of the traffic flow (an example of an inverse problem). Before going too far down this road, let us see what happens when we put a simple U into the model. As heavy traffic generally moves more slowly than light traffic, we want $U(\rho)$ to be a decreasing function of ρ. We may assume a maximum car speed u_{max} and that cars are driven at this speed on an empty road, when $\rho = 0$. Conversely, we can assume a maximum bumper-to-bumper density ρ_{max} at which the traffic comes to a complete halt, so that $u = 0$. This suggests that the speed–density law

Observation suggests that u_{max} is greater than the speed limit . . .

It is an implicit assumption of the model that all drivers behave in the same way, and it is also assumed that they drive as fast as is consistent with the ambient traffic density.

$$u = u_{max} \left(1 - \frac{\rho}{\rho_{max}} \right)$$

should be a reasonable qualitative description.

We can make an immediate and interesting observation. The flux of cars is

$$Q = u\rho = u_{max}\rho_{max} \left(1 - \frac{\rho}{\rho_{max}} \right) \frac{\rho}{\rho_{max}},$$

and it is greatest when $\rho = \frac{1}{2}\rho_{max}$, so that $u = \frac{1}{2}u_{max}$. In this model the assumed free-market individual desire of drivers to minimise their journey time by always driving as fast as possible does not necessarily deliver the maximum-flux solution for drivers as a whole.

Leaving this aside, let us see whether we still have blow-up. Making the trivial scalings $u = u_{max}u'$, $\rho = \rho_{max}\rho'$, with suitable scalings for x and t, and dropping the primes, we have the dimensionless equation

$$\frac{\partial \rho}{\partial t} + \frac{\partial}{\partial x}(\rho(1 - \rho)) = \frac{\partial \rho}{\partial t} + (1 - 2\rho)\frac{\partial \rho}{\partial x} = 0 \qquad (8.1)$$

(this is, of course, just a conservation law). The characteristic equations are

$$\frac{dt}{d\tau} = 1, \qquad \frac{dx}{d\tau} = 1 - 2\rho, \qquad \frac{d\rho}{d\tau} = 0,$$

so the characteristics are again straight, as ρ is constant on them. However, bearing in mind that $0 < \rho < 1$, we see that we can easily prescribe initial data for ρ that will again lead to finite-time blow-up: the characteristic projections can have slopes of either sign and they can easily cross. Indeed, the substitution $v = 1 - 2\rho$ reduces (8.1) to $\partial v / \partial t + v \partial v / \partial x = 0$, so blow-up is inevitable.

Clearly, we must either tinker further with the model so that blow-up is forbidden or face up to the fact that it *will* happen in realistic models, and decide what to do about it.

Note that the characteristic speed, $dx/dt = 1 - 2\rho$, is *not* equal to the car speed – that is $u = 1 - \rho$. Information always propagates more slowly than the cars and can indeed move backwards, if $\rho > \frac{1}{2}$.

8.2 Traffic jams and other discontinuous solutions

Red lights and shocks

We saw in Section 7.3 that the notion of a solution to the conservation law

$$\frac{\partial \rho}{\partial t} + \frac{\partial Q}{\partial x} = 0$$

can be extended to allow jump discontinuities across curves $x = S(t)$, provided that $S(t)$ satisfies the Rankine–Hugoniot relation

$$\frac{dS}{dt} = \frac{[Q]_-^+}{[\rho]_-^+}.$$

Shocks can originate spontaneously, when characteristic projections cross, but a situation in which it is easy to see them is if a stream of traffic with speed u_0 and density ρ_0 is brought to a halt by a traffic light at, say, $x = 0$ and at time $t = 0$.

First, let us look at the cars that did not get through the light, the ones that are in $x < 0$ at $t = 0$; their density is $\rho(x, t)$, which satisfies (8.1). At the moment the light goes red, they are all travelling towards the light with speed u_0 and density ρ_0. These cars therefore see the initial condition $\rho(x, 0) = \rho_0$, $x < 0$. Because $u = 0$ at $x = 0$, the density there takes its maximum value, so $\rho(0, t) = 1$. There are two families of characteristics to consider. Those starting from the initial data on $t = 0$ have characteristic speed $1 - 2\rho_0$ (which may be negative), and they carry the value $\rho = \rho_0$. Those starting from the light at $x = 0$ have speed $1 - 2 = -1$ and carry the value $\rho = 1$. The two families therefore cross immediately and, as shown in Figure 8.1(a), a shock must originate at

Remember that dx/dt is the reciprocal of the gradient in the xt-plane.

Figure 8.1 (a) Shock formation in traffic arriving at a red light. (b) A shock in traffic ahead of a red light. (c) Traffic density profile after the light turns red. In all three cases $\rho_0 = \frac{2}{3}$.

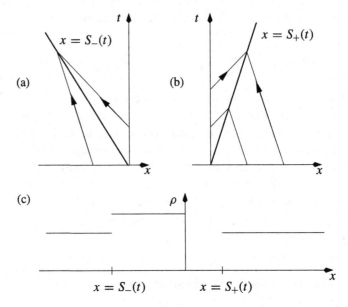

$x = 0, t = 0$. Its speed is given by

$$\frac{\mathrm{d}S_-}{\mathrm{d}t} = \frac{[\rho(1-\rho)]_-^+}{[\rho]_-^+}$$
$$= \frac{0 - \rho_0(1-\rho_0)}{1 - \rho_0}$$
$$= -\rho_0.$$

So, for $x < 0$ there is a shock on $x = -\rho_0 t$ that propagates backwards into the oncoming traffic, bringing it to a halt.

Now consider the traffic that gets through the light before it turns red. We expect these cars to continue on their way at speed u_0, leaving a stretch of empty road behind them. The density satisfies (8.1) for $x > 0$, $t > 0$, with

$$\rho(x, 0) = \rho_0, \qquad \rho(0, t) = 0.$$

As shown in Figure 8.1(b), the characteristics starting from the lights at $x = 0$ all have $\mathrm{d}x/\mathrm{d}t = 1$ and they carry the value $\rho = 0$, while those starting from the initial data on $t = 0$ all have $\mathrm{d}x/\mathrm{d}t = 1 - 2\rho_0 < 1$ and $\rho = \rho_0$. Again, a shock must originate at $x = 0$, $t = 0$. Its speed is given by

$$\frac{\mathrm{d}S_+}{\mathrm{d}t} = \frac{[\rho(1-\rho)]_-^+}{[\rho]_-^+}$$
$$= 1 - \rho_0.$$

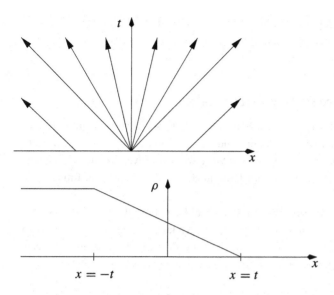

This is just the speed of the last car to get through the lights, and our intuition is confirmed. (Note that both our shocks are causal, as defined in Exercise 7 at the end of Chapter 7.)

Green lights and expansion fans

What happens when the light turns green and a queue of stationary traffic ($\rho = 1$) moves off? In this case the initial data $\rho(x, 0)$ is discontinuous, being equal to 1 for $x < 0$ and 0 for $x > 0$. If we were to smooth out this discontinuity, say with a tanh function, we would see characteristics with all speeds between -1 (corresponding to $\rho = 1$) and $+1$ (corresponding to $\rho = 0$ as in Figure 7.3 on p. 88. This motivates the idea of an *expansion fan*, a collection of characteristic projections all emanating from a single point, as shown in Figure 8.2. It allows the solution to make a continuous transition from $\rho = 1$ to $\rho = 0$.

For our problem, the characteristic projections are all straight and ρ is constant along them. This means that ρ is a function of x/t alone (a similarity solution). On any characteristic, ρ is a constant equal to α, say, where $0 < \alpha < 1$, and the equation of the characteristic projection is $x = (1 - 2\alpha)t$. We can write this as $\rho = \frac{1}{2}(1 - \beta)$ on $x = \beta t$, or explicitly as

In general, the characteristic projections are only locally straight, and the fan curves over.

$$
\rho(x, t) = \begin{cases} 1, & x < -t, \\ \frac{1}{2}(1 - x/t), & -t \le x \le t, \\ 0, & x > t. \end{cases}
$$

Notice that discontinuities in the derivative of ρ propagate along the characteristics $x = -t$, $\rho = 1$ and $x = t$, $\rho = 0$.

It is, in principle, possible to construct the solution for most initial-value problems using a mixture of shocks and expansion fans, but it can become complicated, as we will see in Exercise 3.

8.3 More sophisticated models

The models above have their appeal, but they are rather limited in ambition and realism; we may indeed see fairly abrupt changes in traffic density, but never true jumps. Moreover, the drivers in these models are very myopic: they do not look ahead at all to anticipate future traffic developments.

There are several things we could do about this. One way to go is to introduce a non-local speed–density law, so that $u(x, t)$ depends on $\rho(x + h, t)$ as well as $\rho(x, t)$, where the driver looks a distance h ahead.[1] A limiting case (as $h \to 0$) of this law is the model

$$u = 1 - \rho - \epsilon \frac{\partial \rho}{\partial x},$$

where ϵ is a small positive constant. This says that drivers take into account whether traffic density is increasing or decreasing and slow down if it is increasing. Putting this into $\partial \rho / \partial t + \partial Q / \partial x = 0$ leads to the equation

$$\frac{\partial \rho}{\partial t} + \frac{\partial}{\partial x} \left(\rho(1 - \rho) \right) = \epsilon \frac{\partial}{\partial x} \left(\rho \frac{\partial \rho}{\partial x} \right),$$

a nonlinear diffusion equation with some interesting properties. Because it is parabolic when $\epsilon > 0$, albeit nonlinear (and degenerate because the 'diffusion coefficient' $\epsilon \rho$ vanishes when $\rho = 0$), its solutions may be smoother than is the case when $\epsilon = 0$. In Exercise 5 you are asked to show that travelling-wave solutions of this equation are consistent with the Rankine–Hugoniot conditions that apply when $\epsilon = 0$.

Further models involve an evolution equation for u rather than just a constitutive equation. For example, we might replace $u = U(\rho)$ by an equation such as

$$\frac{\partial u}{\partial t} + u \frac{\partial u}{\partial x} = -\frac{u(\rho, t) - U(\rho)}{\tau},$$

which says that the rate of change of u following a car is proportional to the difference between u and the equilibrium speed U; here τ represents the time over which a driver changes speed to reach equilibrium. To this we might also add an anticipatory term $-\epsilon \, \partial \rho / \partial x$ as above, modelling drivers' tendency to speed up if they see light traffic ahead, and slow down

---

[1] One could also introduce a 'reaction time' delay in the t-variable.

if the traffic is getting worse. All these, and many more, possibilities have been discussed in the traffic literature (see [25] for more details). Roughly speaking, many of the models do a good job in describing generic features such as jams and abrupt changes in traffic density, but they are less successful in forecasting the evolution of traffic from a given starting density (which is, of course, the big question). Only recently has reliable empirical data, gathered by induction loops buried in roads, become available, and I have no doubt that there are many interesting developments to come.

8.4 Sources and further reading

The kinematic wave model for traffic flow is usually attributed to Whitham and Lighthill. An excellent survey of a huge variety of approaches to traffic modelling can be found in [25]; see also www.trafficforum.org.

8.5 Exercises

1 Blinkered cars. Consider the kinematic wave equation

$$\frac{\partial u}{\partial t} + u \frac{\partial u}{\partial x} = 0,$$

where $u(x, 0) = u_0(x)$ is a smooth decreasing function of x. Find the solution in parametric form. Look at the relevant Jacobian to show that the earliest time at which the characteristics cross is

$$t_{\min} = -\frac{1}{\min_{-\infty < x < \infty} u_0'(x)}.$$

Show that the rate at which neighbouring cars get closer to each other is $\partial u / \partial x$ and interpret the blow-up result mentioned on p. 108 in this light.

2 Traffic jams. Consider the traffic model

$$\frac{\partial \rho}{\partial t} + \frac{\partial (u\rho)}{\partial x} = 0,$$

where $u = 1 - \rho$ for $0 \leq \rho \leq 1$.

(a) A tractor is travelling along the road at a quarter of the maximum speed and a very long queue of cars travelling at the same speed has built up behind it. At time $t = 0$ the tractor passes the origin $x = 0$ and immediately turns off the road. Sketch the characteristic diagram; show that there is an expansion fan for ρ centred at $x = 0$, $t = 0$ and find $\rho(x, t)$ for $t > 0$.

(b) A queue is building up at a traffic light at $x = 1$ so that, when the light turns to green at $t = 0$,

$$\rho(x, 0) = \begin{cases} 0 & \text{for } x < 0 \text{ and } x > 1, \\ x & \text{for } 0 < x < 1. \end{cases}$$

Show that the characteristics, labelled by s and starting from $(s, 0)$, are given by $t = \tau$ and

$$
\begin{array}{lll}
x - s = \tau & \text{in } x < \tau \text{ and } x > \tau + 1, \text{ on which } \rho = 0, \\
x - s = (1 - 2s)\tau & \text{in } \tau < x < 1 - \tau, \text{ on which } \rho = s, \\
x - 1 = (1 - 2\rho_0)\tau & \text{in } 1 - \tau < x < 1 + \tau
\end{array}
$$

(these last characteristics, on which $\rho = \rho_0 = (\tau - x + 1)/(2\tau)$, form an expansion fan starting from the light). Draw the characteristic projections in the xt-plane; show that all those starting with $0 < s < 1$ pass through one point and deduce that a collision first occurs at $x = \frac{1}{2}$ and $t = \frac{1}{2}$.

Harder: Show that thereafter there is a shock $x = S(t)$, starting from $(\frac{1}{2}, \frac{1}{2})$, where

$$\frac{dS}{dt} = \frac{S + t - 1}{2t}.$$

Write $S(t) = 1 + \tilde{S}(t)$ to reduce this equation to one that is homogeneous in $\tilde{S}$ and t, and hence solve it.

3 **Red light, green light.** Continue the solution discussed in Section 8.2 as follows. Suppose that the light turns green after time T. Move the time origin to this moment and neglect the traffic that has already passed the light and is in $x > 0$. Find the solution of (8.1) with the initial data (at the new time origin)

$$\rho(x, 0) = \begin{cases} \rho_0, & x < -\rho_0 T, \\ 1, & -\rho_0 T < x < 0, \\ 0, & x > 0. \end{cases}$$

Show that the shock that is initially at $x = -\rho_0 T$ continues to propagate at speed ρ_0 until it is caught up by the characteristic projection $x = -t$ at time $t = \rho_0 T/(1 - \rho_0)$. Show that thereafter there is a shock at $x = S(t)$ where

Switching to the variable $v = \rho - \frac{1}{2}$ helps you to spot the simplification.

$$
\begin{aligned}
\frac{dS}{dt} &= \frac{\frac{1}{4}(1 - S^2/t^2) + \rho_0(1 - \rho_0)}{\frac{1}{2}(1 - S/t) - \rho_0} \\
&= \frac{(\rho_0 - \frac{1}{2})t + S}{2t}.
\end{aligned}
$$

Solve this equation by the substitution $S(t) = ts(t)$.

Harder: Show that the shock initially propagates to the right if $\rho_0 < \frac{1}{2}$ and to the left if $\rho_0 > \frac{1}{2}$. Calculate ρ for $x = S(t)+$ and show that the jump in ρ across the shock is equal to $(\rho_0(1 - \rho_0))^{1/2} (T/t)^{1/2}$ and hence decreases as t increases.

4 Two-lane traffic. Explain why the one-lane model above might be extended to a two-lane model in the form

$$\frac{\partial \rho_1}{\partial t} + \frac{\partial}{\partial x} (\rho_1 u_1) = F_{12}(\rho_1, \rho_2, u_1, u_2),$$

$$\frac{\partial \rho_2}{\partial t} + \frac{\partial}{\partial x} (\rho_2 u_2) = -F_{12}(\rho_1, \rho_2, u_1, u_2),$$

and explain where F_{12} comes from. What general properties should F_{12} have, in your opinion? How would it differ for an American freeway, in which overtaking is allowed on the inside lane, and for British case in which (in principle if not in practice) it is not?

5 Smoothed traffic equation. Consider the equation

$$\frac{\partial \rho}{\partial t} + \frac{\partial}{\partial x} (\rho(1 - \rho)) = \epsilon \frac{\partial}{\partial x} \left(\rho \frac{\partial \rho}{\partial x} \right),$$

a model for anticipatory drivers. Suppose we look for a solution $\rho = f(x - Vt)$, $-\infty < x < \infty$, with $\rho \to \rho_{\pm}$ as $x \to \pm\infty$. Show that

$$V = \frac{[\rho(1 - \rho)]_{-\infty}^{\infty}}{[\rho]_{-\infty}^{\infty}}.$$

Compare this with the Rankine–Hugoniot condition. What do you think happens as $\epsilon \to 0$? We will return to this issue in Chapter 16.

Carry out the same procedure for the smoothed kinematic wave equation

$$\frac{\partial u}{\partial t} + u \frac{\partial u}{\partial x} = \epsilon \frac{\partial^2 u}{\partial x^2},$$

known as Burgers' equation. (Amazingly, it can be reduced to the heat equation by the Cole–Hopf transformation $u = -2\partial \log v/\partial x$, taking $\epsilon = 1$ without loss of generality: try it!)

'The mass of this thing is about 1 kilometre.'

9

The delta function and other distributions

9.1 Introduction

In this chapter we give a very informal introduction to *distributions*, also called *generalised functions*. We do two rather amazing things: we see how to differentiate a function with a jump discontinuity, and we develop a mathematical framework for point forces, masses, charges, sources etc. Furthermore, we find that these two ideas find their expression in the same mathematical object: the Dirac delta function.

When I learned proper real analysis for the first time, we spent ages agonising about continuity, left and right limits, one- and two-sided derivatives, and so on. The result was a lingering fear of pathological functions (continuous everywhere differentiable nowhere, that sort of thing) and associated technicalities. It came as a great relief to find, much later on, alas, that by getting away from the pointwise emphasis of introductory analysis one can give a beautifully consistent and holistic definition of the derivative of the *Heaviside function*[1]

$$\mathcal{H}(x) = \begin{cases} 1, & x > 0, \\ 0, & x \leq 0. \end{cases}$$

In pointwise mode, the best we can do with this function is to talk about the left and right limits of its derivative at the origin. Both these are equal

---

[1] The value $\mathcal{H}(0) = 0$ has been assigned for consistency with probability, as we shall see; but for reasons that will shortly become clear it really doesn't matter what value we take for $\mathcal{H}(0)$.

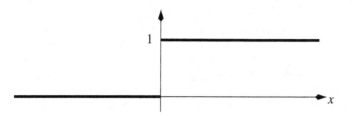

Figure 9.1 The Heaviside function $\mathcal{H}(x)$. Its derivative vanishes for all $x \neq 0$ but it still gets up from 0 to 1. How?

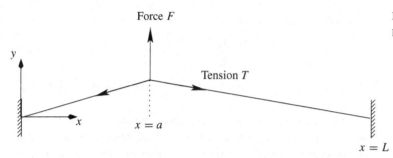

Figure 9.2 A string with a point force.

to zero, but the function nevertheless gets up from 0 to 1. There must be something behind this!

The Heaviside function and its derivative, the delta function (or distribution), are ubiquitous in whole swathes of linear applied mathematics, not to mention discrete probability. They, and other distributions, are invaluable in developing an intuitive framework for modelling and its interaction with mathematics. Don't be inhibited about using them: your mistakes are unlikely to do worse than lead to inconsistencies (for which I hope you are constantly on the look-out) and plainly wrong answers, rather than the deadly 'plausible but fallacious' solution.

9.2 A point force on a stretched string; impulses

Let's start with a couple of motivating physical examples. At some time we have all worked out the displacement of a stretched string under the influence of a point force, shown in Figure 9.2. Under the standard assumptions, that the string is effectively weightless and that the force F (measured upwards, in the same direction as y) can be considered as acting at a point $x = a$ and causes only a small deflection, the equilibrium displacement $y(x)$ of the string satisfies

$$\frac{\mathrm{d}^2 y}{\mathrm{d}x^2} = 0, \qquad 0 < x < a, \quad a < x < L, \tag{9.1}$$

Figure 9.3 Force on an element of a string.

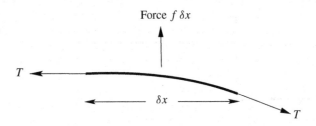

Force $f \, \delta x$

T

δx

T

with the force balance condition

Consistency check on the signs: $F > 0$ and dy/dx is negative to the right of a, positive to the left.

$$\left[T \frac{dy}{dx} \right]_{x=a-}^{x=a+} = -F. \tag{9.2}$$

Notice the implicit assumption that y itself is continuous at x_0 although its derivative is not.

Now we might ask, can we somehow put the force on the right-hand side of (9.1) and have the equilibrium conditions hold at $x = a$ as well? After all, if we have a distributed force per unit length $f(x)$ on the string, the usual force balance on a small element (see Figure 9.3) gives the equation[2]

$$T \frac{d^2 y}{dx^2} = -f(x), \qquad 0 < x < L.$$

For example, when $f = -\rho g$, the gravitational force on a uniform wire of line density ρ, the displacement is a parabola (the small-displacement approximation to a catenary).

Can we devise some limiting process in which all the force becomes concentrated near $x = a$, with the total force $\int_0^L f(x) \, dx$ tending to F?

---

[2] You might wonder why there is a minus sign on the right. If we were to consider the unsteady motion of the string, Newton's second law in the form

$$\text{mass} \times \text{acceleration} = \text{force}$$

gives

$$\rho \frac{\partial^2 y}{\partial t^2} = T \frac{\partial^2 y}{\partial x^2} + f,$$

leading to the minus sign in question. Many mathematicians, writing the wave equation as

$$\rho \frac{\partial^2 y}{\partial t^2} - T \frac{\partial^2 y}{\partial x^2} = f,$$

would write the equilibrium equation for the string as

$$-T \frac{d^2 y}{dx^2} = f(x).$$

Note the absence of minus signs in the impulse example that follows.

A possible way to do this would be to take

$$f(x) = \begin{cases} \dfrac{F}{2\epsilon}, & a - \epsilon < x < a + \epsilon, \\ 0 & \text{otherwise,} \end{cases}$$

and then to let $\epsilon \to 0$. But would we get the same answer if we took the limit of some other concentrated force density, and in any case how, exactly, are we to interpret the result of this limiting process?

In a very similar vein, recall the concept of an impulse in mechanics. In one-dimensional motion, the velocity v of a particle under a force $f(t)$ satisfies Newton's equation

$$m\frac{dv}{dt} = f(t),$$

from which

$$v(t) = v(0) + \frac{1}{m}\int_0^t f(s)\,ds.$$

If the force is very large but only lasts for a short time, say

$$f(t) = \begin{cases} \dfrac{I}{\epsilon}, & 0 < t < \epsilon, \\ 0 & \text{otherwise,} \end{cases}$$

then we can integrate the equation of motion from $t = 0$ to $t = \epsilon$ to find

$$v(\epsilon) = \frac{1}{m}\int_0^\epsilon \frac{I}{\epsilon}\,dt = \frac{I}{m}.$$

Letting $\epsilon \to 0$, we have the result of an *impulse* I: the velocity v changes discontinuously from 0 to I/m. Again, we ask the question, can we put the limiting impulse directly into the equation of motion rather than having to smooth it out and take a limit?

Notice that the wire slope has a jump discontinuity at the position of a point force.

9.3 Informal definition of the delta and Heaviside functions

Obviously the answer to all our questions above is yes. The powerful and elegant theory of distributions allows us to model point forces and much more (dipoles, for example). However, the intuitive view of a point force (mass, charge, ...) as the limit of a distributed force turns out to be technically very cumbersome, and nowadays a more concise and general, but physically less intuitive, treatment is preferred. This oblique approach requires some groundwork, and we defer a brief self-contained description until Chapter 10. You will survive if you don't read it, although I recommend that you do: it is not technically demanding or complex.

Figure 9.4 Three approximations to the delta function: $\epsilon_1 > \epsilon_2 > \epsilon_3 > 0$.

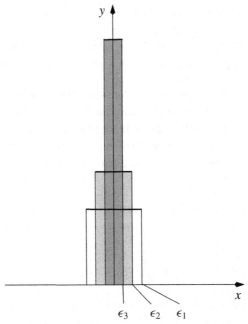

In this chapter we concentrate on the intuitive approach to the delta function. Although this is not how the theory is nowadays developed, *it absolutely is how to visualise this central part of it*. Taking the examples of the previous section and stripping away the physical background, consider the functions

$$f_\epsilon(x) = \begin{cases} \dfrac{1}{2\epsilon}, & -\epsilon < x < \epsilon, \\[2mm] 0 & \text{otherwise.} \end{cases}$$

They are shown in Figure 9.4 for three values of ϵ. The following facts are obvious:

- $\displaystyle\int_{-\infty}^{\infty} f_\epsilon(x)\,dx = 1 \qquad$ for all $\epsilon > 0$;

- for $x \neq 0, \qquad f_\epsilon(x) \to 0 \quad$ as $\quad \epsilon \to 0$.

The limiting 'function' as $\epsilon \to 0$ is very strange indeed. It has a 'mass', or 'area under the graph', of 1, but that mass is all concentrated at $x = 0$. This is just what we need to model a point force, and even though we don't quite know how to interpret it rigorously, we provisionally christen the limit as the *delta function*, $\delta(x)$.

Two extremely useful properties of the delta function are now at least plausible. First, as $\epsilon \to 0$,

$$\int_{-\infty}^{x} f_\epsilon(s)\,ds \to \begin{cases} 1, & x > 0, \\[1mm] 0, & x < 0, \end{cases}$$

and the right-hand side is the Heaviside function $\mathcal{H}(x)$ with its jump discontinuity at $x = 0$. So, we should have

For now, let's not worry what its value is at $x = 0$.

$$\int_{-\infty}^{x} \delta(s)\,\mathrm{d}s = \mathcal{H}(x),$$

at least for $x \neq 0$. Furthermore, fingers crossed and appealing to the fundamental theorem of calculus, we should have, conversely,

$$\frac{\mathrm{d}}{\mathrm{d}x}\mathcal{H}(x) = \delta(x).$$

That is, *delta functions let us differentiate functions with jump discontinuities*. The Heaviside function has a jump up of 1 at $x = 0$ and its derivative is $\delta(x)$; by an obvious extension, the derivative of a function with a jump of A at $x = a$ contains a term $A\delta(x - a)$.

The second vital attribute of $\delta(x)$ is its 'sifting' property. Intuitively, for sufficiently smooth functions $\phi(x)$,

A proof is requested in the exercises.

$$\int_{-\infty}^{\infty} f_\epsilon(x)\phi(x)\,\mathrm{d}x \to \phi(0) \quad \text{as} \quad \epsilon \to 0,$$

simply because all the mass of $f_\epsilon(x)$, and hence of the product $f_\epsilon(x)\phi(x)$, becomes concentrated at the origin. So, we conjecture that we can make sense of the statement

$$\int_{-\infty}^{\infty} \delta(x)\phi(x)\,\mathrm{d}x = \phi(0) \tag{9.3}$$

and, by a simple change of variable,

$$\int_{-\infty}^{\infty} \delta(x - a)\phi(x)\,\mathrm{d}x = \phi(a)$$

for any real a.

These assertions are eminently plausible. However, if you stop to think how you might make them mathematically acceptable, difficulties start to appear. Would we get the same results if we used a different approximating sequence $g_\epsilon(x)$? Do we need to worry about the value of $\mathcal{H}(0)$? Having differentiated $\mathcal{H}(x)$, can we define $\mathrm{d}\delta/\mathrm{d}x$? Clearly this last runs a big risk of being very dependent on the approximating sequence we use.

For example,

$$g_\epsilon(x) = \frac{1}{\epsilon\sqrt{2\pi}}e^{-x^2/(2\epsilon^2)}$$

as discussed in Section 9.4.

For all these reasons, and more, the theory is best developed slightly differently, without the 'epsilonology'.[3] The clue lies in the sifting property. Using the fact that integration is a smoothing process, we can get away from the 'pointwise' view of functions, which is so troublesome, and instead define distributions via *averaged* properties. An example is

---

[3] See [44, p. 97], for this neologism.

the integral (9.3), which leads to the definition of $\delta(x)$.[4] Before looking at this idea in more detail, we consider some examples.

9.4 Examples

9.4.1 A point force on a wire revisited

All our discussion suggests that we should model the point force F acting at $x = a$ by a term $F\delta(x - a)$ in the equilibrium equation for the displacement, and we assume that the latter now holds for *all* x, so that

$$T\frac{d^2y}{dx^2} = -F\delta(x - a), \qquad 0 < x < L.$$

Assuming we believe that differentiation still makes sense. We now know that this means that the left-hand side is the derivative of a function that jumps by F at $x = a$. But the left-hand side is also the derivative of $T\,dy/dx$. Thus, putting the delta function into the equilibrium equation leads *automatically* to the force balance

$$\left[T\frac{dy}{dx}\right]_{a-}^{a+} = -F,$$

and there is no need to state this separately.

9.4.2 Continuous and discrete probabilities

We can interpret each of the approximations $f_\epsilon(x)$ of Figure 9.4 as the probability density of a random variable X_ϵ whose value is uniformly distributed on the interval $(-\epsilon, \epsilon)$. The mean of this distribution is 0 and its standard deviation is $\epsilon/\sqrt{3}$. As $\epsilon \to 0$, the random variable becomes equal to 1 with certainty, because its standard deviation tends to zero, and any random variable with zero standard deviation must be a constant. This suggests that we can interpret the delta function as the probability density 'function' of a variable whose probability of being equal to zero is 1. Likewise, the cumulative density function (distribution function) $F_{X_\epsilon}(x) = P(X_\epsilon < x)$ tends to the Heaviside function.[5]

---

[4] The process of generalisation by looking at a weaker (smoother) definition using an integral, rather than a pointwise definition, is common in analysis. A famous example in applied mathematics is the definition of weak solutions to hyperbolic conservation laws, which leads to the Rankine–Hugoniot relations for a shock.

[5] In this case the strict inequality in the definition of F_{X_ϵ} suggests that we should take $\mathcal{H}(0) = 0$. Looking in the books on my shelf, I find that there is no consensus in the probability world about whether to use $P(X < x)$ or $P(X \le x)$ to define the distribution function (no wonder I can never remember which it is). It is a matter of convention only and would lead to corresponding conventional definitions of $\mathcal{H}(0)$. Another highly

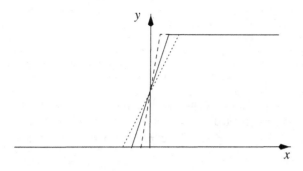

Figure 9.5 Cumulative density functions for the distributions of Figure 9.4.

In a similar vein, we can take approximations

$$g_\epsilon(x) = \frac{1}{\epsilon\sqrt{2\pi}} e^{-x^2/(2\epsilon^2)}$$

which are the density functions of normal random variables with mean zero and standard deviation ϵ. These also clearly tend to the delta function as $\epsilon \to 0$.

Now suppose we have a coin-toss random variable X taking values ± 1 with equal probability $\frac{1}{2}$. As X can only equal 1 or -1, all its probability mass is concentrated at these values: its density function is zero for $x \neq \pm 1$. The density of this random variable is

What is its distribution function?

$$f_X(x) = \frac{1}{2} (\delta(x+1) + \delta(x-1)).$$

In this way, we can unify continuous and discrete probability – at least when the number of discrete events is finite. The extension to infinitely many discrete events is much more difficult and may require the tools of measure theory.

9.4.3 The fundamental solution of the heat equation

If we set $\epsilon = 2t$ in the functions g_ϵ of the previous section, we get the function

$$g(x, t) = \frac{1}{2\sqrt{\pi t}} e^{-x^2/(4t)}.$$

Direct differentiation shows that $g(x, t)$ satisfies the heat equation. As we saw above, for $t \downarrow 0$ we have $g(x, t) \to \delta(x)$. In summary, $g(x, t)$

plausible definition is $\mathcal{H}(0) = \frac{1}{2}$, on the grounds that any Fourier series or transform inversion integral for a function with a jump converges to the average of the values on either side. This sort of hair splitting is one reason why the pointwise view of distributions is not really workable.

satisfies the initial-value problem

$$\frac{\partial g}{\partial t} = \frac{\partial^2 g}{\partial x^2}, \qquad t > 0, \quad -\infty < x < \infty,$$

$$g(x, 0) = \delta(x).$$

Note the infinite propagation speed of the heat: $t = 0$ is a (double) characteristic of the heat equation. Note also the very rapid decay in the solution as $|x|$ increases.

This solution represents the evolution of a 'hot spot', a unit amount of heat that at $t = 0$ is concentrated at $x = 0$.

With this solution, we can solve the more general initial-value problem

$$\frac{\partial u}{\partial t} = \frac{\partial^2 u}{\partial x^2}, \qquad t > 0, \quad -\infty < x < \infty,$$

$$u(x, 0) = u_0(x).$$

We first note that the initial data $u_0(x)$ can be written as

$$u_0(x) = \int_{-\infty}^{\infty} u_0(\xi)\,\delta(x - \xi)\,\mathrm{d}\xi$$

by the picking-out property of the delta function. Now the evolution of a solution with initial data $\delta(x - \xi)$ is just $g(x - \xi, t)$, where g is as above. The integral over ξ amounts to superposing the initial data for these solutions, so that each point contributes a delta function weighted by $u_0(\xi)\,\mathrm{d}\xi$. Because the heat equation is linear we can superpose for $t > 0$ as well, so we have

Confirm that $u(x, t)$ satisfies the heat equation by differentiating under the integral sign.

$$u(x, t) = \int_{-\infty}^{\infty} u_0(\xi)g(x - \xi, t)\,\mathrm{d}\xi$$

$$= \frac{1}{2\sqrt{\pi t}} \int_{-\infty}^{\infty} u_0(\xi)e^{-(x-\xi)^2/4t}\,\mathrm{d}\xi.$$

This solution has a physical interpretation as the superposition of elementary 'packets' of heat evolving independently.[6]

9.5 Balancing singularities

If we have an equation involving 'ordinary' functions and there is a singularity on one side, then there must be a balancing singularity somewhere else. For example, we could never find coefficients a_n such that

$$\frac{1}{\sin x} = a_0 + a_1 x + a_2 x^2 + \cdots$$

---

[6] There is also an interpretation in terms of random walkers following Brownian motion: see Exercise 9.

because the left-hand side clearly has a $1/x$ (simple pole) singularity at $x = 0$. However, there *is* an expansion

This is just the Laurent expansion.

$$\frac{1}{\sin x} = \frac{a_{-1}}{x} + a_0 + a_1 x + a_2 x^2 + \cdots,$$

and furthermore we know that $a_{-1} = 1$, because $1/\sin x \sim 1/x$ as $x \to 0$. Thus both sides have this singularity in their leading-order behaviour as $x \to 0$.

This is a simple but powerful idea, and it applies to distributions as well. In our naive approach, a delta function is a 'function' with a particular singularity at $x = 0$. Thus, if part (for example the right-hand side) of an equation contains a delta function as its 'most singular' term, there must be a balancing term somewhere else. For instance, when we write

$$\frac{dv}{dt} = \frac{I}{m}\delta(t),$$

for the motion of a particle subject to a point force, there must be another singularity to balance the delta function. It can only be in dv/dt, so we know straightaway that v has a jump at $t = 0$; furthermore, we know that the magnitude of the jump is I/m, by 'comparing coefficients' of the delta functions. In this case it is trivial to find the balancing term, because there is only one candidate. Suppose, though, that the equation has a linear damping term:

Go back and look at the point force on a string in this light.

$$m\frac{dv}{dt} = -mkv + I\delta(t),$$

where $k > 0$ is the damping coefficient. The balancing singularity is still in the derivative dv/dt, simply because dv/dt always has worse singularities than v itself. Going back, we can check: if dv/dt has a delta function then v has a jump, which is indeed less singular.

Differentiation makes matters worse, integration makes them better.

9.5.1 The Rankine–Hugoniot conditions

In Chapter 7 we looked briefly at the Rankine–Hugoniot conditions for a first-order conservation law

$$\frac{\partial P}{\partial t} + \frac{\partial Q}{\partial x} = 0,$$

where, for example, P is the density ρ of traffic and Q the flux $u\rho$. We saw that we can construct solutions in which P and Q have jump discontinuities across a shock at $x = S(t)$, provided that

$$\frac{dS}{dt} = \frac{[Q]}{[P]}.$$

We can interpret this condition as a balance of delta functions. If P has a (time-dependent) jump of magnitude $[P](t)$ at $x = S(t)$, we can (very informally) write

$$P(x, t) = [P](t)\mathcal{H}(x - S(t)) + \text{smoother part},$$

and similarly for $Q(x, t)$. Differentiating, we find

$$\frac{\partial P}{\partial t} = -[P](t)\delta(x - S(t))\frac{dS}{dt} + \text{less singular terms},$$

$$\frac{\partial Q}{\partial x} = [Q](t)\delta(x - S(t)) + \text{less singular terms}.$$

Adding these and balancing the coefficients of the delta functions, the Rankine–Hugoniot condition drops out.

9.5.2 Case study: cable-laying

In Chapter 4, we wrote down the model

$$\frac{dF_x}{ds} = -B_x, \qquad \frac{dF_y}{ds} = -B_y + \rho_c g A = 0, \tag{9.4}$$

$$EAk^2\frac{d^2\theta}{ds^2} - F_x \sin\theta + F_y \cos\theta = 0, \tag{9.5}$$

where

$$(B_x, B_y) = \left(\rho_w g A \cos\theta + pA\frac{d\theta}{ds}\right)(-\sin\theta, \cos\theta). \tag{9.6}$$

for a cable being laid on a sea bed; θ is the angle between the cable and the horizontal. We stated, on a rather intuitive basis, that the boundary conditions at $s = 0$ are $\theta = 0$ (no worries about this one) and $d\theta/ds = 0$, namely the continuity of θ and $d\theta/ds$, since $\theta = 0$ for $s < 0$. We can now see why this is necessary. If $d\theta/ds$ is not continuous, $d^2\theta/ds^2$ has a delta function discontinuity at $s = 0$. But then there is no balancing term in (9.5) since, loosely, (9.4) shows that both F_x and F_y are at least as continuous as B_x and B_y, and so from (9.6) they are no worse than $d\theta/ds$ with its jump discontinuity, assumed for a contradiction; we have duly obtained the said contradiction.

Because there is a reaction force between the sea bed and the cable, and maybe some friction, we do not expect the right-hand sides of (9.4) to be continuous at $s = 0$.

9.6 Green's functions

9.6.1 Ordinary differential equations

The two-point boundary-value problem[7]

$$\mathcal{L}_x y(x) = \frac{d}{dx}\left(p(x)\frac{dy}{dx}\right) + q(x)y = f(x), \qquad 0 < x < 1, \quad (9.7)$$

$$y(0) = y(1) = 0, \qquad (9.8)$$

is standard, often arising in a separation of variables calculation using an exotic coordinate system. As a matter of terminology, we call the combination of $\mathcal{L}_x$, the interval on which it is applied and the boundary conditions at the ends of this interval the *differential operator* for this problem. Changing any of these will change the differential operator. The operator (9.7), (9.8) is called *self-adjoint*, a term that will be made clearer later.

One of the first things that one does with problems of this kind is to show that they can be solved with the help of a *Green's function*. Provided that the homogeneous problem ($f(x) \equiv 0$) has no non-trivial solutions, the Green's function is the function $G(x, \xi)$ that satisfies

$$\mathcal{L}_\xi G(x, \xi) = 0, \qquad 0 < \xi < x, \qquad x < \xi < 1, \qquad (9.9)$$

$$G(x, 0) = G(x, 1) = 0, \qquad (9.10)$$

with some rather opaque-seeming conditions at $\xi = x$:

$$[G]_{\xi=x-}^{\xi=x+} = 0, \quad \left[p(\xi)\frac{dG}{d\xi}\right]_{\xi=x-}^{\xi=x+} = 1 \quad \left(\text{or} \quad \left[\frac{dG}{d\xi}\right]_{\xi=x-}^{\xi=x+} = \frac{1}{p(x)}\right). \qquad (9.11)$$

If we can solve this problem then we have a representation for $y(x)$:

$$y(x) = \int_0^1 G(x, \xi) f(\xi) \, d\xi.$$

The elementary proof of this proceeds by direct construction of the Green's function via variation of parameters, assuming the existence of appropriate solutions of the homogeneous equation, and we do not describe it here. The point is that we need only calculate G once, and then we have the solution whatever we take for $f(x)$.[8] In this way, we

Just like inverting a matrix $\mathbf{A}$ to solve $\mathbf{A}x = \mathbf{b}$; see Exercise 7.

---

[7] The subscript to $\mathcal{L}$ tells you which variable to use. Strictly speaking, in much of the discussion to follow all the derivatives should be partial, but it seems to be conventional to stick to ordinary derivatives.

[8] A very common use of the Green's function is to turn a differential equation into an integral equation as a prelude to an iteration scheme to prove existence, uniqueness and regularity. Often the equation has a linear part and some nonlinearity as well, and we use

can think of the operation of multiplying by the Green's function and integrating as the inverse of the differential operator $\mathcal{L}$.

This is all very well, but I don't think it gives a good intuitive feel for what the Green's function really *does*. Suppose, though, that we take the solution

$$y(x) = \int_0^1 G(x, \xi) f(\xi) \, d\xi \tag{9.12}$$

and apply $\mathcal{L}_x$ to it. Assuming that we can differentiate under the integral, we get

$$\mathcal{L}_x y(x) = \int_0^1 \mathcal{L}_x G(x, \xi) f(\xi) \, d\xi$$
$$= f(x).$$

We recognise this: it is the sifting property. Whatever f we take, when we multiply $f(\xi)$ by $\mathcal{L}_x G(x, \xi)$ and integrate, we get $f(x)$. Thus, as a function (actually, a distribution) of x, $G(x, \xi)$ satisfies

$$\mathcal{L}_x G(x, \xi) = \delta(x - \xi),$$

that is

$$\frac{d}{dx} \left(p(x) \frac{dG}{dx} \right) + q(x) G = \delta(x - \xi).$$

Also, the boundary conditions $y(0) = y(1) = 0$ mean that we need to take

$$G(0, \xi) = G(1, \xi) = 0,$$

so that (9.12) satisfies the boundary conditions whatever $f(x)$ we take. In summary, as a function of x the Green's function satisfies the differential equation with a delta function on the right-hand side and with the homogeneous version of the original boundary conditions.

This calculation tells us several things. Thinking physically, it tells us that *the Green's function is the response of the system to a point stimulus (force, charge, ...) at $x = \xi$*. The solution (9.12) is then just

the Green's function for the linear part. A simple example of this procedure is Picard's theorem for the local existence and uniqueness of the solution to $dy/dx = f(x, y)$, $y(0) = y_0$ for a set of first-order equations, where the first step is to write

$$y(x) = y_0 + \int_0^x f(\xi, y(\xi)) \, d\xi;$$

the only modification needed is to adapt the Green's function methodology to cater for initial-value problems, as described in Exercise 4.

the response to $f(x)$, regarded as a superposition of point stimuli (the delta function at $x = \xi$) weighted by $f(\xi)\,d\xi$.

Looking more mathematically, if we expand $\mathcal{L}_x G(x, \xi)$ as

$$\mathcal{L}_x G(x, \xi) = p(x)\frac{d^2 G}{dx^2} + \text{lower-order derivatives,}$$

we can see by balancing the most singular terms (the highest derivatives) that $d^2 G/dx^2$ must have a delta function, scaled by $p(x)$, at $x = \xi$. That is,

$$[G]_{x=\xi-}^{x=\xi+} = 0, \quad \left[p(x)\frac{dG}{dx} \right]_{x=\xi-}^{x=\xi+} = 1 \quad \left(\text{or} \quad \left[\frac{dG}{dx} \right]_{x=\xi-}^{x=\xi+} = \frac{1}{p(\xi)} \right).$$

These conditions should ring a bell. They are the same as the 'opaque' jump conditions (9.11), except that they refer to the x-dependence of $G(x, \xi)$ instead of the ξ-dependence. Indeed, comparing the original definition of G given in (9.9)–(9.11) and recalling that $G(0, \xi) = G(1, \xi) = 0$, we see that the two formulations are identical except that x and ξ have been swapped. That is, we have established that, for self-adjoint problems,

Note that $\delta(x - \xi) = \delta(\xi - x)$.

$$G(x, \xi) = G(\xi, x)$$

and that

$$\mathcal{L}_\xi G(x, \xi) = \delta(\xi - x).$$

We are now in a position to tie together the x- and ξ-dependences of $G(x, \xi)$. Consider the integral

$$\int_0^1 y(\xi)\mathcal{L}_\xi G(x, \xi) - G(x, \xi)\mathcal{L}_\xi y(\xi)\,d\xi.$$

On the one hand, inserting the right-hand sides of the differential equations for G and y, we get

$$\int_0^1 y(\xi)\mathcal{L}_\xi G(x, \xi) - G(x, \xi)\mathcal{L}_\xi y(\xi)\,d\xi$$
$$= \int_0^1 y(\xi)\delta(\xi - x) - G(x, \xi)f(\xi)\,d\xi$$
$$= y(x) - \int_0^1 G(x, \xi)f(\xi)\,d\xi.$$

On the other hand, integrating the same expression by parts, we get

$$\int_0^1 y(\xi)\mathcal{L}_\xi G(x,\xi) - G(x,\xi)\mathcal{L}_\xi y(\xi)\,d\xi$$

$$= \int_0^1 \left\{ y(\xi)\left(\frac{d}{d\xi}\left(p(\xi)\frac{dG}{d\xi}\right) + q(\xi)G(x,\xi)\right) \right.$$

$$\left. -G(x,\xi)\left(\frac{d}{d\xi}\left(p(\xi)\frac{dy}{d\xi}\right) + q(\xi)y(\xi)\right)\right\}d\xi$$

$$= \left[y(\xi)p(\xi)\frac{dG}{d\xi} - G(x,\xi)p(\xi)\frac{dy}{d\xi} \right]_0^1$$

$$- \int_0^1 p(\xi)\frac{dy}{d\xi}\frac{dG}{d\xi} - p(\xi)\frac{dG}{d\xi}\frac{dy}{d\xi}\,d\xi$$

$$= 0.$$

Thus we retrieve the solution

$$y(x) = \int_0^1 G(x,\xi)f(\xi)\,d\xi.$$

As a function of x the differential equation for G is still $\mathcal{L}_x G = 0$ with the same boundary conditions as those for y.

This calculation is really the key to the whole procedure. It tells us that the differential equation and boundary conditions (that is, the differential operator) for G as a function of ξ must be such that we can integrate by parts and get zero (so, after integrating by parts as above, we must have zero multiplying dy/dx, about which we know nothing at the endpoints).

Non-self-adjoint problems. For a self-adjoint problem, such as those discussed thus far, G is symmetric and the two operators, for y and G, are the same. Now suppose that we have a more general problem, such as

$$\mathcal{L}_x y(x) = a(x)\frac{d^2 y}{dx^2} + b(x)\frac{dy}{dx} + c(x)y = f(x),$$

with the boundary conditions, sometimes called *primary boundary conditions*,

$$\alpha_0 y(0) + \beta_0 y'(0) = 0, \qquad \alpha_1 y(1) + \beta_1 y'(1) = 0,$$

where $y' = dy/dx$. We aim to find a differential operator for G which allows us to follow the calculation above as closely as possible. That is, we want to find a combination of derivatives $\mathcal{L}^*$ such that, as a function of ξ, $G(x,\xi)$ satisfies

$$\mathcal{L}^*_\xi G(x,\xi) = \delta(x-\xi),$$

with appropriate boundary conditions. We can then integrate by parts as above, and, provided that

$$\int_0^1 y(\xi)\mathcal{L}_\xi^* G(x,\xi) - G(x,\xi)\mathcal{L}_\xi y(\xi)\,d\xi = 0,$$

we have the answer

$$y(x) = \int_0^1 G(x,\xi)f(\xi)\,d\xi.$$

For the general operator just introduced, the new operator, called the *adjoint operator*, is given by

$$\mathcal{L}^* v(x) = \frac{d^2}{dx^2}\,(a(x)v(x)) - \frac{d}{dx}\,(b(x)v(x)) + c(x)v(x),$$

with the *adjoint boundary conditions*

$$a(0)\big(\alpha_0 v(0) + \beta_0 v'(0)\big) + \beta_0\big(a'(0) - b(0)\big)v(0) = 0,$$
$$a(1)\big(\alpha_1 v(1) + \beta_1 v'(1)\big) + \beta_1\big(a'(1) - b(1)\big)v(1) = 0,$$

as you will find out by doing Exercise 5.

You might very reasonably ask why we bother with the adjoint when all we need to do is differentiate the answer

$$y(x) = \int_0^1 G(x,\xi)f(\xi)\,d\xi$$

under the integral sign to show that

$$\mathcal{L}_x y = \int_0^1 \mathcal{L}_x G(x,\xi)f(\xi)\,d\xi$$
$$= f(x),$$

so that

$$\mathcal{L}_x G(x,\xi) = \delta(x - \xi)$$

with no mention of adjoints at all. An aesthetic reason is the mathematical structure uncovered (compare vector spaces and their duals), but a compelling practical reason is that if the primary boundary conditions are *inhomogeneous*, for example $y(0) = y_0 \neq 0$, $y(1) = y_1 \neq 0$, then only the adjoint calculation works (try it!).

One can take all this a great deal further, both by making it more rigorous and also by looking at more general problems; I recommend reading the relevant parts of [33] or [57] if you want to do this. We move now on to a brief look at partial differential equations.

9.6.2 Partial differential equations

Much of the theory we have just seen can be generalised to linear partial differential equations. This is so much vaster a topic that it is only feasible to discuss one example in detail, the Green's function for Poisson's equation, which is probably the closest in spirit to the two-point boundary value problems we have been discussing so far. We then mention briefly two other canonical problems, for the heat equation and the wave equation.

We first have to generalise the delta function to the multidimensional case. In our informal style, this is easy: we just say that, for $\mathbf{x} \in \mathbb{R}^n$, the delta function $\delta(\mathbf{x})$ is such that

$$\int_{\mathbb{R}^n} \delta(\mathbf{x})\phi(\mathbf{x})\,\mathrm{d}\mathbf{x} = \phi(\mathbf{0})$$

for all smooth functions $\phi(\mathbf{x})$. As before, we can think of this as a limiting process in which the delta function is the limit of a family of functions whose mass becomes more and more concentrated near the origin.[9] Thinking about how the integral is calculated, say in two dimensions with $\mathbf{dx} = \mathrm{d}x\,\mathrm{d}y$, we may also write

$$\delta(\mathbf{x}) = \delta(x)\delta(y),$$

and similarly in three or more variables.

Think of some physical interpretations for u, and then for the Green's function G.

Now suppose that we have to solve the problem

$$\mathcal{L}_x u(\mathbf{x}) = \nabla^2 u(\mathbf{x}) = f(\mathbf{x})$$

in some region D, with the homogeneous Dirichlet boundary condition

$$u(\mathbf{x}) = 0 \quad \text{on} \quad \partial D.$$

We choose the Green's function to satisfy

$$\mathcal{L}_\xi G(\mathbf{x}, \boldsymbol{\xi}) = \delta(\boldsymbol{\xi} - \mathbf{x})$$

The Laplacian is self-adjoint $(\mathcal{L} = \mathcal{L}^*)$ ...

and look at the integral

$$\int_D u(\boldsymbol{\xi})\mathcal{L}_\xi G(\mathbf{x}, \boldsymbol{\xi}) - G(\mathbf{x}, \boldsymbol{\xi})\mathcal{L}_\xi u(\mathbf{x})\,\mathrm{d}\boldsymbol{\xi}$$

$$= \int_D u(\boldsymbol{\xi})\nabla_\xi^2 G(\mathbf{x}, \boldsymbol{\xi}) - G(\mathbf{x}, \boldsymbol{\xi})\nabla_\xi^2 u(\mathbf{x})\,\mathrm{d}\boldsymbol{\xi}$$

$$= \int_D u(\boldsymbol{\xi})\delta(\boldsymbol{\xi} - \mathbf{x}) - G(\mathbf{x}, \boldsymbol{\xi})f(\boldsymbol{\xi})\,\mathrm{d}\boldsymbol{\xi}$$

$$= u(\mathbf{x}) - \int_D G(\mathbf{x}, \boldsymbol{\xi})f(\boldsymbol{\xi})\,\mathrm{d}\boldsymbol{\xi}. \qquad (9.13)$$

---

[9] They might, but need not, be radially symmetric; we might, but won't, worry about how to define integrals in n dimensions.

However, using Green's theorem we have

$$\int_D u(\boldsymbol{\xi})\nabla_{\boldsymbol{\xi}}^2 G(\mathbf{x}, \boldsymbol{\xi}) - G(\mathbf{x}, \boldsymbol{\xi})\nabla_{\boldsymbol{\xi}}^2 u(\mathbf{x})\,d\boldsymbol{\xi}$$

$$= \int_{\partial D} u(\boldsymbol{\xi})\mathbf{n}\cdot\nabla_{\boldsymbol{\xi}} G(\mathbf{x}, \boldsymbol{\xi}) - G(\mathbf{x}, \boldsymbol{\xi})\mathbf{n}\cdot\nabla_{\boldsymbol{\xi}} u(\mathbf{x})\,dS$$

$$= 0, \tag{9.14}$$

> ... because $u\nabla^2 G - G\nabla^2 u$ is a divergence and can be integrated (a generalisation of integration by parts).

provided that we take $G(\mathbf{x}, \boldsymbol{\xi}) = 0$ for $\boldsymbol{\xi} \in \partial D$, where we do not know the normal derivative of u. Putting these together, we have

$$u(\mathbf{x}) = \int_D G(\mathbf{x}, \boldsymbol{\xi})f(\boldsymbol{\xi})\,d\boldsymbol{\xi}.$$

The case of non-zero Dirichlet data $u(\mathbf{x}) = g(\mathbf{x})$ on ∂D is an easy generalisation: we just get an extra known term in (9.14).

Two more things should be noted about this calculation. The first is that we have not yet said anything about the nature of the singularity of $G(\mathbf{x}, \boldsymbol{\xi})$ at $\mathbf{x} = \boldsymbol{\xi}$ (in one space dimension, as we saw above, the first derivative of G has a jump and G itself is continuous). Knowing as we do that line charges (in two dimensions) or point charges (in three dimensions) generate potentials that, away from the charges, are solutions of Laplace's equation, we should not be surprised to see logs in two dimensions and inverse distances in three. This is confirmed by a simple version of the calculation we have just done.[10] In $\mathbb{R}^3$ for example, take $\boldsymbol{\xi} = \mathbf{0}$ and suppose that

> Line or point masses and their gravitational potentials, fluid sources and their velocity potentials, or heat sources and their steady-state temperature fields are all comparable.

$$\nabla^2 G = \delta(\mathbf{x}) \tag{9.15}$$

in the whole space. Clearly, then, G is radially symmetric: $G = G(r)$ where $r = |\mathbf{x}|$. That means that

$$G(r) = A + \frac{B}{r}$$

and if we want $G \to 0$ as $r \to \infty$ then we take $A = 0$. Now use the divergence theorem on the left-hand side of (9.15), integrating over a sphere of radius r centred at $\mathbf{x} = \mathbf{0}$. The left-hand side gives a surface integral equal to $-4\pi B/r$ and the volume integral of the delta function on the right is equal to 1. We conclude that the singular behaviour of $G(\mathbf{x}, \boldsymbol{\xi})$ near $\mathbf{x} = \boldsymbol{\xi}$ is given by

> The meanings of $\sim$ and $O(1)$ are explained in Chapter 12.

$$G(\mathbf{x}, \boldsymbol{\xi}) \sim -\frac{1}{4\pi |\mathbf{x} - \boldsymbol{\xi}|} + O(1),$$

---

[10] In the more classical treatment of Green's functions, you see essentially this calculation when you integrate $u\nabla^2 G - G\nabla^2 u$ over a region consisting of D but with a sphere of radius ϵ around $\mathbf{x} = \boldsymbol{\xi}$ removed. In that case the singular behaviour of G is prescribed (and looks mysterious: why this form?), whereas here it emerges naturally.

Can you now answer the marginal question after (2.3) in Exercise 4 at the end of Chapter 3?

and in two dimensions the corresponding result is

$$G(\mathbf{x}, \boldsymbol{\xi}) \sim \frac{1}{2\pi} \log |\mathbf{x} - \boldsymbol{\xi}| + O(1).$$

The second point to make about the Green's function for the Laplacian is that it has a natural physical interpretation. The singular part, which we have just discussed, gives us the electric potential due to a point charge in the absence of boundaries. The remaining part, $G + 1/(4\pi |\mathbf{x} - \boldsymbol{\xi}|)$, is known as the *regular part* of the Green's function and gives the potential due to the image charge system induced by the boundary condition $G = 0$ on ∂D. Indeed, almost all the Green's functions for which explicit formulas are available are constructed by the method of images (possibly with the help of conformal maps).

You can safely ignore this section, but have a look if you have seen the classical treatments of these problems.

The heat and wave equations.   To round off, let's look quickly at two other equations, the heat and wave equations in two space variables. Let us look at the simplest initial value problem for the heat equation, that defined on the whole line, namely

$$\mathcal{L}_{x,t} u = \frac{\partial u}{\partial t} - \frac{\partial^2 u}{\partial x^2} = 0, \qquad -\infty < x < \infty, \qquad t > 0,$$

$$u(x, 0) = u_0(x).$$

By any of a variety of methods (for example, the Fourier transform in x), we obtain the solution in the form

$$u(x, t) = \frac{1}{2\sqrt{\pi t}} \int_{-\infty}^{\infty} u_0(\xi) e^{-(x-\xi)^2/(4t)} \, d\xi.$$

It is no surprise that this is closely related to the Green's function. The adjoint to the forward heat equation is the backward heat equation and, as a function of ξ and τ (the analogue here of $\boldsymbol{\xi}$ above) $G(x, t; \xi, \tau)$ satisfies

$$\mathcal{L}_{\xi, \tau}^* G = \frac{\partial G}{\partial \tau} + \frac{\partial^2 G}{\partial \xi^2} = \delta(\xi - x)\delta(\tau - t).$$

Two minus signs from the exponent cancel.

Remembering the fundamental solution of the forward heat equation (see Exercise 8) and reversing time,

$$G(x, t; \xi, \tau) = \frac{1}{2\sqrt{\pi(t - \tau)}} e^{-(x-\xi)^2/(4(t-\tau))}.$$

The usual integration in the form

$$\int_{-\infty}^{\infty} \int_0^t u \mathcal{L}_{\xi, \tau} G - G \mathcal{L}_{\xi, \tau} u \, d\tau \, d\xi$$

then yields precisely the solution we derived earlier. It is an exercise to generalise this result to the heat equation with a source term,

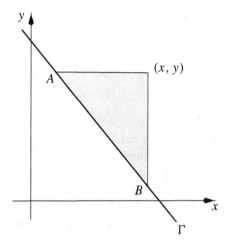

Figure 9.6 Domain of integration for the Riemann function for the wave equation.

$\mathcal{L}u = f(x, t)$; you will get a double integral involving the product of G and f that has the simple physical interpretation of being a superposition of solutions of initial value problems starting at different times. Do it and see.

For the inhomogeneous wave equation in the canonical form

$$\mathcal{L}u = \frac{\partial^2 u}{\partial x \partial y} = f(x, y),$$

with Cauchy data u and $\partial u/\partial n$ given on a non-characteristic curve Γ, we proceed in the same spirit but differently in detail. One of the differences of detail is that the Green's function is now usually called a Riemann function, and we denote it by $R(x, y; \xi, \eta)$. The differential operator $\partial^2/\partial x \partial y$ is self-adjoint, but we have to consider the direction of information flow carefully (see Figure 9.6). When we solve

$$\mathcal{L}^* R = \frac{\partial^2 G}{\partial \xi \partial \eta} = \delta(\xi - x)\delta(\eta - y),$$

we look for a solution valid for $\xi < x, \eta < y$. Then the 'usual' integral

$$\int u\mathcal{L}^* R - R\mathcal{L}u$$

is taken over the characteristic triangle shaded in Figure 9.6, and after use of Green's theorem yields the solution in terms of an integral along Γ from A to B and an integral over the shaded triangle.

The Riemann function for the wave operator $\mathcal{L}$ is particularly simple: Differentiate it and see.

$$R(x, y; \xi, \eta) = \mathcal{H}(x - \xi)\mathcal{H}(y - \eta),$$

i.e. it is equal to 1 in the quadrant $\xi < x, \eta < y$ and zero elsewhere. It yields the familiar D'Alembert solution (see [44]). Unfortunately this is a rare explicit example. Although it is not hard to prove that the Riemann function exists, only for a very few hyperbolic equations can it be found in closed form.

9.7 Sources and further reading

The material on Green's functions is just a small step into Sturm–Liouville and Hilbert–Schmidt theory, eigenfunction expansions and transform methods. If you want to explore further, [26] gives a straightforward account of the theory for ordinary differential equations, [44] presents an informal introduction to the corresponding material for partial differential equations and the excellent [57] contains a more thorough account.

9.8 Exercises

1 Truncated random variables. Suppose that X is a continuous random variable taking values in $(-\infty, \infty)$, for example, Normal. The *truncated* variable Y is defined by

$$Y = \begin{cases} X & \text{if } X < a, \\ a & \text{if } X \geq a. \end{cases}$$

What are its distribution and density functions?

2 A useful identity. Interchange the order of integration (draw a picture of the region of integration) to show that

$$\int_0^x \int_0^\xi f(s)\, ds\, d\xi = \int_0^x (x - \xi) f(\xi)\, d\xi.$$

Generalise to reduce an n-fold repeated integral of a function of a single variable to a single integral.

3 Green's function for a stretched string. Integrate twice to find the solution of the two-point boundary value problem

$$\frac{d^2 y}{dx^2} = f(x), \qquad 0 < x < 1, \qquad y(0) = y(1) = 0,$$

in the form

$$y(x) = \int_0^1 G(x, \xi) f(\xi)\, d\xi.$$

Verify that if you differentiate twice under the integral sign and use the jump conditions at $\xi = x$ you recover the original problem.

4 Green's function for an initial value problem. Use the result of Exercise 2 to show that the solution of the initial value problem

$$\frac{d^2y}{dx^2} = f(x), \quad 0 < x < 1, \qquad y(0) = \frac{dy}{dx}(0) = 0, \qquad (9.16)$$

is

$$y(x) = \int_0^x (x - \xi) f(\xi) \, d\xi.$$

Now pick $X > x$ and write this answer in the form

$$y(x) = \int_0^X G(x, \xi) f(\xi) \, d\xi.$$

What is G? Show that it satisfies

$$\frac{d^2G}{d\xi^2} = \delta(x - \xi), \quad 0 < \xi < X, \qquad G = \frac{dG}{d\xi} = 0 \quad \text{at} \quad x = X.$$

Verify by differentiating under the integral sign that your answer satisfies the original problem. What is the adjoint problem (differential equation and boundary conditions) corresponding to the original problem (9.16)?

This kind of Green's function is the ordinary differential equation analogue of the Riemann function for a hyperbolic equation.

5 Adjoint of a differential operator. Suppose that

$$\mathcal{L}_x y = a(x)\frac{d^2y}{dx^2} + b(x)\frac{dy}{dx} + c(x)y,$$

with

$$\alpha_0 y(0) + \beta_0 y'(0) = 0, \qquad \alpha_1 y(1) + \beta_1 y'(1) = 0.$$

Show that the adjoint is

$$\mathcal{L}_x^* v = \frac{d^2}{dx^2}(a(x)v) - \frac{d}{dx}(b(x)v) + c(x)v$$

with

$$a(0)\big(\alpha_0 v(0) + \beta_0 v'(0)\big) + \beta_0\big(a'(0) - b(0)\big)v(0) = 0,$$
$$a(1)\big(\alpha_1 v(1) + \beta_1 v'(1)\big) + \beta_1\big(a'(1) - b(1)\big)v(1) = 0,$$

in either or both of the following ways:

(a) show that $y\mathcal{L}_x^* v - v\mathcal{L}_x y$ can be integrated by parts as in the text; *The easy way if you know the answer.*
(b) write

$$\mathcal{L}_x^* v = A(x)\frac{d^2v}{dx^2} + B(x)\frac{dv}{dx} + C(x)v$$

This is what you might do if you don't know the answer and can't guess it.

and hack away at the integration by parts (start by integrating the highest derivatives) until everything has been integrated. Whenever terms crop up that can't be integrated, set them equal to zero to find A, B and C and determine the adjoint boundary conditions similarly.

Hence verify that, for self-adjoint operators, $\mathcal{L}_x y$ is of the form

$$\mathcal{L}_x y = \frac{\mathrm{d}}{\mathrm{d}x}\left(p(x)\frac{\mathrm{d}y}{\mathrm{d}x}\right) + q(x)y$$

for some functions $p(x)$ and $q(x)$, while the boundary conditions are as above. Also show that periodic boundary conditions, $y(0) = y(1)$, $y'(0) = y'(1)$, give a self-adjoint operator as long as $p(0) = p(1)$.

What is the adjoint operator if $\mathcal{L}y = \mathrm{d}^2y/\mathrm{d}x^2$, $0 < x < 1$, and the boundary conditions for y are $y(0) = y(1) + y'(1)$, $y'(0) = 0$?

6 The Fredholm alternative: linear algebra and two-point boundary value problems. Suppose that $\mathbf{A}$ is an $n \times n$ matrix, and we want to solve the linear equations

$$\mathbf{A}\mathbf{y} = \mathbf{f}$$

for the vector $\mathbf{y}$ given $\mathbf{f}$. Show that if $\mathbf{y}_1$ and $\mathbf{y}_2$ are two solutions then their difference is an eigenvector of $\mathbf{A}$ with eigenvalue 0.

We know that if the rank of $\mathbf{A}$ is n then $\mathbf{A}$ is invertible, its determinant (equal to the product of the eigenvalues) is nonzero and the solution $\mathbf{y}$ exists and is unique. Suppose now that the rank of $\mathbf{A}$ is $n - 1$, so that the null space of $\mathbf{A}$ has dimension 1 and precisely one eigenvalue of $\mathbf{A}$ is zero. That is, there are vectors $\mathbf{v}$ and $\mathbf{w}$, unique up to multiplication by a scalar, such that

If $\mathbf{A}$ is symmetric, then $\mathbf{v} = \mathbf{w}$.

$$\mathbf{A}\mathbf{v} = \mathbf{0}, \qquad \mathbf{w}^{\mathsf{T}}\mathbf{A} = \mathbf{0}^{\mathsf{T}};$$

they are the right and left eigenvectors of $\mathbf{A}$ with eigenvalue 0. Put another way, the corresponding homogeneous system $\mathbf{A}\mathbf{y} = \mathbf{0}$ has the non-trivial solution $c\mathbf{v}$ for any scalar c.

Premultiply $\mathbf{A}\mathbf{y} = \mathbf{f}$ by $\mathbf{w}^{\mathsf{T}}$ to show that

- **either $\mathbf{w}^{\mathsf{T}}\mathbf{f} = \mathbf{0}$**, in which case the solution exists but is only unique up to addition of scalar multiples of $\mathbf{v}$,
- **or $\mathbf{w}^{\mathsf{T}}\mathbf{f} \neq \mathbf{0}$**, in which case no solution exists at all.

Illustrate this by finding the value of f_2 for which the equations

$$\begin{pmatrix} 1 & -1 \\ 2 & -2 \end{pmatrix} \begin{pmatrix} y_1 \\ y_2 \end{pmatrix} = \begin{pmatrix} 1 \\ f_2 \end{pmatrix}$$

have any solution at all; interpret geometrically.

This result is known as the *Fredholm Alternative*. It applies, *mutatis mutandis*, to two-point boundary value problems. For example, consider

$$\mathcal{L}_x y = \frac{d^2 y}{dx^2} + \alpha^2 y = f(x), \qquad 0 < x < 1, \qquad y(0) = y(1) = 0,$$
$$(9.17)$$

the analogue of $\mathbf{A}\mathbf{y} = \mathbf{f}$. Show that the corresponding homogeneous problem $\mathcal{L}_x y = 0$ has only the trivial solution $y = 0$ unless $\alpha = m\pi$ for integer m (the analogue of the case where $\mathbf{A}$ has zero for an eigenvalue). Find the corresponding eigenfunctions, which are analogous to $\mathbf{v}$ and $\mathbf{w}$ and here are equal as $\mathcal{L}_x$ is self-adjoint. Suppose that $\alpha = \pi$. Multiply (9.17) by the corresponding eigenfunction and integrate by parts to show that there is a solution to (9.17) only if

$$\int_0^1 f(x) \sin \pi x \, dx = 0,$$

the analogue of $\mathbf{w}^\top \mathbf{f} = \mathbf{0}$. Generalise to the case of any (not necessarily self-adjoint) second-order differential operator.

Of course, this is not a coincidence. One could take a two-point boundary value problem and discretise it using finite difference approximations to the derivatives; the result would be a set of linear equations whose solvability or otherwise should be the same, as $n \to \infty$, as that of the original continuous problem.

7 Matrix inversion. In this exercise you will develop the matrix analogue of the calculation of subsection 9.6.1 involving the Green's function for a two-point boundary value problem for an ordinary differential equation. For clarity, we use the summation convention (see p. 9) throughout.

Suppose that the matrix equation $\mathbf{A}\mathbf{y} = \mathbf{f}$ (in which $\mathbf{A}$ is not necessarily symmetric) is written in component form as

$$A_{ij} y_j = f_i \qquad \text{(identify this with } \mathcal{L}_x y = f).$$

Let the inverse matrix $\mathbf{A}^{-1}$ have components $(A^{-1})_{ij} = G_{ij}$, so that from $\mathbf{y} = \mathbf{A}^{-1}\mathbf{f}$ we have

$$y_i = G_{ij} f_j \qquad \text{(identify with } y(x) = \int_0^1 G(x, \xi) f(\xi) \, d\xi).$$

Let δ_{ij} be the Kronecker delta, the discrete analogue of the delta function. Show that $\mathbf{A}^{-1}\mathbf{A} = \mathbf{I}$ and $\mathbf{A}\mathbf{A}^{-1} = \mathbf{I}$ are written

That is, $\delta_{ij} = 0$ if $i \neq j$ and $\delta_{ij} = 1$ if $i = j$. What is δ_{ii}?

$$G_{ij} A_{jk} = \delta_{ik} \qquad \text{(identify with } \mathcal{L}_x G = \delta(x - \xi)),$$
$$A_{ij} G_{jk} = \delta_{ik}.$$

Note that, just as
$\delta(x - \xi) = \delta(\xi - x)$, so
$\delta_{ij} = \delta_{ji}$.

Take the transpose of the last of these equations to identify it with
$\mathcal{L}_\xi^* G = \delta(\xi - x)$. Lastly, take the dot product with the vector (y_k) to
show that

$$0 = A_{ij} G_{jk} y_k - G_{ij} A_{jk} y_k = y_i - G_{ij} f_j;$$

identify this with the calculation involving $\int y \mathcal{L}^* G - G \mathcal{L} y$.

8 The fundamental solution of the heat equation.   Show that the
heat equation

$$\frac{\partial u}{\partial t} = \frac{\partial^2 u}{\partial x^2}$$

has similarity solutions of the form $u(x, t) = t^\alpha f(x/\sqrt{t})$ for all α
and find the ordinary differential equation satisfied by f. Show that

$$\int_{-\infty}^{\infty} u(x, t) \, dx$$

is independent of t when $\alpha = -\frac{1}{2}$, use the result of Exercise 1 of the
next chapter to show that in this case $u(x, 0) \propto \delta(x)$ and hence find
the fundamental solution of the heat equation.

9 Brownian motion.   A particle performs the standard drunkard's
random walk on the real line, in which in timestep i, of length δt,
it moves by $X_i = \pm \delta x$ with equal probability $\frac{1}{2}$. It starts from the
origin and the increments are independent. Define

$$W_n = \sum_1^n X_i.$$

This scaling is the simplest that
allows proper time variation yet
keeps the variance of the limit
finite.

Show that $\mathbb{E}[W_n] = 0$, $\text{var}[W_n] = n \delta x^2 / \delta t$. Now let $n \to \infty$ with
$n \delta t = t$ fixed and $\delta x = \sqrt{\delta t}$. Call the limiting process (assuming it
exists!) W_t. Use the central limit theorem to show that

- for each $t > 0$, W_t has the normal distribution with mean zero and
  variance t.

Show also that

- $W_0 = 0$;
- for each $0 \leq s < t$, $W_t - W_s$ is independent of W_s.

The resulting stochastic process is called *Brownian motion* and it is
central to the modern analysis of financial markets. Give a heuristic
argument that the sample paths (realisations, graphs of the random
walk) are continuous in t but not differentiable.

Now let $p(x, t)$ be the probability density function of many such
random walks (as a function of position x for each t). Go back to the

discrete random walk and, as in the discussion of Poisson processes in Chapter 7, condition on one step to write down

$$p(x, t + \delta t) = \frac{1}{2} \left(p(x - \delta x, t) + p(x + \delta x, t) \right).$$

Expand the right-hand side in a Taylor series and use $\delta x = \sqrt{\delta t}$ to show that

$$\frac{\partial p}{\partial t} = \frac{1}{2} \frac{\partial^2 p}{\partial x^2}.$$

Explain why $p(x, 0) = \delta(x)$ and hence find $p(x, t)$ (see Exercise 8).

The factor in front of the second derivative in the heat equation is a diagnostic feature that the equation is being used by a probabilist as distinct from a 'physical' applied mathematician.

10 **Regular part of the Green's function for the Laplacian.** A horizontal membrane stretched over a region D is stretched to tension T, and a normal force f per unit area is then applied. The displacement (which, like the force, is measured vertically upwards) is zero on the boundary ∂D. Show that the displacement $u(x, y)$ of the membrane satisfies

$$T\nabla^2 u = -f \quad \text{in} \quad D, \qquad u = 0 \quad \text{on} \quad \partial D.$$

Suppose that $f(x, y) = \delta(\mathbf{x} - \boldsymbol{\xi})$, where $\mathbf{x} = (x, y)$ and $\boldsymbol{\xi} = (\xi, \eta)$ is known. How is $u(x, y; \xi, \eta)$ related to the Green's function for the Laplacian in D? *(Do not worry about the infinite displacement!)*

Now suppose that the force is due to a very heavy ball which is free to roll around is in equilibrium at $\boldsymbol{\xi}$. Suppose that we model its effect by that of a point force. Take a small square centred on $\mathbf{x} = \boldsymbol{\xi}$ and resolve forces in the x- and y- directions to show that the gradient of the regular part of G vanishes at $\mathbf{x} = \boldsymbol{\xi}$. Do you think there is always just one such equilibrium point? If not, when might you have one and when more than one? *(Can you find a dimensionless parameter to quantify this modelling assumption?)*

'What's the word beginning with 'd' which means distribution? Oh, distribution.'

10

Theory of distributions

The time has come to look at the theoretical underpinning of the delta function and its relatives. You may choose not to read this chapter, but I promise that you will not find it complex or technically demanding. We begin with a few (as few as we can get away with) necessary definitions.

10.1 Test functions

We noted earlier that the proper way to approach $\delta(x)$ was by thinking of the result of multiplying a suitably smooth function $\phi(x)$ and integrating to get $\phi(0)$. The first step in setting up a robust framework is to define a class of 'suitably smooth' functions, called *test functions*. We say that $\phi(x)$ is a test function if

Because every derivative of ϕ is itself differentiable, the derivatives are all continuous and bounded.

- $\phi(x)$ is a C^∞ function. That is, it has derivatives of all orders at each point $x \in \mathbb{R}$.
- $\phi(x)$ has *compact support*. That is, it vanishes outside some interval (a, b). (The support is the closure of the set where ϕ is non-zero.)

The first requirement makes these functions very smooth indeed.[1] This high degree of regularity guarantees a trouble-free ride for the theory, the reason being that if $\phi(x)$ is a test function, then so are all its derivatives.

---

[1] Roughly speaking, only real analytic functions (defined as equal to the sum of a convergent Taylor series) are smoother, and they can never be test functions because they cannot have compact support (why not?).

We should note that test functions do exist (and that we never need to know much more than this: they are a background tool). The easiest way to see this is to construct one, using the famous example of a function which has derivatives of all orders, and hence a Taylor series, at $x = 0$, but which is not equal to the sum of its Taylor series. That is, look at

See Exercise 5. This standard pathological example from real analysis is more useful than you might imagine on first meeting it.

$$\Phi(x) = \begin{cases} 0, & x \leq 0, \\ e^{-1/x}, & x > 0, \end{cases}$$

which vanishes for $x \leq 0$, is positive for $x > 0$, and is C^∞. The only thing wrong with this function is that it does not have compact support. To rectify this, just multiply by, say, $\Phi(1 - x)$, to give

$$\phi(x) = \Phi(x)\Phi(1 - x),$$

which is a perfectly good test function with support on the interval $(0, 1)$.

We also need a definition of convergence for a sequence of test functions $\{\phi_n(x)\}$. We say that $\phi_n(x) \to 0$ as $n \to \infty$ if

- $\phi_n(x)$ and all its derivatives $\phi_n^{(m)}(x)$ tend to zero, uniformly in both x and m;
- there is an interval (a, b) containing the support of all the ϕ_n.

The first of these is an incredibly strong form of convergence: the ϕ_n have no room to wriggle at all. The second stops them from running away to infinity as n increases.

The only other thing to say about test functions is that we shall denote them by lower-case Greek letters, usually ϕ or ψ.

10.2 The action of a test function

Suppose that $f(x)$ is an integrable[2] function (we denote such functions by lower-case Roman letters f, g, etc.). We define the *action* of f on a test function $\phi(x)$ by

$$\langle f, \phi \rangle = \int_{-\infty}^{\infty} f(x)\phi(x)\,dx.$$

So, this action is a kind of weighted average of $\phi(x)$. If we know the action of f on all test functions then we should know all about f itself (a bit like recovering a probability distribution from its moments). The action, regarded as a map from the space of test functions to $\mathbb{R}$, satisfies

It's also a bit like an inner product: but note that f and ϕ lie in different spaces.

---

[2] We sidestep the question of what exactly we mean by this. Piecewise continuous will do for now, or locally Lebesgue integrable.

the usual linearity properties, such as

$$\langle f, a\phi + b\psi \rangle = a\langle f, \phi \rangle + b\langle f, \psi \rangle$$

for real constants a, b. Also, if $\phi_n(x) \to 0$ in the sense defined above then $\langle f, \phi_n \rangle \to 0$ as a sequence of real numbers.

10.3 Definition of a distribution

In defining distributions, we use the very mathematical idea of taking things about which we already know, here functions, and dropping some of their properties while retaining others in order to obtain something broader or more general. In this way, we see that distributions are indeed 'generalised functions', although this term is not universally used.

As foreshadowed above, the properties that we want to keep are those to do with the action of a function on a test function; that is, we keep the 'smoothing' idea of averaging while quietly dropping all worries about pointwise definition. We do this is such a way that all the properties of distributions are *consistent* with the corresponding properties of (say) piecewise continuous functions. Then, all such functions are subsumed within the larger class of distributions.

Measurable functions would be better, but that requires too much machinery.

The two properties that we keep are those we reached at the end of the previous section: linearity and continuity. We define a distribution $\mathcal{D}$ as *a continuous linear map from the space of test functions to $\mathbb{R}$, denoted by*

$$\mathcal{D} : \phi \mapsto \langle \mathcal{D}, \phi \rangle \in \mathbb{R}.$$

The result of the map, $\langle \mathcal{D}, \phi \rangle$, is known as the *action* of $\mathcal{D}$ on ϕ. We say that two distributions are equal if their action is the same for all test functions.

The properties of linearity and continuity are as above:

$$\langle \mathcal{D}, a\phi + b\psi \rangle = a\langle \mathcal{D}, \phi \rangle + b\langle \mathcal{D}, \psi \rangle,$$

for real constants a, b, and

$$\text{if} \quad \phi_n(x) \to 0 \quad \text{as} \quad n \to \infty \quad \text{then} \quad \langle \mathcal{D}, \phi_n \rangle \to 0.$$

Evidently any piecewise continuous function $f(x)$ corresponds to a distribution $\mathcal{D}_f$ with the obvious action $\langle \mathcal{D}_f, \phi \rangle = \langle f, \phi \rangle$. Indeed, we normally don't bother to write $\mathcal{D}_f$, but just use f itself. This is an example of the consistency referred to above.

We shall mostly use the font style of $\mathcal{D}$, $\mathcal{H}$ to denote distributions, unless they already have another name. The set of test functions is often called $\mathcal{D}$ and the set of distributions is then written $\mathcal{D}'$. Sometimes we write $\mathcal{D}(x)$ to emphasise the dependence on x; the dependence is of course in the test functions, but it's quite OK, and indeed a good idea, to think of distributions as depending on x as well.

Example: the delta function. There could be no better example than the delta distribution, δ or $\delta(x)$. It is defined as a distribution by its action on a test function $\phi(x)$:

We could also have written

$$\langle \delta, \phi \rangle = \phi(0).$$

$$\langle \delta(x), \phi(x) \rangle = \phi(0).$$

You should check carefully that this action does indeed define a distribution satisfying the properties above. Again, it is OK, and indeed a good idea, to think intuitively of the action of the delta function as

$$\langle \delta, \phi \rangle = \int_{-\infty}^{\infty} \delta(x)\phi(x)\,dx.$$

However, for safety you should always use the formal definition to prove anything about $\delta(x)$ or any other distribution.

10.4 Further properties of distributions

If our distributions are to be useful, we need to give them some more properties. We assume that if $\mathcal{D}$ and $\mathcal{E}$ are distributions, a is a real constant, $\phi(x)$ is a test function and $\Phi(x)$ is a C^{∞} function (not necessarily a test function) then there are new distributions $\mathcal{D} + \mathcal{E}$, $a\mathcal{D}$, $\mathcal{D}(x - a)$ and $\mathcal{D}(ax)$ such that:

- $\langle \mathcal{D} + \mathcal{E}, \phi \rangle = \langle \mathcal{D}, \phi \rangle + \langle \mathcal{E}, \phi \rangle$;
- $\langle a\mathcal{D}, \phi \rangle = a\langle \mathcal{D}, \phi \rangle$;
- $\langle \mathcal{D}(x - a), \phi(x) \rangle = \langle \mathcal{D}(x), \phi(x + a) \rangle$;
- $\langle \mathcal{D}(ax), \phi(x) \rangle = \dfrac{1}{|a|}\langle \mathcal{D}, \phi(x/a) \rangle$.
- $\langle \Phi(x)\mathcal{D}(x), \phi(x) \rangle = \langle \mathcal{D}(x), \Phi(x)\phi(x) \rangle$.

Note how we slip in and out of stating the x-dependence explicitly.

Watch out for the modulus sign.

Note that $\Phi(x)\phi(x)$ is a test function even if $\Phi(x)$ is not.

You should check all these when $\mathcal{D}$ corresponds to an integrable function $f(x)$; it will give you some intuition as to why the definitions have been made in this way. Note in particular that from the third definition we have

$$\langle \delta(x - a), \phi(x) \rangle = \langle \delta(x), \phi(x + a) \rangle$$
$$= \phi(a).$$

As expected, we have recovered the sifting property of the delta function.

10.5 The derivative of a distribution

One more idea completes our introduction to the distributional framework. If we want to make sense of ideas such as $d^2 y/dx^2 = \delta(x - \xi)$,

What properties of test functions do we use here?

we'd better define the derivative of a distribution. Again, consistency with ordinary functions provides the way in. If $f(x)$ is differentiable, with derivative $f'(x)$, then integrating by parts we calculate the action of $f'(x)$:

$$
\langle f'(x), \phi(x) \rangle = \int_{-\infty}^{\infty} f'(x)\phi(x)\,\mathrm{d}x
$$
$$
= f(x)\phi(x)\big|_{-\infty}^{\infty} - \int_{-\infty}^{\infty} f(x)\phi'(x)\,\mathrm{d}x
$$
$$
= -\langle f(x), \phi'(x) \rangle.
$$

Notice how the compact support of the test function takes care of $f(x)\phi(x)\big|_{-\infty}^{\infty}$.

We define the *derivative $\mathcal{D}'$* (also written $\mathrm{d}\mathcal{D}/\mathrm{d}x$) *of a distribution $\mathcal{D}$* in terms of its action by

$$
\langle \mathcal{D}', \phi \rangle = -\langle \mathcal{D}, \phi' \rangle
$$

(note that $\phi'(x)$ is also a test function). The point is that although we do not know about $\mathcal{D}'$, we do know about $\mathcal{D}$, so we can calculate $\langle \mathcal{D}, \phi' \rangle$ and hence $\langle \mathcal{D}', \phi \rangle$.

For example, let us show that $\mathcal{H}'(x) = \delta(x)$. We define the Heaviside function $\mathcal{H}(x)$ by its action:

$$
\langle \mathcal{H}, \phi \rangle = \int_{0}^{\infty} \phi(x)\,\mathrm{d}x;
$$

this is entirely consistent with our view of $\mathcal{H}(x)$ as the unit step function since

$$
\mathcal{H}(x)\phi(x) = \begin{cases} 0, & x \le 0, \\ \phi(x), & x > 0. \end{cases}
$$

Now consider the action of $\mathcal{H}'(x)$:

$$
\langle \mathcal{H}'(x), \phi(x) \rangle = -\langle \mathcal{H}(x), \phi'(x) \rangle
$$
$$
= -\int_{0}^{\infty} \phi'(x)\,\mathrm{d}x
$$
$$
= \phi(0)
$$
$$
= \langle \delta(x), \phi(x) \rangle.
$$

Since their actions are identical, we conclude that $\mathcal{H}'(x) = \delta(x)$ (as distributions).

We can extend this definition recursively, to give the action of the mth derivative of $\mathcal{D}$ as

$$
\langle \mathcal{D}^{(m)}(x), \phi(x) \rangle = (-1)^m \langle \mathcal{D}, \phi^{(m)}(x) \rangle
$$

for $m = 1, 2, 3, \ldots.$ Because every derivative of a test function is a test function, we see that distributions have derivatives of all orders too, an example of the technical simplicity of this theory.

10.6 Extensions of the theory of distributions

We conclude with an overview (a glimpse, really) of two vital extensions of the theory just outlined.

10.6.1 More variables

It is a very straightforward business to define multidimensional distributions, in the context of functions of several variables. We first define test functions to have compact support and to be C^∞ in all their arguments. Then, we define distributions as continuous linear maps from this space of test functions to $\mathbb{R}$. In particular, the delta function satisfies

$$\langle \delta(\mathbf{x}), \phi(\mathbf{x}) \rangle = \phi(\mathbf{0}).$$

The partial derivatives of a distribution $\mathcal{D}(\mathbf{x})$ are defined recursively using the formula

$$\left\langle \frac{\partial \mathcal{D}}{\partial x_i}, \phi \right\rangle = -\left\langle \mathcal{D}, \frac{\partial \phi}{\partial x_i} \right\rangle.$$

Again, D has derivatives of all orders and, because the mixed partial derivatives of the test functions are always equal, so also are the mixed partials of $\mathcal{D}$. Thus identities such as $\nabla \wedge \nabla \mathcal{D} \equiv \mathbf{0}$ are automatically true for distributions. The whole theory is splendidly robust, and we need have no qualms at all about writing down equations such as $\nabla^2 G = \delta(\mathbf{x} - \boldsymbol{\xi})$.

10.6.2 Fourier transforms

Space does not permit a full description of the theory of Fourier transforms of distributions in one or more variables. Nevertheless, here is an outline. For technical reasons, we use a slightly different class of test functions, which are still C^∞ but no longer have compact support. Instead, they and all their derivatives decay faster than any power of x as $x \to \pm\infty$. In principle, this defines a different class of distributions (known as *tempered distributions* – the compact support ones are *Schwartz*[3] *distributions*), but we won't notice the difference.

---

[3] Rather to my surprise, Schwartz, who invented the theory in 1944, died as recently as the time of writing. A fearless opponent of political and military oppression and a great mathematician, his support was the interval (1915, 2002).

The new test functions can be shown to have the nice property that if $\phi(x)$ is a test function then so is its Fourier transform; this is why we use this class of test functions. We write the transform as[4]

$$\hat{\phi}(k) = \int_{-\infty}^{\infty} \phi(x) e^{ikx} dx.$$

See Exercise 12 to see why this would not be so for compact support test functions.

This is just the usual Fourier transform; we write the inverse as

$$\check{\psi}(x) = \frac{1}{2\pi} \int_{-\infty}^{\infty} \psi(k) e^{-ikx} dk,$$

and we recall the standard results

$$\widehat{\frac{d\phi}{dx}} = -ik\hat{\phi}, \qquad \widehat{x\phi} = -i\frac{d\hat{\phi}}{dk},$$

the first of which is established by integration by parts and the second by differentiation under the integral sign.

Let's examine the action of the Fourier transform of an ordinary function on a test function. The Fourier transform of a tempered distribution $\mathcal{D}$ is then defined to be consistent with this; as ever, we look at its action and transfer the work to the test function. Formally,

You might want to write this out, swapping the dummy variables x and k in the second line.

$$\langle \hat{f}, \phi \rangle = \int_{-\infty}^{\infty} \left(\int_{-\infty}^{\infty} f(x) e^{ikx} dx \right) \phi(k) dk$$

$$= \int_{-\infty}^{\infty} \left(\int_{-\infty}^{\infty} \phi(k) e^{ikx} dk \right) f(x) dx$$

$$= \langle f, \hat{\phi} \rangle.$$

We therefore define

$$\langle \hat{\mathcal{D}}, \phi \rangle = \langle \mathcal{D}, \hat{\phi} \rangle,$$

Check this one for an ordinary function.

and similarly we define the inverse by

$$\langle \check{\mathcal{D}}, \phi \rangle = \langle \mathcal{D}, \check{\phi} \rangle.$$

Notice how important it is that $\hat{\phi}$ should be a test function too. Unless it is, we cannot be confident that some of these actions are properly defined. Notice too that the factors of 2π don't appear here: they are all hidden in the inverse of ϕ.

---

[4] Beware: notations differ, both in the signs in the exponent and in the placement of the 2π, which can appear in the exponent or symmetrically as a factor $1/\sqrt{2\pi}$ multiplying both the transform and its inverse. The definition here is probably the commonest among applied mathematicians since a factor 2π in the exponent would prove a nuisance in the formulae for the transforms of derivatives. However, this factor is often used in signal processing, when x is replaced by time t and k by angular frequency ω; $2\pi\omega t$ is a frequency in Hz multiplied by t. The factor 2π is also used in pure mathematics.

Using these deceptively simple formulas, we can prove that the Fourier transform of the derivative $\mathcal{D}' = \mathrm{d}\mathcal{D}/\mathrm{d}x$ is $-\mathrm{i}k\hat{\mathcal{D}}$:

$$\langle \widehat{\mathcal{D}'}, \phi \rangle = \langle \mathcal{D}', \hat{\phi} \rangle$$
$$= -\langle \mathcal{D}, \mathrm{d}\hat{\phi}/\mathrm{d}k \rangle$$
$$= -\langle \mathcal{D}, \widehat{\mathrm{i}x\phi} \rangle$$
$$= \langle -\mathrm{i}k\hat{\mathcal{D}}, \phi \rangle$$

Line 1 is the definition of the transform; line 2 is the distributional derivative; line 3 is a standard identity; in line 4 we swap x for k and shift it to the first argument of the action.

as required. It is an exercise for you to prove that the transform of $x\mathcal{D}$ is $-\mathrm{i}\mathrm{d}\hat{\mathcal{D}}/\mathrm{d}k$.

We end this section by finding the transforms of $\delta(x)$ and 1. (Yes, 1 has a Fourier transform in this theory; so do x, $|x|$, etc.).[5] The transform of $\delta(x)$ must surely be 1: informally,

$$\int_{-\infty}^{\infty} \delta(x)e^{\mathrm{i}kx}\,\mathrm{d}x = e^{\mathrm{i}k0} = 1.$$

Very informally, because $e^{\mathrm{i}kx}$ is not a test function although one could 'truncate' it by multiplying by a test function that is small for $|x| > R$ and taking $R \to \infty$.

Formally,

$$\langle \hat{\delta}, \phi \rangle = \langle \delta, \hat{\phi} \rangle$$
$$= \hat{\phi}(0)$$
$$= \int_{-\infty}^{\infty} \phi(x)\,\mathrm{d}x$$
$$= \langle 1, \phi \rangle,$$

so we do indeed have

$$\hat{\delta}(k) = 1.$$

For the inverse, we have

$$\check{\delta} = \frac{1}{2\pi} \int_{-\infty}^{\infty} \delta(k)e^{-\mathrm{i}kx}\,\mathrm{d}k$$
$$= \frac{1}{2\pi},$$

so taking the transform of both sides, remembering that $(\check{\delta})\widehat{} = \delta$, we get

$$\hat{1}(k) = 2\pi\,\delta(k).$$

You may like to show this from the formal definitions alone, using the fact that for test functions $\langle 1, \check{\phi} \rangle = 2\pi \langle 1, \hat{\phi} \rangle$.

---

[5] The transforms of sums of delta functions are the characteristic functions of discrete random variables.

Example: the heat equation.   We conclude with an example: it's one we have seen before but we will do it in a different way. Consider the following initial value problem for the heat equation:

$$\frac{\partial u}{\partial t} = \frac{\partial^2 u}{\partial x^2}, \qquad -\infty < x < \infty, \qquad t > 0,$$

$$u(x, 0) = \delta(x).$$

This time we'll take a Fourier transform in x. The equation for $\hat{u}(k, t)$ is

$$\frac{\partial \hat{u}}{\partial t} = -k^2 \hat{u}, \qquad -\infty < k < \infty, \qquad t > 0,$$

$$u(x, 0) = \hat{\delta}(k) = 1.$$

The solution is

$$\hat{u}(k, t) = e^{-k^2 t},$$

and inversion by any of a number of methods (see Exercise 14) yields the answer

$$u(x, t) = \frac{1}{2\sqrt{\pi t}} e^{-x^2/(4t)}.$$

10.7 Sources and further reading

If the idea of extending our definition of functions to make sense of the result

$$\int_{-1}^{1} \frac{dx}{x^2} = -2$$

appeals to you then you should definitely read [52].

The theory of distributions in its modern form was developed by Schwartz [55]; the epsilonological approach is exemplified by Lighthill's book [39]. My description of the modern theory is heavily based on the very approachable book by Richards & Youn [52].

10.8 Exercises

1 **Constructing delta functions from continuous functions by the Lebesgue dominated convergence theorem.**   Suppose that $f(x) \in L^1$ is continuous and $\int_{-\infty}^{\infty} f(x)\,dx = 1$. Take a test function $\phi(x)$ and show that, as $\epsilon \to 0$,

$$I_\epsilon = \int_{-\infty}^{\infty} \frac{1}{\epsilon} f\left(\frac{x}{\epsilon}\right) \phi(x)\,dx \to \phi(0),$$

as follows. First show that

$$I_\epsilon = \int_{-\infty}^{\infty} f(s)\phi(\epsilon s)\,ds.$$

Next, show that

$$|f(s)\phi(\epsilon s)| < M\,|f(s)|$$

for some constant $M > 0$, that if $f(s) \in L^1$ then $f(s)\phi(\epsilon s) \in L^1$ and that, for each s, $f(s)\phi(\epsilon s) \to f(s)\phi(0)$ as $\epsilon \to 0$. Deduce from the dominated convergence theorem that interchanging the limit and the integral can be justified:

$$\lim_{\epsilon \to 0} \int_{-\infty}^{\infty} f(s)\phi(\epsilon s)\, ds = \phi(0).$$

2 Constructing delta functions from continuous functions by splitting the range of integration. If you don't know about Lebesgue integration, derive the following slightly weaker result. Suppose that $f(x)$ is any continuous function with

$$\int_{-\infty}^{\infty} f(x)\, dx = 1, \quad \int_{-\infty}^{\infty} |f(x)|\, dx < \infty, \quad \int_{-\infty}^{\infty} |xf(x)|\, dx < \infty.$$

Take a test function $\phi(x)$ and show that, as $\epsilon \to 0$,

$$I_\epsilon = \int_{-\infty}^{\infty} \frac{1}{\epsilon} f\left(\frac{x}{\epsilon}\right) \phi(x)\, dx \to \phi(0),$$

as follows. First write $x = \epsilon s$ in the integral and split up the range of integration to get

$$I_\epsilon = \int_{-\infty}^{-1/\sqrt{\epsilon}} + \int_{-1/\sqrt{\epsilon}}^{1/\sqrt{\epsilon}} + \int_{1/\sqrt{\epsilon}}^{\infty} f(s)\phi(\epsilon s)\, ds.$$

Noting that $|\phi(x)|$ is bounded and using the idea that if $|h| < c$ then $|\int gh| \le \int |gh| \le c \int |g|$, show that the first and third integrals tend to zero as $\epsilon \to 0$, because f is integrable. For the inner integral, expand $\phi(\epsilon s)$ using Taylor's theorem to get

$$\int_{-1/\sqrt{\epsilon}}^{1/\sqrt{\epsilon}} f(s)\big(\phi(0) + \epsilon s\phi'(\xi(s))\big)\, ds$$

where $\xi(s)$ lies between 0 and s. Show that the first term in this integral tends to what we want and, noting that $|\phi'|$ is bounded, that the second tends to zero as $\epsilon \to 0$.

3 Delta sequences. Consider the functions

$$f_n(x) = \frac{n}{\pi \left(1 + n^2 x^2\right)} \quad \text{and} \quad g_n(x) = \frac{\sin nx}{\pi x}.$$

Sketch them and show that $f_n(x)$ tends to $\delta(x)$ as $n \to \infty$, in the distributional sense, so that, for any test function $\phi(x)$,

$$\langle f_n, \phi \rangle \to \phi(0)$$

as $n \to \infty$. Use the method of Exercise 2, but be careful when estimating the integrals as $f_n(x)$ does not satisfy all the conditions of that exercise. Repeat for $g_n(x)$.

This might suggest that if $\delta_n(x)$ is a sequence tending to $\delta(x)$ then $\delta_n(0) \to \infty$. Construct a piecewise constant example to show that this is false.

4 Discrete and continuous sources. Suppose that $u(\mathbf{x})$ is a classical solution of $\nabla^2 u = f(\mathbf{x})$ in $\mathbb{R}^n$, $n \geq 2$, where $f(\mathbf{x})$ is smooth and has compact support, and where appropriate growth conditions at infinity are assumed. Let $\phi(\mathbf{x})$ be a test function. Use Green's theorem in the form

$$\int_D v\nabla^2 w - w\nabla^2 v = \int_{\partial D} v\frac{\partial w}{\partial n} - w\frac{\partial v}{\partial n}, \qquad (10.1)$$

where D is a region containing the support of f, to show that

$$\langle u, \nabla^2\phi \rangle = \langle f, \phi \rangle.$$

Now suppose that we approximate $f(\mathbf{x})$ by delta functions, defining the sequence of distributions

$$\mathcal{F}_n = \sum_1^n \alpha_n\delta(\mathbf{x} - \mathbf{x}_n)$$

and taking the limit $n \to \infty$ in such a way that all the weights α_n tend to zero but

$$\langle \mathcal{F}_n, \phi \rangle \to \langle f, \phi \rangle$$

for all test functions ϕ. Also let u_n be the solution of $\nabla^2 u_n = \mathcal{F}_n$. Show that

$$\langle u_n, \nabla^2\phi \rangle = \langle \mathcal{F}_n, \phi \rangle,$$

and deduce that $u_n \to u$ (as a distribution). Interpret this result in terms of the gravitational potential due to a finite mass distribution (or in electrostatic terms).

5 The function $e^{-1/x}$. Consider

$$\Phi(x) = \begin{cases} 0, & x \leq 0, \\ e^{-1/x}, & x > 0. \end{cases}$$

Remember that $X^n e^{-X} \to 0$ as $X \to \infty$ for all N.

Show that for $x > 0$ its nth derivative $\Phi^{(n)}(x)$ is a polynomial in $1/x$ times $e^{-1/x}$, and hence that $\lim_{x\downarrow 0} \Phi^{(n)}(x) = 0$. Deduce that the Taylor coefficients of this function are all zero. Does the complex function $e^{-1/z}$ have a Taylor series at $z = 0$? If not, what does it have?

6 The distribution $\delta(ax)$. Show from its interpretation as an integral that

$$\delta(ax) = \frac{1}{|a|}\delta(x).$$

7 Derivatives of the delta function. Show carefully, using the definition of a distributional derivative, that if $\Psi(x)$ is a smooth (C^∞)
function and $\mathcal{D}$ a distribution then $(\mathcal{D}\Psi)' = \mathcal{D}'\Psi + \mathcal{D}\Psi'$ (Leibniz).
Deduce that

$$x^n \delta^{(m)}(x) = \begin{cases} 0, & m < n, \\ \dfrac{(-1)^n m!}{(m-n)!} \delta^{(m-n)}(x), & m \geq n, \end{cases}$$

($\delta^{(m)} = m$th derivative). What is $x\delta(x)$? Show that $\delta(x) = -x\delta'(x)$.

8 Convergence of a series of distributions. We say that a sequence
$\{\mathcal{D}_n\}$ of distributions converges to $\mathcal{D}$ if

$$\langle \mathcal{D}_n, \phi \rangle \to \langle \mathcal{D}, \phi \rangle$$

for all test functions $\phi(x)$. This is a remarkably tolerant form of
convergence, because our definition of the convergence of a sequence
of test functions is so stringent: show that if $\mathcal{D}_n \to \mathcal{D}$ then the same
applies to all the derivatives, so that $\mathcal{D}_n^{(m)} \to \mathcal{D}^{(m)}$. Show also that
you can differentiate a convergent series of distributions term by
term.

Find the Fourier series of the sawtooth function

$$f(x) = \begin{cases} \dfrac{1}{2} - \dfrac{x}{2\pi} & 0 < x < \pi, \\ -\dfrac{1}{2} - \dfrac{x}{2\pi} & -\pi < x < 0. \end{cases}$$

Now differentiate both sides, noting that the jumps of 1 in $f(x)$
at $x = 2n\pi$ contribute delta functions $\delta(x - 2n\pi)$, to establish the
result

$$\sum_{n=-\infty}^{\infty} \delta(x - 2n\pi) = \frac{1}{2\pi} \sum_{m=-\infty}^{\infty} \cos mx,$$

an identity that makes classical nonsense but perfect distributional
sense.

Note: It can be shown that every distribution $\mathcal{D}$ is the distributional limit of a sequence of test functions, which are C^∞. So the set
of distributions is not unboundedly diverse.

9 Derivative of a distribution. Let $\mathcal{D}(x)$ be a distribution. Show (by
considering its action) that

$$\mathcal{D}'(x) = \lim_{h \to 0} \frac{\mathcal{D}(x+h) - \mathcal{D}(x)}{h}.$$

Remember that

$$\langle \mathcal{D}(x+h), \phi(x) \rangle$$
$$= \langle \mathcal{D}(x), \phi(x-h) \rangle.$$

Use the right-hand side of this equation to confirm (again by considering the action) that $\delta(x) = \mathcal{H}'(x)$.

10 Dipoles. The derivative of the delta function, $\delta'(x)$, is known as a (one-dimensional) *dipole*, which you can think of as the limit as $\epsilon \to 0$ of a positive delta function at $x = \epsilon$ and a negative one at $x = 0$ (see Exercise 9). What is its action on a test function $\psi(x)$?

In hydrodynamics, a mass dipole aligned with the x-axis is obtained as the limit of point (in two dimensions, line) sources of strength q at $(\pm\epsilon, 0, 0)$, keeping the product $m = 2\epsilon q$ constant as $\epsilon \to 0$. Explain why the velocity potential ϕ for inviscid irrotational flow with a point source at the origin satisfies

$$\nabla^2 \phi = q\delta(\mathbf{x}),$$

and deduce that if there is a dipole as above at the origin then the potential satisfies

$$\nabla^2 \phi = m\frac{\partial \delta}{\partial x}.$$

(The right-hand side may also be written as $\delta'(x)\delta(y)\delta(z)$ in three dimensions, or $\delta'(x)\delta(y)$ in two.) Hence calculate the potential for a dipole and sketch the streamlines in two dimensions. Show that the potential $U(r\cos\theta + a^2\cos\theta/r)$ for flow past a cylinder is that for uniform flow plus that for a dipole.

Interpret these results in terms of electric charges. (Whereas point charges generate electric fields, because there are no magnetic monopoles the basic generator of magnetic fields is the infinitesimal current loop, giving a dipole field with lines of force similar to those of a bar magnet. Higher-order derivatives, called multipoles, are important in, for example, analysis of the far field of radio transmitters.)

11 Vector distributions. Develop the following two ways of defining vector-valued distributions in $\mathbb{R}^3$. In both cases aim to establish the identities $\nabla \cdot \nabla \wedge \boldsymbol{\mathcal{D}} \equiv 0$, $\nabla \wedge \nabla \mathcal{D} \equiv \mathbf{0}$ for vector and scalar distributions $\boldsymbol{\mathcal{D}}$ and $\mathcal{D}$ respectively. You will need to establish variants of Green's theorem in order to define the action of the operators div and curl using integration by parts.

(a) Take scalar test functions $\phi(\mathbf{x})$ and define their action on a vector function $\mathbf{v}(\mathbf{x})$ as the vector

$$\langle \mathbf{v}, \phi \rangle = \int_{\mathbb{R}^3} \mathbf{v}(\mathbf{x})\phi(\mathbf{x})\,d\mathbf{x}.$$

Then define a vector-valued distribution $\boldsymbol{\mathcal{D}}$ as a continuous linear

Notation clash! ϕ is not a test function here.

map from the space of test functions to $\mathbb{R}^3$, consistently with this action.

(b) Use vector test functions ϕ and the action

$$\langle \mathbf{v}, \phi \rangle = \int_{\mathbb{R}^3} \mathbf{v} \cdot \phi \, d\mathbf{x}.$$

12 Open support test functions. To get an idea of the reason why compact support test functions do not lead to a good theory for the distributional Fourier transform, work out the Fourier transform of

It's not hard: just integrate.

$$f(x) = \begin{cases} 1, & -1 < x < 1, \\ 0 & \text{otherwise,} \end{cases}$$

and observe that, unlike $f(x)$, $\hat{f}(k)$ does not have compact support. (Although $f(x)$ is not a test function, a similar result would hold if it were.) Now look at the definition of the Fourier transform to see why compact support test functions are not useful here.

13 Commutation of the Fourier transform and its inverse. Show directly from the definitions that if $\mathcal{D}$ is a distribution with Fourier transform $\hat{\mathcal{D}}$ then

$$(\hat{\mathcal{D}})\check{} = (\check{\mathcal{D}})\hat{} = \mathcal{D},$$

assuming that this holds for test functions.

14 The inverse of $e^{-k^2 t}$. Find the inverse of $\hat{u}(k, t) = e^{-k^2 t}$ in the following two ways.

(a) Write down the inversion integral and complete the square in the exponent; then, thinking of the integral as a contour integral in the complex k-plane, move the integration contour to the line $\operatorname{Im} k = -x/(2t)$ (check that the endpoint contributions vanish) and evaluate a standard real integral, using the result $\int_{-\infty}^{\infty} e^{-s^2} ds = \sqrt{\pi}$.

(b) Show that $\partial \hat{u}/\partial k = -2kt\hat{u}$, then use the standard identities for the transforms of $\partial u/\partial x$ and xu to obtain a similar ordinary differential equation for u; solve this and choose the 'constant of integration' (which is actually a function of t) so as to set $\int_{-\infty}^{\infty} u(x, t) \, dx = 1$ for all t (which is easy to show from the original problem).

15 The pseudofunction $1/x$. Obviously, $1/x$ is defined for $x \neq 0$ as an ordinary function. Its definition for all $x \in \mathbb{R}$ is achieved by defining its action on a test function $\phi(x)$:

$$\langle 1/x, \phi(x) \rangle = \lim_{\epsilon \to 0} \langle 1/x, \phi(x) \rangle_\epsilon,$$

where

$$\langle 1/x, \phi(x) \rangle_\epsilon = \int_{-\infty}^{-\epsilon} + \int_{\epsilon}^{\infty} \frac{\phi(x)}{x} \, dx;$$

the limiting integral, denoted by

$$\fint_{-\infty}^{\infty} \frac{\phi(x)}{x} \, dx,$$

is called a *Cauchy principal value integral*. Note that the small interval $(-\epsilon, \epsilon)$, which we remove before integrating and taking the limit $\epsilon \to 0$, is symmetric about $x = 0$.

Show that the limit exists for all test functions $\phi(x)$. Show directly from the distributional definitions that

$$\frac{1}{x} = \frac{d}{dx} \log |x|;$$

that is, show that

$$\langle d \log |x|/dx, \phi(x) \rangle = -\langle 1/x, d\phi/dx \rangle$$

by considering the same statement with $\langle \cdot, \cdot \rangle$ replaced by $\langle \cdot, \cdot \rangle_\epsilon$ and then letting $\epsilon \to 0$.

Show also (for future reference) that

$$\fint_{-1}^{1} \frac{dx}{x} = 0. \tag{10.2}$$

16 The Fourier transform of $\mathcal{H}(x)$. A distribution $\mathcal{D}(x)$ is called *odd* if the result of its action gives $\mathcal{D}(-x) = -\mathcal{D}(x)$ and *even* if $\mathcal{D}(-x) = \mathcal{D}(x)$. Show that $\delta(x)$ is even and that $x\delta(x) = 0$. If $\mathcal{H}(x)$ is the Heaviside function, show that $\widetilde{\mathcal{H}}(x) = \mathcal{H}(x) - \frac{1}{2}$ is odd.

Show that the Fourier transform of a real-valued odd function is a purely imaginary odd function of k, and deduce (or assert) that the same applies to distributions.

Since $\mathcal{H}'(x) = \delta(x)$, taking the Fourier transform gives

$$-ik\widehat{\mathcal{H}} = \hat{\delta} = 1.$$

However, before dividing through by k we must realise that we can add $ck\delta(k) \, (= 0)$ to the right-hand side, where c is an as yet unspecified complex constant. By considering instead the transform of the odd distribution $\widetilde{\mathcal{H}}(x)$ and recalling that $\hat{1} = 2\pi\delta(k)$, show that

$$\widehat{\mathcal{H}}(k) = -\frac{1}{ik} + \pi\delta(k).$$

Note that $\widehat{\mathcal{H}}$ requires the definition of $1/k$ introduced in the previous exercise.

17 The Fourier transform of $\mathcal{H}(x)$ again. Here are two more ways of calculating $\widehat{\mathcal{H}}(k)$.

(i) Consider

$$\int_0^{1/\epsilon} e^{ikx}\,dx = -\frac{1}{ik} + \frac{e^{ik/\epsilon}}{ik}.$$

The first part is already in the answer, so the second part must tend to $\pi\delta(k)$ as $\epsilon \to 0$. Write

$$\frac{e^{ik/\epsilon}}{ik} = \frac{\sin(k/\epsilon)}{k} - i\frac{\cos(k/\epsilon)}{k}$$

and note that the real part has been shown (in Exercise 3) to give $\pi\delta(k)$. It remains to show that the principal value integral

$$\fint_{-\infty}^{\infty} \frac{\cos(k/\epsilon)\phi(k)}{k}\,dk \to 0$$

as $\epsilon \to 0$ for any test function ϕ. Write ϕ as the sum of its even and odd parts and note that we need only consider the odd part of ϕ, since the integral of the even part vanishes by symmetry. Now proceed as in earlier exercises, splitting the range of integration into $|k| > \sqrt{\epsilon}$ and $|k| < \sqrt{\epsilon}$ and dealing with each separately. Alternatively, don't bother with the odd/even split, and just use (10.2) for the inner integral.

They are $\frac{1}{2}(\phi(k) \pm \phi(-k))$; show that both of these are test functions.

Use the decay properties of the test function to justify the use of the Riemann–Lebesgue lemma for the outer integrals, and expand ϕ in a Taylor series for the inner one.

(ii) Consider the Fourier transform of

$$\mathcal{H}_\epsilon(x) = \mathcal{H}(x)e^{-\epsilon x},$$

which clearly exists for $\epsilon > 0$; show that it is

$$\widehat{\mathcal{H}}_\epsilon(k) = \frac{1}{\epsilon - ik}.$$

Note that $\mathcal{H}_\epsilon(x)$ 'does the right sort of thing' as $\epsilon \to 0$: it tends to $-1/(ik)$ for $k \neq 0$ and to infinity for $k = 0$.

Writing

$$\frac{1}{\epsilon - ik} + \frac{1}{ik} = \frac{\epsilon}{\epsilon^2 + k^2} - \frac{i\epsilon^2}{k(\epsilon^2 + k^2)},$$

show that as $\epsilon \to 0$ the action of the right-hand side on a test function tends to that of $\pi\delta(k)$. (You will need to interpret the second term as a principal value integral; use the results of Exercise 3.)

Remember that $\widehat{xf} = -\mathrm{i}\,\mathrm{d}\hat{f}/\mathrm{d}k$.   **18 More Fourier transforms.**   What are the Fourier transforms of

$$x, \quad x^n, \quad |x|,$$

for integral $n > 0$?

'You can always make infinity smaller by multiplying by h.'

11

Case study: the pantograph

11.1 What is a pantograph?

In the late 1960s, British Rail was planning a new generation of high-speed electric trains. One question was flagged as a potential problem area: could the waves generated in the overhead cable by the current-collecting device, called a *pantograph*,[1] build up and cause interruptions in the current flow, in particular when the train passes a support? At about the same time the US Air Force developed a facility in which a rocket slung from a taut cable was accelerated along the cable, to allow flight characteristics to be tested and to enable precise targeting for impact tests. In one such test, the rocket was accelerated to 1.04 times the wavespeed in the wire, with the dramatic result indicated in Figure 11.1: a large disturbance propagates forward of the rocket.

The pantograph problem was one of the first to be discussed at an Oxford Study Group with Industry meeting and has become a minor classic of mathematical modelling since the original paper [47]. As so often with industrial problems, it provoked a strand of theoretical research, into the so-called pantograph equation, which is still active.

---

[1] From the Greek, 'universal writer'. The original pantograph was a mechanical device whereby a series of linkages allowed a drawing to be copied exactly. The theory of linkages was at one time intensively studied because of their importance in machinery, the prototype problem being how to transform the up-and-down motion of a piston into circular motion of a wheel. It is related to Ptolemy's epicycle model of the heavens, in which the apparently irregular motion of the planets was accounted for by the assumption that they move around the earth in an arrangement of small circular orbits mounted on larger ones, much like various fairground rides (teacups, waltzer, cyclone).

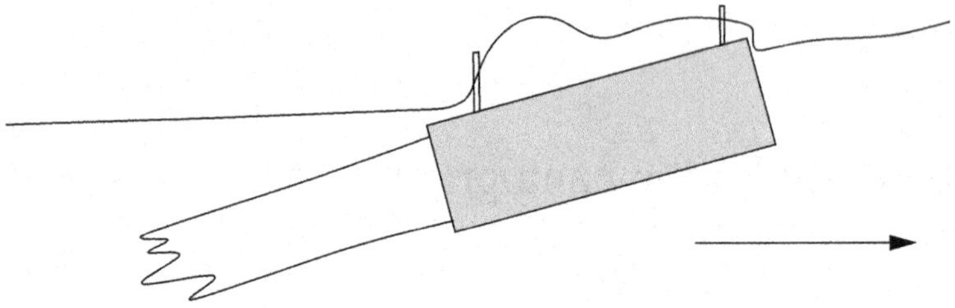

Figure 11.1 A rocket slung from a cable and accelerated to 1.04 times the wavespeed in the cable. The diagram is based on an indistinct photograph in [54]; the cable is fairly accurately reproduced but the rest of the diagram is schematic.

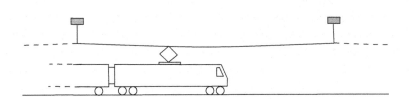

Figure 11.2 A locomotive and its power supply.

11.2 The model

Suppose, then, that a locomotive moves with constant speed U below a cable stretched to tension T between supports at $x = 0, \pm L, \pm 2L, \ldots$, as in Figure 11.2. We can distinguish three main parts of the system that we should model: the motion of the cable, the dynamics of the pantograph and the dynamics of the supports. Some modelling assumptions are immediately reasonable.

- The cable can be modelled as a uniform string of line density ρ, because (based on typical values) its bending stiffness is small. Also its displacement from equilibrium is small. Thus, apart from the static displacement due to gravity, the vertical wire displacement $y(x, t)$ satisfies the wave equation away from the pantograph and the supports. (In a more sophisticated treatment, we may have to reconsider this assumption near to the pantograph and supports.)
- The contact area between the pantograph and the wire is small compared with L. This means that we can represent the effect of the pantograph by a point force at the position $x = Ut$, with a suitable choice of time origin.

We also make some simplifying assumptions, the removal of which leads to unedifying complications.

- Although the dynamics of the supports can be surprisingly complex (see [47]), we assume that the wire is rigidly attached to them. This has the great advantage that what happens in one span of the cable does not affect what happens in the others. In our discussion below, we focus on the span $0 < x < L$, into which the train enters at time $t = 0$.

- The pantograph itself can be modelled as a linear system, so that the force $F(t)$ that it exerts on the wire depends linearly on its vertical displacement $Y(t) = y(Ut, t)$. More specifically, we may expect
  - a spring force, intended to keep the pantograph in contact with the wire; for a linear spring this would contribute a term

$$F_0 - F_1 Y(t)$$

  to the force $F(t)$, where both F_0 and F_1 are positive and we expect that the combination $F_0 - F_1 Y$ is also positive for reasonable values of Y;
  - a damping force, which in the linear case has the form

$$-F_2 \frac{\mathrm{d}Y}{\mathrm{d}t}.$$

In the case of a rocket, a separate calculation of its dynamics adds a term proportional to $\mathrm{d}^2 F/\mathrm{d}t^2$; see Exercise 8.

Just as in our earlier examples, the point force is modelled by a delta function; the difference now is that the force is moving. The motion of the wire is described by the inhomogeneous wave equation

$$\rho \frac{\partial^2 y}{\partial t^2} - T \frac{\partial^2 y}{\partial x^2} = F(t)\delta(x - Ut) - \rho g,$$

where the last term models the gravitational force. Including both spring and damping forces, the pantograph dynamics are modelled by

$$F(t) = F_0 - F_1 Y(t) - F_2 \frac{\mathrm{d}Y}{\mathrm{d}t},$$

where $Y(t) = y(t, t)$. The initial and boundary conditions are these: the wire starts at rest in its equilibrium shape $y_0(x) = -\rho g x(L - x)/(2T)$, and its displacement vanishes at $x = 0, L$.

Let us make this problem dimensionless. There are two velocities, U and the wavespeed $c = \sqrt{T/\rho}$, and we use the latter for scaling purposes, which with the length scale L for x gives a timescale L/c; also, we write $U = cu$ (you might think of u as a Mach number for the train). In order to scale y, either we can use the maximum wire displacement under gravity, $y^* \rho g L^2/(8T)$, or we can use the displacement caused by a typical pantograph force. As we want to focus on the pantograph, let us use the latter and scale y with $F^* L/T$, where F^* is a typical size for the pantograph force (it might, for example, be equal to the constant force F_0). With these scalings, you should check that, the primes having

It hardly matters which we use. The choice of c was governed by the desire to keep u visible – we might want to look at a range of values of u.

In doing the scalings, remember that $\delta(ax) = |a|^{-1}\delta(x)$.

been dropped, the dimensionless problem is

$$\frac{\partial^2 y}{\partial t^2} - \frac{\partial^2 y}{\partial x^2} = f(t)\delta(x - ut) - \alpha, \qquad 0 < x < 1, \qquad (11.1)$$

where, retaining the notation $Y(t) = y(ut, t)$, the dimensionless force has the form

As an exercise work out the dimensionless coefficients f_0 etc. in terms of their dimensional parents.

$$f(t) = f_0 - f_1 Y - f_2 \frac{dY}{dt} \qquad (11.2)$$

and $\alpha = \rho g L / F^*$ is a dimensionless parameter measuring the ratio of the weight of the wire to the force exerted by the pantograph.

11.2.1 What happens at the contact point?

Looking ahead to when we calculate the displacement of the wire, clearly we are going to rely heavily on the general solution of the wave equation in the usual form. That means that we will need to join up solutions of this type across the train path $x = ut$, so we need to know what happens to the gradient of y across this line. We should proceed with caution when we see something unfamiliar such as $\delta(x - ut)$: it is not immediately obvious what it means. One fairly safe way to proceed is to change coordinates so as to reduce the train to rest. That is, we replace x by $\xi = x - ut$ and use ξ and t as independent variables. When we do this, a straightforward chain rule calculation shows that (11.1) becomes

We started Chapter 9 by writing down jump conditions on the (static) wire slope and went from there to the delta function; now the boot is on the other foot as we are confident that the delta function should be there but we don't know how to interpret it!

Since the right-hand side of this equation is a distribution, so also is the left-hand side. However, the chain rule still applies for smooth coordinate changes.

$$\frac{\partial^2 y}{\partial t^2} - 2u \frac{\partial^2 y}{\partial \xi \partial t} - (1 - u^2)\frac{\partial^2 y}{\partial \xi^2} = f(t)\delta(\xi) - \alpha.$$

Now we are in a position to balance the most singular terms. We know that y is continuous at $\xi = 0$ and, assuming smoothness in t, $\partial^2 y/\partial t^2$ should also be continuous. The finger points at $\partial^2 y/\partial \xi^2$ as the most singular term, and we see that this has for its leading-order singular behaviour a delta function of magnitude $-f(t)/(1 - u^2)$. That is, $\partial y/\partial \xi$, which is the same as $\partial y/\partial x$, has a jump of this magnitude,

At least for $U < c$; Figure 11.1 suggests otherwise for $U > c$!

Please check for consistency that the mixed partial derivative is less singular than the leading order terms.

$$\left[\frac{\partial y}{\partial x} \right]_{x=ut-}^{x=ut+} = -\frac{1}{1 - u^2} f(t), \qquad (11.3)$$

which is the time-dependent generalisation of the static condition (9.2) on p. 116. A physical derivation of this condition is given in Exercise 1.

Sign consistency check: We expect the pantograph to push the cable up, giving a negative jump in $\partial y/\partial x$; this is fine if $u < 1$.

11.3 Impulsive attachment for an undamped pantograph

The simplest situation to consider is one in which gravity is neglected ($\alpha = 0$), so that the cable is initially straight, and at $t = 0$ the train

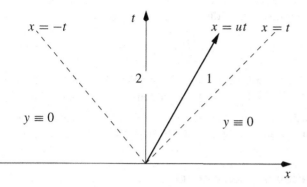

Figure 11.3 Characteristic
diagram for impulsive
attachment.

attaches to the cable impulsively at $x = 0$. In this case we expect disturbances to propagate ahead of and behind the train with (dimensionless) speed 1 so that the cable displacement is only nonzero for $-t < x < t$, as shown in the characteristic diagram of Figure 11.3.

Our strategy is to join together general solutions of the wave equation, of the form $g(t - x) + h(t + x)$, finding the arbitrary function involved from the conditions at the pantograph. Clearly the wire displacement is identically zero except in region 1, $ut < x < t$, and region 2, $-t < x < ut$, shown in Figure 11.3. Otherwise, information would have to travel faster than the wavespeed. Across the characteristics $x = \pm t$ we expect to see discontinuities in the derivatives of y, as we know that these can only propagate along characteristics.

Perhaps you usually write $g(x - t)$; I do. It turns out that $g(t - x)$ is more convenient later, as we don't then get negative arguments for the function g_1.

Bearing in mind that all the information comes from the train, the solution must have the form

$$y(x, t) = \begin{cases} g_1(t - x) & \text{in region 1,} \\ h_2(t + x) & \text{in region 2,} \end{cases}$$

representing forward and backward travelling waves respectively. The functions g_1 and h_2 are as yet unknown, except that we can say that $g_1(0) = h_2(0) = 0$.

At the train $x = ut$, we first express the continuity of the cable:

$$g_1(t - ut) = h_2(t + ut). \tag{11.4}$$

Next, from (11.3), we have

$$-g_1'(t - ut) - h_2'(t + ut) = -\frac{1}{1 - u^2} f(t).$$

Using (11.2) to express $f(t)$ in terms of $Y(t) = g_1(t - ut)$ and eliminating $h_2(t + ut)$ by differentiating (11.4), we have

$$g_1'(t - ut) = \frac{1}{2(1 - u)} f(t)$$

$$= \frac{1}{2(1 - u)} \big(f_0 - f_1 g_1(t - ut) - (1 - u) f_2 g_1'(t - ut) \big).$$

That is, $g_1(\xi)$ satisfies the ordinary differential equation

$$(1 - u)(2 + f_2)\frac{dg_1}{d\xi} + f_1 g_1 - f_0 = 0,$$

whose solution is easily found as a constant plus a decaying exponential. (The large-time behaviour has the pantograph displacement tending to the value f_0/f_1 at which the spring force vanishes; in practice this value would be very large and a support would intervene before it was reached.)

11.4 Solution near a support

A rather more surprising thing occurs when we look at what happens shortly after the train passes a rigid support. As the characteristic diagram Figure 11.4 shows, there are again only two regions where the cable displacement is not zero. Let us again neglect the static displacement of the cable (this is even more realistic near a support, where it must be very small). The difference between this configuration and impulsive attachment is that in region 2 waves can be reflected off the rigid support. Thus the cable displacement has the form

$$y(x, t) = \begin{cases} g_1(t - x) & \text{in region 1,} \\ g_2(t - x) + h_2(t + x) & \text{in region 2.} \end{cases}$$

Continuity of the cable at $x = ut$ now gives

$$g_1(t - ut) = g_2(t - ut) + h_2(t + ut), \tag{11.5}$$

and the pantograph force balance is

$$-g_1'(t - ut) + g_2'(t - ut) - h_2'(t + ut) \tag{11.6}$$

$$= -\frac{1}{1 - u^2} f(t)$$

$$= -\frac{1}{1 - u^2} \left(f_0 - f_1 g_1(t - ut) - f_2(1 - u)g_1'(t - ut) \right). \tag{11.7}$$

Figure 11.4 Characteristic diagram for a train passing a support.

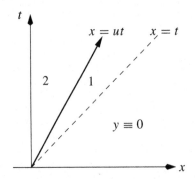

Lastly, we have the rigid-support condition

$$g_2(t) + h_2(t) = 0.$$

With three equations for three unknown functions, we can proceed confidently. We first eliminate h_2 to find from (11.5)

$$g_1(t - ut) = g_2(t - ut) - g_2(t + ut)$$

and from (11.7)

$$-g_1'(t - ut) + g_2'(t - ut) + g_2'(t + ut)$$
$$= -\frac{1}{1 - u^2} \left(f_0 - f_1 g_1(t - ut) - (1 - u) f_2 g_1'(t - ut) \right).$$

Then we observe that we can eliminate $g_1(t - ut)$ throughout, to give (after tidying up)

$$(1 + u)(2 + f_2)g_2'(t + ut) + f_1 g_2(t + ut)$$
$$= -f_0 + f_1 g_2(t - ut) + (1 - u) f_2 g_2'(t - ut).$$
$$(11.8)$$

If we can solve this equation then we will have found g_2 and hence g_1 and the force on the pantograph. The left-hand side of (11.8) is as expected, but the right-hand side is not. Because it contains the function g_2 evaluated at an *earlier* time than on the left-hand side, we have arrived not at an ordinary differential equation but at a kind of *delay differential equation* for g_2. This kind of equation has come to be known as a *pantograph equation* and has given rise to a substantial literature in the last two decades.

Let us consider (11.8) in the special case $f_1 = 0$. Then it can immediately be integrated once, to give

$$(2 + f_2)g_2(t + ut) = -f_0 t + f_2 g_2(t - ut).$$

Writing $\tau = t(1 + u)$ and

$$\mu = \frac{1 - u}{1 + u} < 1,$$

we have

$$(2 + f_2)g_2(\tau) = -\frac{f_0}{1 + u}\tau + f_2 g_2(\mu\tau).$$

We can spot the particular solution $g_2(\tau) = a\tau$, where a is easily found, and we claim that this is the only solution. To show this, consider the difference between two solutions, which satisfies the homogeneous equation

$$g(\tau) = \frac{f_2}{2 + f_2}g(\mu\tau), \qquad g(0) = 0;$$

note that the fraction on the right is less than 1. Suppose that, for a fixed τ, $g(\tau) = g_0 \neq 0$. This means that $g(\mu\tau)$ is greater in modulus than g_0 and, iterating, that

$$g(\mu^n\tau) = \left(\frac{2 + f_2}{2}\right)^n g_0.$$

But as $n \to \infty$, $\mu^n\tau \to 0$ and $|g(\mu^n\tau)| \to \infty$, a contradiction. Hence $g_0 = 0$ and the solution is unique.

It is possible to transform (11.8) to a delay differential equation with a constant time lag by making an exponential substitution; see Exercise 4.

11.5 Solution for a whole span

Let us look briefly at the cable motion in a whole span. This is the case we really need to analyse, because of the possibility that the solution will 'pile up' at the far end of the span $x = 1$. The characteristic diagram is now much more complicated, as seen in Figure 11.5. The initial disturbance propagates as a gradient discontinuity across the characteristic $x = t$, as in the previous section, but then it reflects off the support at $x = 1$. We therefore have a reflected characteristic, $x = 2 - t$, with another gradient discontinuity; this in turn is reflected off the train path and so on, to generate an infinite series of characteristics separating regions in which the solution is smooth.

We have already looked at the regions near the first support, and we know from the previous section that when the pantograph force is a constant plus a linear damping term the pantograph displacement $Y(t) = y(ut, t)$ is linear in t. This solution holds until the time $t_1 = 2/(1 + u)$

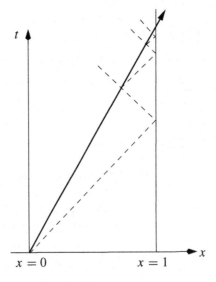

Figure 11.5 Characteristic diagram for a whole span. If, as here, $\frac{1}{3} < u < 1$, no reflections of characteristics from $x = 0$ reach the train.

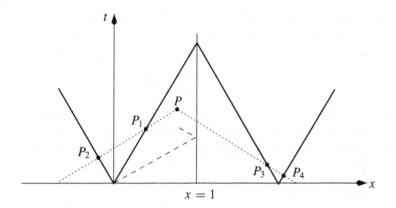

Figure 11.6 A characteristic triangle for $t > 2/(1 + u)$, formed by the dotted lines, which are characteristics, and a segment of the x-axis. The dashed line is the leading characteristic and its reflection. The thick solid line represents the train path and its images in the supports.

at which the reflection ($x = 2 - t$) of the leading characteristic ($x = t$) meets the train path. It is a reasonable guess that the pantograph displacement is a piecewise linear function of t at later times, and we can show that this is the case.

We will adopt a slightly different, and more sophisticated, approach to the wave equation than simply writing down its general solution. In Figure 11.6, we see the train path with its images in the supports. We intend to extend the domain of definition from $0 < x < 1$ to the whole real line in the usual way, so as to satisfy the conditions at the support; this means that the (fictitious) pantograph force from alternate, downward-sloping, parts of the image train path is minus that from the upward-sloping parts, and the same is true for the pantograph displacement. The extended train path seen in Figure 11.6, which also shows a characteristic triangle, which we call $\triangle$, for a point P above the reflection of the leading characteristic (recall that we already know the solution below this line).

Now suppose that we integrate the wave equation (11.1) (again with $\alpha = 0$ for simplicity) over the interior of the characteristic triangle $\triangle$, in the spirit of Exercise 6. We had better be careful about the jumps across the train path, so for safety we integrate separately over each of the polygons that make up the triangle and consider integrals along both sides of the train path. Using Green's theorem, this gives

$$\int_{\triangle} \frac{\partial y}{\partial x}\, dt + \frac{\partial y}{\partial t}\, dx + \int_{\text{train path}} \left[\frac{\partial y}{\partial x}\right] dt + \left[\frac{\partial y}{\partial t}\right] dx = 0,$$

where as before the square brackets denote the jump in their contents. The first integral gives us $2y(P) = 2y(x, t)$. On the train path, we know that y itself has no jump, so

$$\left[\pm u \frac{\partial y}{\partial x} + \frac{\partial y}{\partial t}\right] = 0,$$

A factor of $1 - u^2$ cancels.

the plus sign being taken on the upward-sloping parts of the path and the minus sign on the others. This lets us eliminate $[\partial y/\partial t]$ and then use the fact that $dx = \pm u \, dt$ and the pantograph jump condition to show that

$$2y(P) = \int_{\text{train path}} \pm f(s) \, ds,$$

where again the plus and minus signs take account of the image forces.

The next step is to let P approach the train path and calculate the t-coordinates of the points P_2, P_3 and P_4 (P_1 coincides with P). This results in

$$2y(ut, t) = 2Y(t)$$

$$= -\int_0^{\mu t} + \int_0^{t} - \int_0^{t/\mu - 2/(1-u)} + \int_0^{t - 2/(1+u)} f(s) \, ds,$$

where $\mu = (1 - u)/(1 + u)$ is as defined at the end of Section 11.4. If we differentiate this with respect to t, we can eliminate dY/dt on the left-hand side in favour of the pantograph force, using (11.2) (without the linear spring term: otherwise we get a genuine pantograph differential equation). This gives us

$$\frac{2(f_0 - f(t))}{f_2} = -\mu f(\mu t) + f(t) - \frac{1}{\mu} f\left(\frac{t}{\mu} - \frac{2}{1 - u}\right)$$

$$+ f\left(t - \frac{2}{1 + u}\right), \tag{11.9}$$

again a delay equation but now with *three* delays.

Fortunately, there is some structure to the solution. It is easy to show that the time coordinates of the points at which the reflected characteristics in Figure 11.5 meet the train path are given by

$$t = t_n = \frac{1}{u}(1 - \mu^n).$$

These are the only places at which any sort of discontinuity in $f(t)$ can occur. Furthermore, for $n > 1$,

$$t_n - \frac{2}{1 + u} = \mu t_{n-1}, \qquad \frac{t_n}{\mu} - \frac{2}{1 - u} = t_{n-1}.$$

When you write out the right-hand side, you'll get f evaluated at $t_{n-1}\pm$, $\mu t_n\pm$ and $\mu t_{n-1}\pm$. Only in exceptional cases could μt_n be equal to t_m for some $m < n$, so f is continuous at these points and contributes nothing to the jump.

So, if t in (11.9) is equal to one of the t_n then all the terms on the right-hand side occur in the same equation at $t = t_n$ or $t = t_{n-1}$. With all this going for us, we need only look for piecewise smooth functions between these points and join them up across $t = t_n$, using (11.9) to relate the discontinuity at t_n to that at t_{n-1}. Simply writing (11.9) at $t = t_n\pm$ and subtracting the two expressions gives

$$[f(t)]_{t_n-}^{t_n+} = \frac{f_2}{\mu(2 + f_2)}[f(t)]_{t_{n-1}-}^{t_{n-1}+},$$

and differentiating (11.9) gives the corresponding relations

$$\left[\frac{\mathrm{d}f}{\mathrm{d}t}\right]_{t_n-}^{t_n+} = \frac{f_2}{\mu^2(2+f_2)}\left[\frac{\mathrm{d}f}{\mathrm{d}t}\right]_{t_{n-1}-}^{t_{n-1}+}.$$

These relations are sufficient to determine a piecewise linear solution for values of t between the t_n, and they show that the solution is well behaved as $n \to \infty$ provided that $f_2/(\mu(2+f_2)) < 1$.

11.6 Sources and further reading

The description of the pantograph problem closely follows that of [58] but be careful, the notation is slightly different.

11.7 Exercises

1 **Conservation of momentum.** A string of line density ρ and tension T is pulled with speed $U < c = \sqrt{T/\rho}$ through a small frictionless ring, as shown in Figure 11.7. Remembering that

force = rate of change of momentum,

show that for small displacements the force on the ring is

$$F = -\rho(c^2 - U^2)\left[\frac{\partial y}{\partial x}\right]_-^+.$$

2 **Removing the static displacement.** Show that if we calculate the static displacement $y_s(x)$ of the cable and subtract it from $y(x,t)$ then the difference $\bar{y}(x,t)$ satisfies the $\alpha = 0$ version of the problem but with an additional known time-dependent term in the force relation (11.2).

3 **Impulsive attachment of a point force.** A wire of density ρ is stretched to tension T. At time $t = 0$, the wire is straight and motionless; a constant point force is implied impulsively at $x = 0$ and thereafter it moves with speed $U < c$. Draw a characteristic diagram; show that, of the four regions in it, the wire displacement is only nonzero in region 1 $(Ut < x < ct)$ and region 2 $(-ct < x < Ut)$ and that the displacements there are of the form $g_1(t - x/c)$, $h_2(t + x/c)$ respectively. Apply the pantograph conditions (with a constant force) at

Figure 11.7 Conservation of momentum.

$x = Ut$ to find these functions and sketch the wire displacement at a later time t. Repeat for $U > c$ and comment on the results.

Repeat the exercise in the case where the wire is also subject to gravity, so that its initial (static) displacement is $y_s(x) = \frac{1}{2}\rho g x^2$. (Of course the wire is held up by distant supports.)

4 Delay differential equations.   Consider the equation

$$y'(t) = \alpha_0 - \alpha_1 y(t - \tau), \qquad t > 0,$$

where $\tau > 0$ is a constant, with the initial condition that $y = 0$ for $-\tau \le t < 0$ (this generalises the 'point' initial equation with no delay). Show that, with this initial condition, the Laplace transform of $y(t - \tau)$ is $e^{-p\tau} \bar{y}(p)$. Hence show that

$$\bar{y}(p) = \frac{\alpha_0}{p(p + \alpha_1 e^{-p\tau})}.$$

Put the right-hand side into partial fractions to deduce that the solution can be found as a constant plus a series of exponentials in t involving the roots of $p + \alpha_1 e^{-p\tau} = 0$ (when $\alpha_1 > 0$ it can be shown that these have negative real parts, so the associated exponentials decay). Confirm (for quality control) that you get the expected answer when $\tau = 0$.

Show that the substitution $t = e^s$ reduces the pantograph equation (11.8) to a more complicated constant-delay equation, defined on the whole real line.

5 The solution for one span.   Complete the details of the solution of Section 11.5, finding the coefficients in the linear expression for $f(t)$ in each interval $t_n < t < t_{n+1}$. Include the static displacement of the wire (this generates a particular solution of the delay equation, which you can subtract out).

6 D'Alembert's solution to the wave equation.   Consider the initial value problem

$$\frac{\partial^2 y}{\partial t^2} - \frac{\partial^2 y}{\partial x^2} = G(x, t),$$

with initial conditions

$$y(x, 0) = y_0(x), \qquad \frac{\partial y}{\partial t}(x, 0) = v_0(x), \qquad -\infty < x < \infty.$$

Write the left-hand side of the wave equation in divergence form and integrate over a characteristic triangle to derive the D'Alembert solution

$$y(x, t) = \frac{1}{2}(y_0(x - t) + y_0(x + t)) + \frac{1}{2}\int_{x-t}^{x+t} v_0(s)\, ds$$
$$+ \frac{1}{2}\iint G(\xi, \tau)\, d\xi\, d\tau,$$

where the double integral is taken over the characteristic triangle. Show that this solution is unique by considering the energy

$$E(t) = \frac{1}{2}\int_{-\infty}^{\infty} \left(\frac{\partial y}{\partial x}\right)^2 + \left(\frac{\partial y}{\partial t}\right)^2 dx.$$

7 Pantograph with variable velocity.

(a) Suppose that a point force moves along the wire at $x = X(t)$, so that there is a jump in $\partial y/\partial x$ given by

$$\left[\frac{\partial y}{\partial x}\right] = -\frac{1}{1 - (X'(t))^2} f(t).$$

Bearing in mind that y is continuous, what is the corresponding jump in $\partial y/\partial t$? Modify the argument of Exercise 6 to show that, if the wire is initially at rest,

$$y(x, t) = \frac{1}{2}\int f(\tau)\frac{1 - (X'(\tau))^2}{1 - (X'(\tau))^2}\, d\tau,$$

where the integral is along the part of $x = X(t)$ lying inside the characteristic triangle. (Thank you, Claudia. I know that the fraction is equal to 1. It's written like that because the two bits of it come from different places.)

    Why is this solution not valid if $X'(t) > 1$?

(b) If $X = \frac{1}{2}at^2$ and $f(t) = 1$, show that for $0 < t < 1/a$ there is a region of the x, t-plane in which

$$y(x, t) = \frac{1}{2}t^*(x, t)$$

where $t^*(x, t)$ is the appropriate root of

$$x + (t - t^*) = \frac{1}{2}at^{*2}.$$

(c) Define a distribution $\mathcal{D}(x, y) = \delta(x)f(y)$, where $f(y)$ is smooth. Explain why it is reasonable that

$$\int_0^1 \int_0^1 \mathcal{D}(x, y)\, dx\, dy = \int_0^1 f(y)\, dy.$$

How do you generalise this to $\mathcal{D}(x, y) = f(y)\delta(ax - by)$ for constant a and b, if the integral is over a general region? Use these ideas and Exercise 6 to derive the result in (a) from the equation of motion with a delta function on the right hand side.

8 Dynamics of a rocket. Consider a rocket of mass m slung from a long horizontal cable in the pantograph framework. Let its horizontal position be $x = X(t)$. Ignoring the static displacement of the cable, derive the dimensional model

$$\rho\frac{\partial^2 y}{\partial t^2} - T\frac{\partial^2 y}{\partial x^2} = F(t)\delta(x - X(t)), \qquad 0 < x < \infty,$$

$$y(0, t) = 0, \qquad y(x, 0) = \frac{\partial y}{\partial t}(x, 0) = 0,$$

where

$$F(t) = -mg - m\frac{d^2 Y}{dt^2}, \qquad Y(t) = y(t, t).$$

In the case $X(t) = Ut$, where U is constant, derive and solve an equation for the rocket displacement. When the rocket accelerates at a constant rate a, draw the characteristic diagram, indicating all the significant characteristics.

'One side of a triangle is always shorter than the sum of the other three sides.'

Part III
Asymptotic techniques

12
Asymptotic expansions

12.1 Introduction

The rest of this book deals with systematic procedures to exploit small or large parameters in a dimensionless problem, a collection of ideas grouped together under the umbrella of *asymptotic analysis*. In this chapter, we open the proceedings with the basics of what an asymptotic approximation is, following which we look at a selection of common techniques.

We start with a very simple example. Consider the quadratic

$$\epsilon x^2 + x - 1 = 0, \tag{12.1}$$

where ϵ is a fixed very small positive number, say $0.000\,000\,1$. Forget for the moment that we know how to solve quadratics exactly: can we exploit the fact that ϵ is small to find approximate values for the roots?

If $\epsilon = 0$ then we have $x = 1$ and, furthermore, if we put $x = 1$ into the equation for small positive ϵ, then the error, namely what remains on the left-hand side, is small; here it equals ϵ. So, a natural first try is to write

$$x = x_0 + \epsilon x_1 + \epsilon^2 x_2 + \cdots$$

where, obviously, $x_0 = 1$ (but it is reassuring to know that, as we see below, we can show this systematically). This kind of assumed form for an expansion is known as an *ansatz*; here we are assuming that ϵ crops up only in positive integral powers. We substitute this into the quadratic, obtaining

$$\epsilon \left(x_0 + \epsilon x_1 + \epsilon^2 x_2 + \cdots \right)^2 + x_0 + \epsilon x_1 + \epsilon^2 x_2 + \cdots - 1 = 0,$$

and then collect terms by powers of ϵ:

$$x_0 - 1 + \epsilon \left(x_1 + x_0^2\right) + \epsilon^2 \left(x_2 + 2x_0 x_1\right) + \cdots = 0.$$

Now we equate coefficients of successive powers of ϵ to zero. From ϵ^0,

$$x_0 - 1 = 0$$

so $x_0 = 1$ as expected. From ϵ^1,

$$x_1 + x_0^2 = 0$$

so $x_1 = -1$; from ϵ^2,

$$x_2 + 2x_0 x_1 = 0$$

so $x_2 = 2$, and so on. We have found x_0, x_1, x_2 recursively, giving us the approximation

$$x = 1 - \epsilon + 2\epsilon^2 + \cdots .$$

It is clear that we can carry on in this way to find as many terms as we like (and you can check the answer we get against the small-ϵ expansion of the quadratic formula).

Hold on! Don't quadratics have *two* roots? Where did we lose the other one? Well, one way to find out, is to look at what we did when we calculated x_0. In effect, we simply put $\epsilon = 0$ in (12.1) and so we lost the x^2 term, which, being of course the term of highest degree, tells us how many roots there are. Put another way, we said that the terms x and -1 must balance each other, leaving ϵx^2 as a small correction which we use to improve our solution iteratively. But is that the only possible balance? We can enumerate the other candidates.

- We might balance all three terms: this is on the face of it implausible, and in any case it is ruled out by our analysis above.
- We might balance ϵx^2 and -1, assuming that x is smaller than these terms. This doesn't look so silly, but take it a bit further: if ϵx^2 balances -1 then the size of $|x|$ is $1/\sqrt{\epsilon}$ (which is large) plus a smaller correction. But then the term x, which was supposed to be small relative to ϵx^2 and -1, is in fact much bigger than either. It stands head and shoulders above the other two, with no counterbalance. We have not made the right choice.
- The only remaining possibility is to balance ϵx^2 and x.

If, then, ϵx^2 and x balance, we see that $|x|$ is of size $1/\epsilon$. So we *rescale*, writing

$$x = \frac{1}{\epsilon} X,$$

after which the quadratic (12.1) becomes

$$X^2 + X - \epsilon = 0,$$

confirming immediately that the third term (which was -1 before) is indeed small compared with the other two. We expand X as above:

$$X = X_0 + \epsilon X_1 + \cdots ;$$

skipping the details (which you should work out), the lowest-order terms give

$$X_0^2 + X_0 = 0,$$

with the *two* roots

$$X_0 = -1 \quad \text{and} \quad X_0 = 0.$$

One root ($X_0 = 0$) reproduces the root we found earlier, while the root $X_0 = -1$ is the first term in the expansion of the root we did not find. It is left to you to calculate a couple more terms and verify that the expansions are correct by comparison with the exact solution.

Our example is mathematically trivial. However, it illustrates some important points about asymptotic approaches.

- When we equate coefficients of powers of ϵ to zero, we are in effect embedding our particular problem, with a given numerical value of ϵ, say 0.000 000 1, into a continuous set of problems for *all* ϵ within a small interval $[0, \epsilon^*)$ containing 0.000 000 1. If, as we hope, the dependence of the roots on ϵ has some smooth 'structure' as $\epsilon \to 0$, we should first be able to extract their general behaviour for all small ϵ and only then reinstate our particular numerical value.
- Systematic approximation procedures start with the identification of the dominant balance(s) in an equation. Physical and mathematical intuition may both help in finding these balances, as may iteration ideas (see Exercise 2) and, once they are found, the remaining terms should be smaller corrections. We can consider ourselves unlucky if more than two or three mechanisms are simultaneously in balance.
- It may be necessary to rescale some of the dependent or independent variables to achieve a balance.

Now it's time for some definitions.

12.2 Order notation

It is useful to have a way of writing down the idea that two functions are 'about the same size' near a point x_0 (usually 0 or ∞) or ϵ_0 (almost always 0, as ϵ is almost always used to denote a small parameter). We say that

$$f(x) = O(g(x)) \quad \text{as} \quad x \to x_0$$

if there is a constant A such that

$$|f(x)| \le A\,|g(x)|$$

for all x sufficiently near x_0. So, for example,

$$3x + x^2 = O(x) \quad \text{as} \quad x \to 0;$$

here any $A > 3$ will do. In our quadratic equation example, the roots $x^{(1)}$ and $x^{(2)}$ satisfy

$$x^{(1)}(\epsilon) = O(1) \quad \text{and} \quad x^{(2)}(\epsilon) = O(1/\epsilon)$$

as $\epsilon \to 0$. Successively more precise estimates for $x^{(1)}$ are

$$x^{(1)}(\epsilon) = 1 + O(\epsilon) \quad \text{and} \quad x^{(1)}(\epsilon) = 1 - \epsilon + O(\epsilon^2).$$

If we want a more specific estimate of the size of $f(x)$, we may try to find a function $g(x)$ whose 'leading order' behaviour is the same as that of $f(x)$. We write

$\sim$ is pronounced 'twiddles', by me at any rate. 'Goes as' is another possibility.

$$f(x) \sim g(x) \quad \text{as} \quad x \to x_0$$

if

$$\lim_{x \to x_0} \frac{f(x)}{g(x)} = 1.$$

So, $3x + x^2 \sim 3x$ as $x \to 0$, and

$$x^{(1)}(\epsilon) \sim 1 - \epsilon + 2\epsilon^2 \quad \text{as} \quad \epsilon \to 0.$$

Lastly there is a compact notation for the idea that one function is much smaller than another. We write

$$f(x) = o(g(x)) \quad \text{as} \quad x \to x_0$$

if

$$\lim_{x \to x_0} \frac{f(x)}{g(x)} = 0;$$

this is often written $f(x) \ll g(x)$. So, for example,

$$e^{-x} = o(x^{-1}) \quad \text{as} \quad x \to \infty$$

and, for any n,

$$x^n \ll e^x \quad \text{as} \quad x \to \infty.$$

In our quadratic example,

$$x^{(1)}(\epsilon) = 1 - \epsilon + o(\epsilon) \quad \text{as} \quad \epsilon \to 0.$$

The order notation is most often used to quantify the error in an approximation, so that we know when we can safely use it. A good example is the remainder of a Taylor approximation (series). If we take $n + 1$ terms of a Taylor series for $f(x)$ about x_0, the error is $o(x - x_0)^n$, and usually $O(x - x_0)^{n+1}$. For $n = 0$,

$$f(x) = f(x_0) + O(x - x_0);$$

the error is also $o(1)$. For $n = 1$,

$$f(x) = f(x_0) + (x - x_0)f'(x_0) + O(x - x_0)^2;$$

the error is also $o(x - x_0)$. For $n = 2$,

$$f(x) \sim f(x_0) + (x - x_0)f'(x_0) + \tfrac{1}{2}(x - x_0)^2 f''(x_0)$$

with an error of $O(x - x_0)^3$ or $o(x - x_0)^2$.

12.2.1 Asymptotic sequences and expansions

Suppose that we are looking at a function of ϵ as $\epsilon \to 0$ (or any other limit point). We may aim to write its asymptotic behaviour in this limit in terms of simple functions of ϵ such as powers. A well-known example here is a power series in ϵ, if one exists, and we note that increasing powers of ϵ have an important property: each one is smaller than its predecessor, so that $\epsilon^{n+1} = o(\epsilon^n)$ as $\epsilon \to 0$. Less specifically, we may have non-integral powers, logs and so on, so we generalise this idea of using powers of ϵ by saying that a set of *gauge functions* $\{\phi_n(\epsilon)\}$, $n = 0, 1, 2, \ldots$, is an *asymptotic sequence* as $\epsilon \to 0$ if

$$\phi_{n+1}(\epsilon) = o\left(\phi_n(\epsilon)\right)$$

for all n. For example, $\{\epsilon^n\}$, $\{\epsilon^{n/2}\}$ are asymptotic sequences as $\epsilon \to 0$, while $\{e^{-nx}\}$ is an asymptotic sequence as $x \to \infty$. Making the right choice of asymptotic sequence for a specific problem is something of an art, albeit one in which common sense and simple iteration ideas play a large part.

Once we have an asymptotic sequence, we can expand functions. We say that a function $f(\epsilon)$ has an *asymptotic expansion* with respect to the asymptotic sequence $\{\phi_n(\epsilon)\}$ if there are constants a_k such that, for each n,

$$f(\epsilon) = \sum_{k=0}^{n} a_k \phi_k(\epsilon) + o\left(\phi_n(\epsilon)\right),$$

or

$$f(\epsilon) \sim \sum_{k=0}^{n} a_k \phi_k(\epsilon)$$

as $\epsilon \to 0$. In our quadratic equation example, we found the expansions

$$x^{(1)}(\epsilon) = 1 - \epsilon + 2\epsilon^2 + o(\epsilon^2)$$

with respect to the sequence $\{1, \epsilon, \epsilon^2, \dots\}$, and

$$x^{(2)}(\epsilon) = \epsilon^{-1} + o(\epsilon^{-1})$$

with respect to the sequence $\{\epsilon^{-1}, 1, \epsilon, \epsilon^2, \dots\}$, although we only calculated one term.

We have very often a function of several independent variables, here represented by a generic $\mathbf{x}$, and a small parameter ϵ. In such a case, we may look for an expansion in the form

$$f(\mathbf{x}; \epsilon) \sim \sum_{k=0}^{n} a_k(\mathbf{x}) \phi_k(\epsilon),$$

and we hope that the problem of calculating the a_k (sequentially) will be easier than finding $f(\mathbf{x}; \epsilon)$ all at one go. This is usually the reason for trying an asymptotic expansion in the first place.

12.3 Convergence and divergence

So what's the big deal: haven't we just found a straightforward generalisation of the Taylor series approach? Well, no, not exactly. The point of a Taylor series is that, for a fixed value of ϵ, or whatever we've called the independent variable, as we take more and more terms the sum of the series gets closer and closer to the function it represents. That is, the series converges as the number of terms in the partial sums, n, tends to infinity. All Taylor series are thus *de facto* asymptotic expansions.

There are, however, *two* limiting processes going on when we write down an an asymptotic expansion, $n \to \infty$ and $\epsilon \to 0$, and they need not commute. On the one hand, when we do a Taylor (or Laurent) series expansion we first take the limit as $n \to \infty$ and only then think what happens as ϵ varies. An asymptotic expansion, on the other hand, is designed to provide an accurate approximation as $\epsilon \to 0$ for each n, and many useful expansions don't converge at all as $n \to \infty$.

A famous example is the incomplete exponential integral:[1] evaluate

$$I(\epsilon) = \int_{1/\epsilon}^{\infty} x^{-1} e^{-x} \mathrm{d}x.$$

Repeated integration by parts shows that

$$I(\epsilon) = \epsilon e^{-1/\epsilon} \left(1 - 1!\epsilon + 2!\epsilon^2 + \cdots + (-1)^n n!\epsilon^n \right) + R_n(\epsilon), \quad (12.2)$$

where $R_n(\epsilon)$ is easily shown to be asymptotically smaller than the last retained term; see Exercise 3.

It is quite clear that the series we have generated does not converge for *any* $\epsilon > 0$ because of the growth of the factorial. That isn't the point, though. What *is* important is that as $\epsilon \to 0$, with n fixed, the series should give us an accurate description of the behaviour of the integral. That is, the smaller we take ϵ to be, the smaller should be the relative error of the approximation. In fact what happens is that if we take a fixed value of ϵ and consider more and more terms in the expansion then at first successive terms get smaller and smaller (as they would for a convergent series); however, starting from values of n that are $O(1/\epsilon)$, they increase again. The best approximation is given by cutting the series off at this optimal truncation point.[2] Even for $\epsilon = 1/4$, which is not particularly small, truncation of the series after four terms gives the reasonable approximation $0.004\,01\ldots$, which compares with $0.003\,78\ldots$ from numerical integration. When $\epsilon = 1/8$, eight terms of the series give $0.112\,434 \times e^{-8} = 0.000\,037\,717$, which may be compared with the true value, $0.000\,037\,666$. The relative error is less than 0.14%.

In most practically generated, as opposed to mathematically generated, asymptotic problems we are unable to calculate enough terms to decide whether the asymptotic series is divergent. Indeed, it's usually next to impossible (or at least a rather strenuous exercise in mathematical weightlifting) trying to prove that the remainder after even one or two terms is small as it should be. We have to live with these lacunae: we proceed knowing that experience tells us where, mostly, things will work out.

What are the gauge functions?

The figures are accurate to the number of decimal places given, always a good practice.

For another asymptotic expansion that works well even when the relevant parameter, here n, is not small, try putting $n = 1$ in Stirling's formula, which says that $n! \sim n^{n+1/2} e^{-n} / \sqrt{2\pi}$ as $n \to \infty$.

---

[1] The approximation of integrals and special functions is a particularly happy hunting ground for asymptoticists, although we shan't be going very far that way.

[2] The location of the optimal truncation point is determined by the value $n(\epsilon)$ at which successive terms have the same size. The series for $I(\epsilon)$ has a 'factorial-power' form for the terms in the expansion, a very general phenomenon, and it is easy to calculate that successive terms are closest in size when n is the integer part of $1/\epsilon$. The precise behaviour of the remaining error, and how to deal with it, is part of the trendy subject of *hyperasymptotics*, also known as *asymptotics beyond all orders* or *exponential asymptotics*.

12.4 Further reading

There are many excellent books on asymptotic expansions. A short but intensive introduction is provided by Hinch's book of [27], the first chapter of which probably put the quadratic example into my mind (if you think about it, $\epsilon x^2 + x - 1 = 0$ is the irreducible minimum of that kind of problem). Keener [33] is another good source. If you are interested in ordinary differential equations try the book by O'Malley [49]; for a more wide-reaching treatment, including partial differential equations, see the books of Kevorkian & Cole [36] or Bender & Orszag [5]. Olver [48] is a good starting point for the analysis of special functions and, as ever, Carrier, Krook & Pearson [6] is very well worth reading.

12.5 Exercises

1 Roots of equations.   Find expansions for the roots of

$$\epsilon x^3 + x - 1 = 0$$

as $\epsilon \to 0$, with at least two and preferably three non-zero terms in each expansion.

Draw graphs to see where the roots are.

Repeat for the real roots of $\epsilon x \tan x = 1$ and then for $x \tan x = \epsilon$. In the latter case you will have to consider the first root separately, as well as rescaling to get the large roots. In addition there is a range of roots you can't get approximations to; where is it?

2 Iteration.   Show that $x \log x \to 0$ as $x \to 0$; draw its graph. Suppose that we want to find an asymptotic expansion for the solution to $x \log x = -\epsilon$, where $0 < \epsilon \ll 1$. In this case, it is not obvious what gauge functions we should use, so we find them by *iteration*. Write

$$x(\epsilon) \sim x_0(\epsilon) + x_1(\epsilon) + \cdots,$$

where all we know is that $x_0 \gg x_1 \gg \cdots$. Take logarithms of the original equation (a key manipulative step, because it replaces multiplication of two small terms by addition of their (large) logarithms) and substitute the expansion for $x(\epsilon)$ in, to find that

$$\log(x_0 + x_1 + \cdots) + \log(-\log(x_0 + x_1 + \cdots)) = \log \epsilon. \quad (12.3)$$

Now ignore x_1 in order to show that taking $x_0 = \epsilon$ gives a *residual* (i.e., left-hand side $-$ right-hand side) that is much smaller than the dominant terms. Put $x_0 = \epsilon$ back into (12.3) and expand the logarithms in powers of x_1/ϵ, which is small, to show that $x_1 = -\epsilon/|\log \epsilon|$. Finally calculate one more term in the expansion.

Repeat the calculation after making the simplifying initial scaling $x = \epsilon X$ (which you might not spot the first time round).

3 **The exponential integral.** Show that the expression (12.2) for the incomplete exponential integral is indeed an asymptotic expansion, as follows. Consider

$$I(\epsilon) = \int_{1/\epsilon}^{\infty} x^{-1} e^{-x} dx.$$

Integrate by parts $n + 1$ times to show that

$$I(\epsilon) = \epsilon e^{-1/\epsilon} \left(1 - 1!\epsilon + 2!\epsilon^2 + \cdots + (-1)^n n! \epsilon^n\right) + R_n(\epsilon),$$

and now integrate by parts once more to get

$$R_n = \epsilon e^{-1/\epsilon}(-1)^{n+1}(n+1)!\epsilon^{n+1} + R'_n,$$

where a simple estimate using $e^{-x} \le e^{-1/\epsilon}$ for $x \ge 1/\epsilon$ shows that R'_n is at most of the same order as the first term on the right-hand side of the expression for R_n. Conclude that as $\epsilon \to 0$ we have the estimate $R_n = o(\epsilon^n e^{-1/\epsilon})$.

4 **Stock market crashes and six-sigma quality control.** The probability density function for the standard normal distribution $N(0, 1)$ is

$$f_X(x) = \frac{1}{\sqrt{2\pi}} e^{-x^2/2}.$$

Integrate by parts to find a one-term approximation for $P(X < x)$ as $x \to -\infty$ and show that it is asymptotically correct (see the previous exercise).

'Six-sigma standards' in manufacturing demand that the probability that an individual component is defective is less than the probability of being six or more standard deviations away from the mean of a standard normal distribution. What is this probability, approximately? ($e^3 \approx 20$, $2\pi \approx 25/4$.) If the manufacturers use this standard, what is the probability that none of the 10 000 components in a computer or the 1 000 000 components in an aeroplane is faulty?

Hint: which well-known limit lets you work out an approximation to $(0.99)^{100}$?

In the standard Black–Scholes model for financial markets, the daily percentage changes in, say, the FTSE-100 or S&P-500 index are independent random variables that are approximately normal with a very small mean value and a standard deviation of about 1%. What is the probability of a fall of 10% or more in one day? What is the probability of two such falls on consecutive days? (In October 1987 the UK stock market fell by more than 10% on both Black Monday and the day after. There have been several other changes of this magnitude in the (roughly) 25 000 days for which stock indices have been calculated.)

'It's a half-plane, so it probably goes on for quite a long distance.'

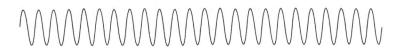

13
Regular perturbation expansions

13.1 Introduction

We begin our tour of asymptotic methods for simplifying complex problems using the most straightforward idea, that of a *regular asymptotic*, or *perturbation*, *expansion*. This is just the plain vanilla common-sense expansion you carry out when it seems that the dominant-balance terms in your model do indeed reflect the dominant physical mechanisms, and everything else is a small correction. For example, the expansion

$$x^{(1)}(\epsilon) = 1 - \epsilon + 2\epsilon^2 + o(\epsilon^2)$$

in the quadratic equation example of the previous chapter is beautifully regular. In some problems, we can characterise a regular expansion by saying that it is expected to be a *uniformly valid approximation* to the solution; for example, when we consider the standard model for waves on a string we hope that, for all times and positions on the string, the wave equation is a good approximation to the fully nonlinear model we could write down for displacements that are not small. Having said this, there probably isn't a watertight definition of when an expansion is 'regular'; it may be safest just to leave it as 'any expansion that is not singular, boundary layer, multiple-scale, . . .' and to let your sense of the meaning of the term grow with experience.

In the models we look at later on, we'll see a variety of scalings and transformations that help us to understand less straightforward situations but, after all these contortions, we end up with a regular expansion. When we've got to a regular expansion, nine times out of ten we've done as much simplifying as we can with asymptotic approximations. As a simple example, to find the other root $x^{(2)}(\epsilon)$ of the quadratic in Section 12.1, we first had to introduce the singular (as $\epsilon \to 0$) scaling $x = X/\epsilon$; only then did we obtain a regular expansion for X.

There is really not much more to say of a general nature. The rest of the chapter consists of a collection of examples of the regular perturbation technique in action. They are necessarily in order, because this is not a hypertext document, but they are not rigidly so: wander as you will.

13.2 Example: stability of a spacecraft in orbit

Many asymptotic techniques have their origin in astronomy. We begin with a simple one: the stability of circular planetary orbits. In the classical Newtonian model for the motion of, say, a satellite or a space station[1] orbiting the earth, the space station's plane polar coordinates $(r(t), \theta(t))$, with origin at the centre of the earth, satisfy

$$\ddot{r} - r\dot{\theta}^2 = -\frac{GM}{r^2}, \qquad \frac{1}{r}\frac{\mathrm{d}}{\mathrm{d}t}\left(r^2\dot{\theta}\right) = 0,$$

where again the dot means $\mathrm{d}/\mathrm{d}t$, G is the universal gravitational constant and M is the mass of the earth. Hence we retrieve Kepler's second law

$$r^2\dot{\theta} = h,$$

'The radius vector sweeps out equal areas in equal times': the element of area is $\frac{1}{2}r^2\,\mathrm{d}\theta$ (draw a diagram to see why).

the constant h being equal to the angular momentum (per unit mass, to be pedantic).

A circular orbit is an obvious solution with

$$r = a, \quad \dot{\theta} = \omega, \qquad \text{where} \qquad a^3\omega^2 = GM,$$

the relation between a and ω being Kepler's third law. Suppose that a booster rocket on the space station gives it a small radial velocity ϵv (note that this does not change h). Is the orbit stable or will the space station plunge to earth or fly off into the depths of space?

Write

$$r(t) = a + \epsilon r_1(t) + \cdots, \qquad \dot{\theta}(t) = \omega + \epsilon\dot{\theta}_1(t) + \cdots, \qquad \epsilon \ll 1,$$

so that

$$\epsilon\ddot{r}_1 + \cdots - (a + \epsilon r_1 + \cdots)\left(\omega + \epsilon\dot{\theta}_1 + \cdots\right)^2 = -\frac{GM}{(a + \epsilon r_1 + \cdots)^2}$$

$$(13.1)$$

---

[1] See www.heavens-above.com for predictions of when to see the International Space Station and other satellites.

and

$$(a + \epsilon r_1 + \cdots)^2 \left(\omega + \epsilon\dot{\theta}_1 + \cdots\right) = h = a^2\omega. \qquad (13.2)$$

Expand the right-hand side of (13.1) by the binomial theorem; then, remembering that $a^3\omega^2 = GM$ and equating the $O(\epsilon)$ terms in (13.1), (13.2) gives

$$\ddot{r}_1 - \omega^2 r_1 - 2a\omega\dot{\theta}_1 = 0, \qquad 2a\omega r_1 + a^2\dot{\theta}_1 = 0,$$

that is,

$$\ddot{r}_1 + \omega^2 r_1 = 0.$$

We see that r_1 oscillates without growing or decaying, so it looks as if the system is neutrally stable; this is not so surprising when we recall that the full system is conservative. The period of oscillation is equal to the original period of orbital rotation, and the perturbed orbit is slightly elliptical with the centre of the earth at one focus; the furthest and nearest distances from earth (apogee and perigee) occur one-quarter and three-quarters of an orbit after the initial thrust. At every half-orbit the space station returns to the original location relative to earth: if the astronauts drop a spanner before applying the thrust, they will have two opportunities per orbit to reach out and grab it.

13.3 Linear stability

Linear stability analysis, of which the orbiter problem is an example, is an archetypal example of a regular perturbation. We take a solution $\mathbf{u}_0$ of a system, often an equilibrium state or a steady state such as a travelling wave, perturb it to $\mathbf{u}_0 + \epsilon\mathbf{u}_1$, write down a regular perturbation expansion to determine $\mathbf{u}_1$ and then see whether $\epsilon\mathbf{u}_1$ is small, or (in a time-dependent problem) remains small. The perturbation may be to the initial and/or boundary data of our problem, or to the geometry, or it may be structural via changes to the parameters or equations of the problem.

We say that a time-dependent system is *linearly stable* if a suitable norm (measure of the size) of the perturbation decays, *linearly unstable* if the norm grows and (linearly) *neutrally stable* if it remains the same size.

In many systems the result of linearising about $\mathbf{u}_0$ is a linear evolution problem in the form

$$\frac{\partial\mathbf{u}_1}{\partial t} = \mathcal{L}\mathbf{u}_1,$$

where $\mathcal{L}$ is a linear differential operator. Often $\mathcal{L}$ has time-independent coefficients, and then the solution has the form

$$\mathbf{u}_1 = e^{\lambda t}\mathbf{U}_1,$$

where $\mathbf{U}_1$, which is independent of t, is an eigensolution of $\mathcal{L}$ with eigenvalue λ:

$$\mathcal{L}\mathbf{U}_1 = \lambda\mathbf{U}_1$$

(we expect an eigenproblem because of the scaling invariance of the linear problem). This is all well illustrated by the very familiar phase-plane analysis, which we now briefly review.

13.3.1 Stability of critical points in a phase plane

The phase-plane analysis of critical points is a classic example of linear stability analysis. Take a two-dimensional autonomous dynamical system

$$\frac{d\mathbf{x}}{dt} = \mathbf{f}(\mathbf{x}), \qquad \mathbf{x} = \begin{pmatrix} x \\ y \end{pmatrix}, \qquad \mathbf{f} = \begin{pmatrix} f_1(x, y) \\ f_2(x, y) \end{pmatrix}.$$

The *critical points* $\mathbf{x}_0 = (x_0, y_0)^\top$ are equilibrium points satisfying $\mathbf{f}(\mathbf{x}_0) = \mathbf{0}$. In order to analyse their stability, we first write down a regular expansion for $\mathbf{x}(t)$ about $\mathbf{x}_0$,

$$\mathbf{x} \sim \mathbf{x}_0 + \epsilon\mathbf{x}_1 + \cdots.$$

You can do it in more than two dimensions, but two dimensions is much easier to analyse, because the dimension of the phase paths is one less than the dimension of the plane; in three or more dimensions the extra degree(s) of freedom make life much more difficult.

Then, we expand $\mathbf{f}(\mathbf{x})$ in a Taylor series about $\mathbf{x}_0$:

$$\mathbf{f}(\mathbf{x}_0 + \epsilon\mathbf{x}_1 + \cdots) = \begin{pmatrix} f_1(x_0 + \epsilon x_1 + \cdots, \; y_0 + \epsilon y_1 + \cdots) \\ f_2(x_0 + \epsilon x_1 + \cdots, \; y_0 + \epsilon y_1 + \cdots) \end{pmatrix}$$

$$\sim \begin{pmatrix} 0 + \epsilon\left(x_1\dfrac{\partial f_1}{\partial x} + y_1\dfrac{\partial f_1}{\partial y}\right) + \cdots \\ 0 + \epsilon\left(x_1\dfrac{\partial f_2}{\partial x} + y_1\dfrac{\partial f_2}{\partial y}\right) + \cdots \end{pmatrix}$$

$$= \mathbf{0} + \epsilon\mathbf{J}\mathbf{x}_1 + \cdots,$$

It is de facto an asymptotic expansion provided that $|\mathbf{x} - \mathbf{x}_0| = O(1)$ and $\mathbf{f}$ is smooth.

where $\mathbf{J}$, a constant matrix, is the Jacobian

$$\frac{\partial(f_1, f_2)}{\partial(x, y)} = \begin{pmatrix} \dfrac{\partial f_1}{\partial x} & \dfrac{\partial f_1}{\partial y} \\ \dfrac{\partial f_2}{\partial x} & \dfrac{\partial f_2}{\partial y} \end{pmatrix}$$

evaluated at $\mathbf{x}_0$.

At $O(1)$ we have $\mathbf{0} = \mathbf{0}$, and at $O(\epsilon)$ we find a linear equation for $\mathbf{x}_1$,

$$\frac{d\mathbf{x}_1}{dt} = \mathbf{J}\mathbf{x}_1.$$

With a constant-coefficient equation, it is natural to look for a solution

$$\mathbf{x}_1 = e^{\lambda t} \mathbf{v}_1,$$

which reveals the eigenvalue problem

$$\mathbf{J}\mathbf{v}_1 = \lambda \mathbf{v}_1.$$

The stability or otherwise of the fixed point is thus determined by the real parts of the eigenvalues of $\mathbf{J}$, since a positive real part for either eigenvalue leads to exponential growth and hence instability. The details of the behaviour are surprisingly complicated, largely because of special cases when the eigenvalues are equal.[2] When they are distinct, things are easier and we have the familiar catalogue of possible behaviours:

Here is a use for the canonical-form reductions of linear algebra: we see a differential-equation interpretation of the difference between algebraic and geometric multiplicity.

- stable (unstable) nodes when both λ_1 and λ_2 are real and negative (positive);
- saddles where there are real eigenvalues of opposite sign;
- stable (unstable) spirals for complex eigenvalues with negative (positive) real parts;
- centres when the eigenvalues are pure imaginary.

The eigenvalues are a conjugate pair: why?

Care is needed with the last case, which, unlike the rest, is clearly structurally unstable to small changes in the entries of $\mathbf{J}$, as the following digressionary example shows.

13.3.2 Side track: example of a system that is linearly neutrally stable but nonlinearly stable or unstable

Consider the two systems (one for $+$, the other for $-$)

$$\frac{dx}{dt} = y \pm x \left(x^2 + y^2 \right), \tag{13.3}$$

$$\frac{dy}{dt} = -x \pm y \left(x^2 + y^2 \right). \tag{13.4}$$

If we look for solutions near the obvious equilibrium point $(0, 0)$, say with $x(0) = \epsilon \xi_0$, $y(0) = \epsilon \eta_0$, we can write $x = \epsilon X$, $y = \epsilon Y$ and then expand in the form

$$X \sim X_0 + \epsilon^2 X_2 + \cdots, \qquad Y \sim Y_0 + \epsilon^2 Y_2 + \cdots$$

(it is fairly clear that the $O(\epsilon)$ terms vanish). Then

$$\frac{dX_0}{dt} = Y_0, \qquad \frac{dY_0}{dt} = -X_0,$$

and we have neutral stability since $X_0^2 + Y_0^2 = \xi_0^2 + \eta_0^2$ is constant.

---

[2] See [32].

However, at $O(\epsilon^2)$ we find

$$\frac{dX_2}{dt} = \pm X_2 \left(\xi_0^2 + \eta_0^2\right), \qquad \frac{dY_2}{dt} = \pm Y_2 \left(\xi_0^2 + \eta_0^2\right),$$

and the 'plus' system is plainly unstable at this order, while the 'minus' system is stable. In fact this is in accordance with the exact result since multiplying equations (13.3) and (13.4) by x and y respectively and adding them gives

$$\frac{d}{dt} \left(x^2 + y^2\right) = \pm \left(x^2 + y^2\right)^2,$$

and it is an exercise to solve this equation and show that there is finite-time blow-up if we have the plus sign and existence for all t if we have the minus sign.

This example shows that linear stability analysis can be the tip of the iceberg. Nevertheless, although it is possible to construct examples that are linearly stable and nonlinearly unstable, and vice versa, nonetheless as a general rule linear stability analysis is a good guide to the overall behaviour.

13.4 Example: the pendulum

Let us have another look at the pendulum model we introduced in Chapter 3. Recall that the dimensionless pendulum model, without the primes on t, is

$$\frac{d^2\theta}{dt^2} + \gamma \frac{d\theta}{dt} + \sin\theta = 0,$$

with

$$\theta = \alpha_0, \qquad \frac{d\theta}{dt} = \beta_0, \qquad \text{at} \quad t = 0.$$

Of course, we can treat this equation via the phase plane (this relatively straightforward procedure is treated in Exercise 3). However, the purpose of this chapter is to see the *modus operandi* of regular perturbations, so let's do this problem from scratch.

Suppose that α_0 is small, say[3] $\alpha_0 = \epsilon a_0$ where $\epsilon \ll 1$, and $\beta_0 = 0$, so that we are releasing the pendulum from rest with only a small initial displacement. Can we retrieve linear theory, and how big is the error?

Write

$$\theta \sim \theta_0 + \epsilon\theta_1 + \cdots . \tag{13.5}$$

---

[3] I could have chosen to expand in terms of α_0, instead of writing $\alpha_0 = \epsilon a_0$, and that might have looked less contrived. However, for continuity of exposition I want the small parameter to be called ϵ wherever possible.

Then

$$\frac{d^2\theta}{dt^2} + \epsilon \frac{d^2\theta_1}{dt^2} + \cdots + \gamma \left(\frac{d\theta}{dt} + \epsilon \frac{d\theta_1}{dt} + \cdots \right)$$

$$+ \sin(\theta_0 + \epsilon\theta_1 + \cdots) = 0, \qquad (13.6)$$

with the initial conditions

$$\theta_0 + \epsilon\theta_1 + \cdots = \epsilon\alpha_0, \qquad \frac{d\theta_0}{dt} + \epsilon \frac{d\theta_1}{dt} + \cdots = 0 \qquad (13.7)$$

at $t = 0$.

Equating coefficients of powers of ϵ, at $O(1)$ (equivalent to putting $\epsilon = 0$) we see immediately that $\theta_0 \equiv 0$: it satisfies

$$\frac{d^2\theta_0}{dt^2} + \gamma \frac{d\theta_0}{dt} + \sin\theta_0 = 0,$$

with

$$\theta_0 = \frac{d\theta_0}{dt} = 0 \quad \text{at} \quad t = 0,$$

and the solution $\theta_0(t) \equiv 0$ is unique, by standard Picard theory.

With more experience, we would have seen this straightaway and accounted for it by using the (regularly) scaled variable $\theta = \epsilon\tilde{\theta}$. However, let's press on. We now know that

$$\theta \sim \epsilon\theta_1 + o(\epsilon),$$

and so

$$\sin\theta \sim \epsilon\theta_1 + o(\epsilon).$$

Putting these two into (13.6) and retaining only the terms of $O(\epsilon)$, we find that

$$\frac{d^2\theta_1}{dt^2} + \gamma \frac{d\theta_1}{dt} + \theta_1 = 0,$$

while the $O(\epsilon)$ terms from the initial conditions (13.7) give

$$\theta_1 = a_0, \qquad \frac{d\theta_1}{dt} = 0, \qquad \text{at} \quad t = 0.$$

As promised, we have retrieved the linear theory.

As we noted in Chapter 12, what we have done is in effect to embed the problem for our particular value of α_0, say 0.001, into a family of problems parametrised by ϵ, and we are looking for an expansion valid for all ϵ in an interval containing $(0, 0.001)$. We hope that the solution depends smoothly on α_0 (and hence on ϵ) when α_0 is small, so that our procedure of expanding in powers of ϵ is justified. Indeed, if the solution is often enough differentiable with respect to ϵ at $\epsilon = 0$ then we are just identifying a function by its Taylor series. Even if this is not the case, we hope that *the asymptotic expansion gives a good approximation to the solution as $\epsilon \to 0$*, the key requirement of such a representation.

13.5 Small perturbations of a boundary

In this section, we look at two problems in which the perturbation to a simple solution is induced by a small irregularity in the boundary of the domain in which we solve rather than in the field equation itself.

13.5.1 Example: flow past a nearly circular cylinder

Suppose that we want to calculate two-dimensional potential flow, with velocity $(U, 0)$ at infinity, past a nearly circular cylinder whose equation in plane polar coordinates is $r = a(1 + \epsilon \cos 2\theta)$, where $\epsilon \ll 1$. The shape of this cylinder is close to that of an ellipse with small eccentricity. The velocity potential ϕ satisfies

$$\nabla^2 \phi = 0, \qquad r > a(1 + \epsilon \cos 2\theta), \tag{13.8}$$

with the boundary conditions

$$\frac{\partial \phi}{\partial n} = \mathbf{n} \cdot \nabla \phi = 0, \qquad r = a(1 + \epsilon \cos 2\theta),$$
$$\phi \sim Ur \cos \theta + o(1), \qquad r \to \infty. \tag{13.9}$$

We know the solution when $\epsilon = 0$, namely

$$\phi_0 = U \left(r \cos \theta + \frac{a^2}{r} \cos \theta \right), \tag{13.10}$$

and it seems very likely that the solution for $0 < \epsilon \ll 1$ is close to this. The only obstacle is that for $\epsilon > 0$ the boundary condition $\partial \phi / \partial n = 0$ is applied in an inconvenient place. We deal with this by *linearising* it onto $r = a$: we replace the exact boundary condition by an approximate one on the more convenient location. This entails two steps. First, we expand the full condition $\mathbf{n} \cdot \nabla \phi = 0$ in powers of ϵ and then discard small terms. Next, we use a second expansion to replace the resulting approximate condition at $r = a(1 + \epsilon \cos 2\theta)$ by one on $r = a$. Again, we discard small terms and, as long as we do so consistently, the accuracy of our approximation will not be degraded. I am going to go through this process in excruciating detail, because this is an important technique and one where it is easy to slip up.

Suppose that we want to calculate the solution correct to $O(\epsilon)$. That means that as we go along we can discard any $O(\epsilon^2)$ terms (as long as we are confident that they won't get divided by ϵ later). The unit normal

Show that this curve is, to $O(\epsilon)$, the same as

$$\frac{x^2}{(1+\epsilon)^2} + \frac{y^2}{(1-\epsilon)^2} = 1.$$

to $r = a\,(1 + \epsilon f(\theta))$, for any smooth function $f(\theta)$, is[4]

$$
\mathbf{n} = \frac{\nabla\big(r - a\,(1 + \epsilon f(\theta))\big)}{\big|\nabla\big(r - a\,(1 + \epsilon f(\theta))\big)\big|}
$$

$$
= \frac{\mathbf{e}_r - \dfrac{\epsilon f'(\theta)}{1 + \epsilon f(\theta)}\,\mathbf{e}_\theta}{\left(1 + \epsilon^2\dfrac{(f'(\theta))^2}{(1 + \epsilon f(\theta))^2}\right)^{1/2}}
$$

$$
= \mathbf{e}_r - \epsilon f'(\theta)\mathbf{e}_\theta + O(\epsilon^2).
$$

In polars,

$$
\nabla = \mathbf{e}_r\frac{\partial}{\partial r} + \frac{1}{r}\mathbf{e}_\theta\frac{\partial}{\partial\theta}.
$$

Note that a has the dimension of length but $\mathbf{n}$ is dimensionless: the a's cancel in the second line.

So, we have

$$
\mathbf{n}\cdot\nabla\phi|_{r=a(1+\epsilon f(\theta))}
$$

$$
= \Big(\mathbf{e}_r - \epsilon f'(\theta)\mathbf{e}_\theta + O(\epsilon^2)\Big)\cdot\left(\mathbf{e}_r\frac{\partial\phi}{\partial r} + \frac{1}{r}\mathbf{e}_\theta\frac{\partial\phi}{\partial\theta}\right)\bigg|_{r=a(1+\epsilon f(\theta))}
$$

$$
= \left(\frac{\partial\phi}{\partial r} - \epsilon\frac{f'(\theta)}{a}\frac{\partial\phi}{\partial\theta}\right)\bigg|_{r=a(1+\epsilon f(\theta))} + O(\epsilon^2). \qquad (13.11)
$$

The next stage is to expand $\partial\phi/\partial r$ and $\partial\phi/\partial\theta$ in (13.11) in Taylor series about $r = a$. That is, we write

$$
\frac{\partial\phi}{\partial r}\bigg|_{r=a(1+\epsilon f(\theta))} = \frac{\partial\phi}{\partial r}\bigg|_{r=a} + \epsilon a f(\theta)\frac{\partial^2\phi}{\partial r^2}\bigg|_{r=a} + O(\epsilon^2), \qquad (13.12)
$$

and, as hindsight shows that we only need one term for $\partial\phi/\partial\theta$,

$$
\frac{\partial\phi}{\partial\theta}\bigg|_{r=a(1+\epsilon f(\theta))} = \frac{\partial\phi}{\partial\theta}\bigg|_{r=a} + O(\epsilon). \qquad (13.13)
$$

We can now substitute for $\partial\phi/\partial r$ and $\partial\phi/\partial\theta$ from (13.12) and (13.13) into (13.11), to find that

Do you now see why we only bother with one term for $\partial\phi/\partial\theta$?

$$
\frac{\partial\phi}{\partial n}\bigg|_{r=a(1+\epsilon f(\theta))} = \left(\frac{\partial\phi}{\partial r} + \epsilon a f(\theta)\frac{\partial^2\phi}{\partial r^2} - \epsilon\frac{f'(\theta)}{a}\frac{\partial\phi}{\partial\theta}\right)\bigg|_{r=a} + O(\epsilon^2).
$$

Check dimensions: all the terms should have dimension $[\phi]/[\mathrm{L}]$, which they do.

In our case, $f(\theta) = \cos 2\theta$ and so, instead of the exact problem (13.8), (13.9), we solve the approximate problem[5] $\nabla^2\phi = 0, r > a$, with the boundary conditions

$$
\frac{\partial\phi}{\partial r} + \epsilon a\cos 2\theta\frac{\partial^2\phi}{\partial r^2} + 2\epsilon\frac{\sin 2\theta}{a}\frac{\partial\phi}{\partial\theta} = 0, \qquad r = a,
$$

$$
\phi \sim Ur\cos\theta + o(1), \qquad r \to \infty.
$$

---

[4] Note that the expansion may be invalid if $f'(\theta)$ is large, specifically $O(1/\epsilon)$.

[5] A pedant says: 'This function ϕ is different from the original one, so you should use a different notation for it.' I reply: 'Go away and leave me alone. There is too much unnecessary notation in the world without adding to it.'

We have done the hard work. The approximate problem yields immediately to the regular expansion

$$\phi(r, \theta) \sim \phi_0(r, \theta) + \epsilon \phi_1(r, \theta) + O(\epsilon^2).$$

The leading-order problem is (as expected) just the standard flow past a circular cylinder with solution ϕ_0 as given in (13.10). The problem for ϕ_1 is then

$$\nabla^2 \phi_1 = 0, \qquad r > a, \qquad \phi_1 = o(1), \qquad r \to \infty,$$

More details to be filled in.

with the approximate condition on $r = a$

$$\frac{\partial \phi_1}{\partial r} = -a \cos 2\theta \, \frac{\partial^2 \phi_0}{\partial r^2} - 2 \sin 2\theta \, \frac{\partial \phi_0}{\partial \theta}$$

$$= -2U(\cos \theta \cos 2\theta - 2 \sin \theta \sin 2\theta)$$

$$= U(\cos \theta - 3 \cos 3\theta).$$

We can look up ϕ_1 in our (mental) library of separable solutions of Laplace's equation, and the solution, to $O(\epsilon)$, is

$$\phi(r, \theta) = U \left(r \cos \theta + \frac{a^2}{r} \cos \theta \right) + \epsilon U \left(-\frac{a^2 \cos \theta}{r} + \frac{a^4 \cos 3\theta}{r^3} \right) + O(\epsilon^2).$$

We can (should) run a couple of consistency checks on this solution. First, the correction ϕ_1 has the right dimensions, velocity × length. Second, look at the velocity correction on the x-axis. Our cylinder sticks out beyond the circle $r = a$ near the downstream end $\theta = 0$ and near the

It helps to draw a picture.

You can do the same argument at the upstream stagnation point, but you are more likely to lose a minus sign because you have to remember that $\partial / \partial r = -\partial / \partial x$ there.

upstream end $\theta = \pi$; the flow has stagnation points on the boundary at $\theta = 0, \pi$. At the downstream stagnation point, the leading-order horizontal velocity $\partial \phi_0 / \partial r$ is small and positive, because the leading-order flow, which is overall left to right, has a stagnation point just to the left. Thus, to get zero horizontal velocity here, $\partial \phi_1 / \partial r$ must be negative, which it is: we have got the right signs.

13.5.2 Example: water waves

For our second example of a boundary perturbation, we look at the very classical problem of two-dimensional small-amplitude surface gravity waves on deep water. The new feature of the problem is that the boundary that is perturbed is itself unknown: it is called a *free boundary* or *free surface*. You may have done this problem previously in an ad hoc way, 'neglecting quadratic terms'. This can be viewed formally as constructing the corresponding linearised problem, an extension of the idea of the derivative in calculus as a locally linear operator that approximates a nonlinear function. Form our point of view, however, it can be regarded as constructing an asymptotic expansion correct to $O(\epsilon)$, neglecting $O(\epsilon^2)$.

Let us build in the fact that the amplitude is small by writing the water surface as $y = \epsilon h(x, t)$, where $\epsilon \ll 1$ and $h = O(1)$ (and there is an implicit assumption that derivatives of h are not large either). Then the full problem to be solved for the velocity is

$$\nabla^2 \phi = 0, \qquad y < \epsilon h(x, t),$$

for which the free-surface conditions are the kinematic condition

$$\frac{D}{Dt}\big(y - \epsilon h(x, t)\big) = 0,$$

namely

$$\frac{\partial \phi}{\partial y} = \epsilon \left(\frac{\partial h}{\partial t} + \frac{\partial \phi}{\partial x}\frac{\partial h}{\partial x} \right), \qquad y = \epsilon h(x, t), \qquad (13.14)$$

and the Bernoulli condition

$$\frac{\partial \phi}{\partial t} + \frac{1}{2}|\nabla \phi|^2 + gy = 0, \qquad y = \epsilon h(x, t). \qquad (13.15)$$

Particles in the surface stay there, so the material derivative of $y - \epsilon h(x, t)$ is zero. It also says that the normal velocity of the water is equal to the normal velocity of the interface.

Woe to those who spell him Bernouilli.

The unknown location of the surface makes this a formidably hard problem, and even after decades of effort there are many open questions. The first step on the road, however, is easy.

We are aiming for an asymptotic expansion in powers of ϵ. A quick look at the kinematic condition (13.14) shows immediately that there is no $O(1)$ term in the velocity potential, and so its expansion has the form

$$\phi(x, y, t) \sim \epsilon \phi_1(x, y, t) + \cdots .$$

It is now clear that the leading-order terms in the kinematic and dynamic boundary conditions are

$$\frac{\partial \phi_1}{\partial y} = \frac{\partial h}{\partial t}, \qquad \frac{\partial \phi_1}{\partial t} + gh = 0, \qquad (13.16)$$

Technically, we should expand $h(x, t)$ as well, but we only need one term so we don't bother. If we want the $O(\epsilon^2)$ term as well, we have to do this.

which apply on $y = \epsilon h(x, t)$ and then, by a trivial linearisation, on $y = 0$ without loss of accuracy to this order. The rest is history: taking a representative wave[6] $h(x, t) = ae^{i(kx - \omega t)}$ with wavenumber k and frequency ω, where a is the constant amplitude, we have from Laplace's equation and the first equation in (13.16) that

$$\phi_1(x, y, t) = -\frac{i\omega a}{|k|}e^{|k|y}e^{i(kx - \omega t)},$$

and then from the second equation in (13.16) we get the *dispersion relation*

$$\omega^2 = g|k|,$$

Waves with $k > 0$ travel to the right, those with $k < 0$ to the left; $|k|$ in ϕ ensures decay as $y \to -\infty$. Of course we take the real part of h and ϕ for the physical quantities.

Check the dimensions.

---

[6] Representative because we can superpose these waves to solve any initial value problem; this is equivalent to taking a Fourier transform in x.

giving us the phase speed $c = \omega/|k| = \sqrt{g/|k|}$ in terms of the wavenumber.

This calculation can be viewed as a prototype of stability analyses of all sorts of free boundary problems, ranging from fluid flow to the solidification of ice or steel. If we think of it in this light, it tells us that the surface of our water is neutrally stable, because the fact that ω is purely real tells us that small disturbances neither grow nor decay. If, however, we take $g < 0$, equivalent to having the water above the air, ω is purely imaginary and the linearised problem always has exponentially growing modes, because then $e^{\pm i\omega t} = e^{\pm |\omega| t}$. Why does water fall out of a glass if you turn it upside down, even though the atmospheric pressure (about 10 m of water) is more than enough to hold it in?

Fill a glass to the brim, slide a piece of card across, and hold it in place while you invert the glass: it will stay there when you take your hand away. Why does this not work if there is an air gap before you put the card on?

13.6 Caveat expandator

Regular expansions don't always work. Sometimes the reasons for this are obvious: the procedure falls flat on its face early on. For example, consider the very easy differential equation

$$\epsilon\frac{dy}{dx} = y - 1, \qquad x > 0, \qquad y(0) = 0,$$

for small ϵ. A straightforward regular expansion in powers of ϵ gives $y \sim 1$, and *all other terms vanish*. The regular expansion completely fails to satisfy the initial condition, and in this case inspection of the exact solution, $y = 1 - e^{-x/\epsilon}$, shows that there is a *boundary layer* near $x = 0$ in which the solution changes too rapidly to be describable by a regular expansion. We will look at problems of this type in Chapter 16. Another example where a regular expansion doesn't even get to first base is

$$\epsilon^2\frac{d^2y}{dx^2} + y = 0,$$

whose solutions $e^{\pm ix/\epsilon}$ oscillate very rapidly. We look at problems of this kind in Chapter 23.

There are, however, some problems where what goes wrong is more subtle. Let us return to the undamped small-displacement pendulum equation of Section 13.4,

$$\frac{d^2\theta}{dt^2} + \sin\theta = 0, \qquad \theta(0) = \epsilon a_0, \qquad \frac{d\theta}{dt} = 0 \quad \text{at} \quad t = 0,$$

for small displacements. We showed that we could recover linear theory as the first term in an expansion

$$\theta \sim \epsilon a_0 \cos t + \epsilon^2\theta_2 + \epsilon^3\theta_3 + O(\epsilon^4).$$

(I've kept two terms after θ_1 because it very soon becomes clear that $\theta_2 = 0$.). If, encouraged by this success, we continue, the problem for θ_3 is

$$\frac{d^2\theta_3}{dt^2} + \theta_3 = \frac{a_0^3}{6}\cos^3 t$$

$$= \frac{a_0^3}{24}\left(3\cos t + \cos 3t\right),$$

with

$$\theta_3 = \frac{d\theta_3}{dt} = 0 \quad \text{at} \quad t = 0.$$

The solution is found after a small effort:

$$\theta_3 = \frac{a_0^3}{16}t\sin t - \frac{a_0^3}{192}\left(\cos 3t - \cos t\right).$$

There's just one problem with this solution: the term $t\sin t$ grows unboundedly as t increases. Eventually, when $\epsilon^3 t\sin t$ and $\epsilon\cos t$ are of the same size, that is when $t = O(1/\epsilon^2)$, the expansion is no longer valid, because successive terms are no longer decreasing in size. Terms like this are known as *secular terms* (the word in this context means 'having an enormous duration'); the origin is in the analysis of planetary orbits and in particular the slowly developing effect of one planet's gravitational field on the motion of another, which is how the outer planets were found.

The nonuniformity arises because we are trying to describe a periodic function of t whose period is not quite the 2π of the base solution θ_1. (Such nonuniformities are always lurking in problems with conserved quantities or similar structure if the functions we use to approximate are not quite compatible with the conserved quantities.) We take a brief look at problems of this kind in Chapter 22.

Exercise.

How to do this:

$$e^{3it} = \cos 3t + i\sin 3t$$

$$= (e^{it})^3 = (\cos t + i\sin t)^3;$$

expand the last and use $\cos^2 t + \sin^2 t = 1$ to get $\cos 3t = \cos^3 t - 4\cos t$ from the real part.

Why no $t\cos t$ in θ_3? Because the forcing function on the right-hand side is even: with a second derivative, an undifferentiated term and zero initial derivative, we are bound to get an even solution. There's no point in making work for ourselves by putting in odd terms. For much the same reason there is no $\sin t$ term either, which makes fitting the initial zero value for θ_3 and $d\theta_3/dt$ a trivial business.

13.7 Exercises

1 Space stations. A space station is in a circular orbit about the earth at a distance a from the centre and with angular speed ω. Its tangential speed is increased from $a\omega$ to $a\omega + \epsilon v$, where $\epsilon \ll 1$. Carry out the linear stability analysis of the orbit (remember that the angular momentum has to be perturbed).

2 Phase planes. Referring back to subsection 13.3.1, suppose that the Jacobian $\mathbf{J}$ is real and symmetric at a critical point. Show that, by diagonalising $\mathbf{J}$, the linearised equations can be reduced to

$$\frac{d}{dt}\begin{pmatrix} X \\ Y \end{pmatrix} = \begin{pmatrix} \lambda_1 & 0 \\ 0 & \lambda_2 \end{pmatrix}\begin{pmatrix} X \\ Y \end{pmatrix},$$

where the axes of the coordinates X_i are along the eigenvectors of $\mathbf{J}$. Deduce that the orbits are locally given by the curves

$$|X|^{\lambda_2}|Y|^{-\lambda_1} = \text{constant}$$

and that they look roughly like hyperbolae when $\lambda_1\lambda_2 < 0$.

Show that when $\mathbf{J}$ is skew-symmetric, $\mathbf{J} = -\mathbf{J}^\top$, the critical point is a centre.

3 Pendulum phase planes. Consider the damped pendulum equation

$$l\ddot{\theta} + k\dot{\theta} + g\sin\theta = 0, \qquad \theta(0) = \theta_0, \qquad \dot{\theta}(0) = \omega_0,$$

made dimensionless with the timescale $t_0 = \sqrt{l/g}$ so that it becomes

$$\ddot{\theta} + \gamma\dot{\theta} + \sin\theta = 0, \qquad \theta(0) = \alpha_0, \qquad \dot{\theta}(0) = \beta_0.$$

First suppose that $\gamma = 0$. Show that there are centres in the phase plane at $(\theta, \dot{\theta}) = (2n\pi, 0)$ and saddles at $((2n+1)\pi, 0)$. Sketch the phase plane, indicating the direction of the trajectories. Indicate the curves for which $\alpha_0 = 0$, $\beta_0 \ll 1$. Find a suitable scaling for θ to show that they are approximately circles.

Put an arrow on one trajectory at, say, $(\alpha_0, 0)$; the rest follow by continuity.

Still with $\gamma = 0$, indicate the curves for which $\beta_0 \gg 1$. What is the pendulum doing on one of these? Take $\alpha_0 = 0$ and rescale time by writing $t = \tilde{t}/\beta_0$; show that this gives

$$\frac{d^2\theta}{d\tilde{t}^2} + \frac{1}{\beta_0^2}\sin\theta = 0, \qquad \theta(0) = 0, \qquad \frac{d\theta}{d\tilde{t}}(0) = 1.$$

Now write

$$\theta = \theta_0 + \frac{1}{\beta_0^2}\theta_1 + \cdots,$$

and find θ_0 and θ_1 by equating terms of $O(1)$ and $O(1/\beta_0^2)$ to zero separately. Interpret these results.

Now suppose that $\gamma > 0$. Show that the saddles remain saddles but the centres become stable spirals for $0 < \gamma < 2$. What happens for $\gamma > 2$? Sketch the phase plane when (a) $0 < \gamma \ll 1$, (b) $\gamma \gg 1$.

4 Satellites. Investigate the linear stability of a satellite orbit using the more general approach of subsection 13.3.1 as follows. Write the equations for the motion of a satellite,

$$\ddot{r} - r\dot{\theta}^2 = -\frac{GM}{r^2}, \qquad \frac{1}{r}\frac{d}{dt}\left(r^2\dot{\theta}\right) = 0,$$

as a first-order system $\dot{x}_i = F_i(x_j)$ for $x_1 = r$, $x_2 = \dot{r}$ and $x_3 = \dot{\theta}$. Show that all points on the curve $x_1^3 x_3^2 = GM$, $x_2 = 0$, are equilibrium points. Taking a representative point $(a, 0, \omega)$ on this curve, show that the Jacobian $(\partial F_i / \partial x_j)$ is given by

$$\begin{pmatrix} 0 & 1 & 0 \\ 3\omega^2 & 0 & 2a\omega \\ 0 & -2\omega/a & 0 \end{pmatrix}.$$

Deduce from the trace and determinant of this matrix, which you can evaluate without detailed calculation, the consistency check that at least one of the eigenvalues vanishes and that the others sum to zero, from which it follows that just one eigenvalue vanishes and the other two sum to zero. Confirm this by finding the eigenvalues to be $0, \pm i\omega$.

Note the presence of a zero eigenvalue, which is due to the existence of a non-isolated set of equilibrium points (why?). What would happen if you wrote the equations as a 4×4 system for $r, \dot{r}, \theta, \dot{\theta}$?

5 **Motion under gravity near the earth.**   Absolutely everybody meets the motion of projectiles early in their mathematical career: $m\ddot{\mathbf{r}} = -m\mathbf{g}$, $\mathbf{r}(0) = (0, 0, h)$, $\dot{\mathbf{r}}(0) = \mathbf{v}$. Expand in terms of $\epsilon = h/R_e$, where R_e is the radius of the earth, to reconcile this with the full Newtonian model in which the force on a particle of mass m is $m\nabla\phi$ and the gravitational potential ϕ is $GM_e/|\mathbf{R}|$, G being the universal gravitational constant, M_e the mass of the earth and $\mathbf{R}$ the position vector measured from the centre of the earth. What restriction on $\mathbf{v}$ is necessary for your approximation to be valid?

6 **Flagpoles again.**   Look up your derivation of the dimensionless flagpole equation oscillated at the base, and write it in the form

$$\frac{\partial^2 y}{\partial t^2} + \alpha^4 \frac{\partial^4 y}{\partial x^4} = 0,$$

with the boundary conditions

$$y_{xx} = y_{xxx} = 0 \quad \text{at} \quad x = 1, \qquad y = \cos t, \quad y_x = 0 \text{ at } \quad x = 0,$$

and a condition of periodicity in time. Suppose that $\alpha \gg 1$, and write $\epsilon = 1/\alpha$. Find the solution correct to $O(\epsilon)$ by a regular perturbation method. What is happening physically in this regime?

7 **The Euler strut (ii).**   Recall from Chapter 4, Exercise 1, that for the Euler strut the angle θ between the strut and the x-axis satisfies

$$\frac{d^2\theta}{ds^2} + \alpha^2 \sin\theta = 0, \qquad \theta(0) = \theta(1) = 0,$$

where $\alpha^2 = FL^2/b$ is the bifurcation parameter. Show that if θ is small then the procedure of that exercise is equivalent to finding the first term in a regular expansion for θ.

Now suppose that α is just above the critical value π, say $\alpha^2 = \pi^2 + \epsilon^2$ where ϵ is small. Seek a solution in which θ is small, say $\theta = \phi\delta$ where $\delta \ll 1$ (so far, we do not know how large δ should be). Show that

$$\frac{d^2\phi}{ds^2} + (\pi^2 + \epsilon^2)\left(\phi - \delta^2\frac{\phi^3}{6} + \cdots\right) = 0, \qquad \phi(0) = \phi(1) = 0.$$

Conclude, provisionally, that a sensible choice for δ is $\delta = \epsilon$ (we return to this below).

Construct a regular expansion

$$\phi \sim \phi_0 + \epsilon^2\phi_1 + O(\epsilon^4),$$

This is the Fredholm Alternative for a two-point boundary value problem; see Exercise 6 on p. 136.

show that $\phi_0 = A \sin \pi s$ for an as yet unknown constant A and write down the problem satisfied by ϕ_1. Multiply by ϕ_0 and integrate by parts to show that it only has a solution if

$$\int_0^1 \left(\pi^2\frac{\phi_0}{6} - \phi_0\right)\sin \pi s \, ds = 0,$$

and conclude that ϕ_1 only exists if $A = 0$ or $\pm 2\sqrt{2}/\pi$.

As α varies, define a measure $M_\theta(\alpha)$ of the size of the solution $\theta(s;\alpha)$ by

$$M_\theta(\alpha) = \max_{0 \leq s \leq 1} \theta(s;\alpha),$$

and show that, for α near π, either $M_\theta(\alpha) = 0$ or

$$M_\theta(\alpha) = \pm 2\sqrt{2}\sqrt{\alpha^2 - \pi^2}/\pi.$$

Plot the *response diagram* $M_\theta(\alpha)$ against α and you will see why this bifurcation is called a *pitchfork bifurcation*.

Finally, go back and convince yourself that other choices for the magnitude of δ do not lead to sensible expansions. Also show (using the analysis above) that if α is slightly *below* the critical value then the only solution is $\phi = 0$.

8 The forced logistic equation.   Explain why the equation

$$\frac{du}{dt} = ku(1 - u)$$

is a crude model for population dynamics supported by a finite resource (what happens to u if it is small, or just above or below 1?). Which term in the equation corresponds to the size of the resource? Now suppose that the resource fluctuates seasonally, so that the population equation is

$$\frac{du}{dt} = ku(1 + \epsilon \cos t - u).$$

Find a periodic solution $u = 1 + \epsilon u_1(t) + \cdots$ correct to $O(\epsilon)$ (with an error of $O(\epsilon^2)$).

Show that the equation can be solved exactly by putting $u = 1/v$. Does this help matters?

9 Electric potential of a nearly circular cylindrical annulus. Find the electric potential ϕ, satisfying $\nabla^2 \phi = 0$ between the two cylinders $r = a$, on which $\phi = 0$, and $r = b > a$, on which $\phi = V$. Suppose that the inner cylinder is perturbed to $r = a(1 + \epsilon \sin n\theta)$. Calculate ϕ correct to $O(\epsilon)$; to build up your arithmetical strength, calculate it correct to $O(\epsilon^2)$. What restriction on n is necessary for your expansion to be valid?

Here (r, θ) are plane polar coordinates. When $n = 1$ the inner cylinder is nearly circular but displaced. When $n = 2$ it is close to elliptical.

'But I want to cross out *all* the terms!'

14

Case study: electrostatic painting (2)

14.1 Small parameters in the electropaint model

When we left this problem (see Chapter 4), we had a dimensionless model with a number of small parameters in it. Let's revisit it in the light of our discussion of regular expansions.

Recall that we have a number density n of particles, with velocity $\mathbf{v}_p$, an electric field $\mathbf{E}$ and gas velocity and pressure $\mathbf{v}_g$ and p. There are several small dimensionless parameters in the model, and we'll leave them all out except the least small, which we call

$$\epsilon = \frac{q_p V_0 L}{K U_g}.$$

The numerical value of ϵ is about 0.1. There is also one $O(1)$ parameter,

$$A = \frac{n_0 K L}{\rho_g U_g},$$

whose numerical value is about 1.

The model consists of an equation of motion for the particles,[1]

$$\mathbf{v}_p - \mathbf{v}_g = \epsilon \mathbf{E} \tag{14.1}$$

an equation for the conservation of particles,

$$\frac{\partial n}{\partial t} + \nabla \cdot (n \mathbf{v}_p) = 0 \tag{14.2}$$

---

[1] Now I need to apologise: how, to an applied mathematician of taste, could $\epsilon \mathbf{E}$ be anything but $\mathbf{D}$?

and an equation for the electric field,

$$\nabla \cdot \mathbf{E} = n. \qquad (14.3)$$

Lastly, we have the equations of motion and conservation of mass for the gas,

$$\frac{D\mathbf{v_g}}{Dt} = -\nabla p + An\left(\mathbf{v_p} - \mathbf{v_g}\right) \qquad (14.4)$$

and

$$\nabla \cdot \mathbf{v_g} = 0. \qquad (14.5)$$

Now expand

$$\mathbf{v_p} \sim \mathbf{v_{p0}} + \epsilon \mathbf{v_{p1}} + \cdots ,$$

with similar expansions for the other variables. It's clear from (14.1) that

$$\mathbf{v_{0p}} = \mathbf{v_{0g}},$$

which confirms that the paint particles follow the gas, to leading order. If we use this on the right-hand side of (14.4), then we see that $\mathbf{v_{0g}}$ just satisfies an ordinary fluid flow problem with no body force from the paint particles. Let's assume that we can solve this, and carry on.

The next thing to do is to calculate the evolution of the number density n. The leading-order terms in (14.2) are

$$\frac{\partial n_0}{\partial t} + \nabla \cdot \left(n_0 \mathbf{v_{p0}}\right) = 0.$$

Bearing in mind that $\mathbf{v_{p0}} = \mathbf{v_{g0}}$ and $\nabla \cdot \mathbf{v_{g0}} = 0$, this simplifies to

$$\frac{\partial n_0}{\partial t} + \mathbf{v_{g0}} \cdot \nabla n_0 = 0,$$

a first-order hyperbolic equation[2] whose characteristics are, not surprisingly, the gas particle paths.

Having found n_0, the last task is to find the leading-order electric field as the solution of

$$\nabla \cdot \mathbf{E_0} = n_0.$$

We can now go round the cycle again, using the equations in the same order. First, (14.1) tells us that

$$\mathbf{v_{p1}} - \mathbf{v_{g1}} = \mathbf{E_0}; \qquad (14.6)$$

If any of the following steps are not clear, take a minute to write them out.

---

[2] If you have only studied first-order partial differential equations in two independent variables, it should be a relief to find that the extension to more independent variables is very straightforward; see [44], Chapter 1.

then (14.4) and (14.5) form a linear system for $\mathbf{v}_{g1}$. Thus, we know the correction to the particle velocity from (14.6); we can also calculate n_1 from (14.2) and lastly $\mathbf{E}_1$ from (14.3). Notice how the asymptotic expansion suggests an order in which to solve the equations, which might also be a sensible basis for an iterative numerical scheme (at least in the steady case).

You can develop this problem further by doing the exercises on it.

14.2 Exercises

1 Electrostatic painting I. In the electrostatic painting model, we wrote down the conservation of particle number,

$$\frac{\partial n}{\partial t} + \nabla \cdot \left(n\mathbf{v}_{p}\right) = 0$$

and the conservation of mass for the gas (assumed incompressible),

$$\nabla \cdot \mathbf{v}_{g} = 0.$$

(Remember that this is dimensionless, and n is scaled with a typical number density $n_0 \approx 10^9$ m^{-3}; what is the average distance between the particles?)

In fact this isn't quite right, because we have a two-phase flow and gas may be displaced by particles and vice versa. If we take a macroscopically small representative volume V (which at the same time is large compared with the cube of the average particle separation), show that we can nevertheless justify it as follows.

(a) Show that the proportion of V that is occupied by particles is ϵn, where $\epsilon = 4\pi n_0 a^3/3$ if all the particles are spherical with the same radius a ($a \approx 10^{-5}$ m). Estimate the numerical size of ϵ.

(b) Deduce that the proportion of V occupied by gas is $1 - \epsilon n$.

(c) Using the general form

$$\frac{\partial(\text{density})}{\partial t} + \nabla \cdot (\text{flux}) = 0$$

for a conservation law, show that the conservation of mass for the gas (remember it's incompressible so its density is constant) is given by

$$\frac{\partial(1 - \epsilon n)}{\partial t} + \nabla \cdot \left((1 - \epsilon n)\mathbf{v}_{g}\right) = 0.$$

Show also that the conservation-of-particle-number equation given above is correct.

(d) Just to confirm, show that

$$\nabla \cdot \left(\epsilon n \mathbf{v}_p + (1 - \epsilon n) \mathbf{v}_g \right) = 0,$$

and interpret this flux in physical terms.

(e) Expand n and $\mathbf{v}_g$ in powers of ϵ, and show that the leading-order equations are those given above.

2 Electrostatic painting II. Complete the derivation of the $O(\epsilon)$ equations for this problem, and verify that all the equations you obtain are linear in $\mathbf{v}_{g1}, n_1$ etc.

3 Electrostatic painting III. Consider steady-state solutions of the leading-order $(O(1))$ equations for this problem. Show that $\mathbf{v}_{g0} \cdot \nabla n_0 = 0$. If the flow is two-dimensional, with stream function ψ, deduce that $n_0 = f(\psi)$ for some function f determined by the inlet conditions.

4 Space charge. A variant of the electropainting problem occurs when the charged particles are so small that they do not exert any significant body force on the gas. The sort of physical situation that can be modelled in this way is the motion of charged ions from a high-voltage DC power cable, or the electrostatic scrubbing used to clean power station emissions.

If there is no imposed gas flow, briefly justify the model

$$\mathbf{v}_p = \mathbf{E}, \qquad \frac{\partial n}{\partial t} + \nabla \cdot \left(n \mathbf{v}_p \right) = 0, \qquad \nabla \cdot \mathbf{E} = n.$$

Given that we can write $\mathbf{E} = -\nabla \Phi$ (since from Maxwell's equations on this timescale we have $\nabla \wedge \mathbf{E} = \mathbf{0}$), show that the system becomes

$$\frac{\partial n}{\partial t} + n^2 - \nabla n \cdot \nabla \Phi = 0, \qquad \nabla^2 \Phi = -n.$$

In the steady state, show that the characteristics of the first of these equations have tangent $-\nabla \Phi$. Deduce that they are orthogonal to the equipotentials and, parametrising them by τ, derive the ordinary differential equation

$$\frac{dn}{d\tau} + n^2 = 0$$

along them. Show also how to model a point source of charged particles by allowing $n \to \infty$ as $\tau \to 0$.

Now suppose that there is an irrotational imposed gas flow, so that there is a velocity potential ϕ with $\mathbf{v}_g = \nabla \phi$, where $\nabla^2 \phi = 0$ (potential flow is often a very reasonable model upstream of an obstacle, less so downstream where the effects of boundary layers, separation and so on are felt). Show that there are parameter ranges where the model

$$\mathbf{v}_p = A\mathbf{v}_g + \mathbf{E}, \qquad \frac{\partial n}{\partial t} + \nabla \cdot (n\mathbf{v}_p) = 0, \qquad \nabla \cdot \mathbf{E} = n$$

is valid with A an $O(1)$ constant. Show that the results of the previous paragraph hold but with the characteristics derived from the modified potential $\Phi - A\phi$.

5 **Paint layer again.** Suppose that a thin layer of paint particles, deposited electrostatically as in the text, is growing on $y = 0$ and that its thickness is $y = h(x, t)$. If $\epsilon = H/L \ll 1$, where H and L are a typical thickness of the layer and lengthscale of the workpiece respectively, justify the approximate boundary condition

$$\frac{\partial h}{\partial t} = \mathbf{v}_p \cdot \mathbf{n}$$

on the workpiece (see Exercise 1 at the end of Chapter 6).

'That means it's the outer end of the length.'

15
Case study: piano tuning

15.1 The notes of a piano: the tonal system of Western music

This section contains a short description of the mathematical structure of the tonal system used for a piano. It can be omitted by those not interested.

The particular sound of a given note of a piano or other musical instrument is characterised reasonably well by its fundamental frequency and a variety of higher harmonics; damping rates also play a role. These harmonics are often (approximately – as we shall see, that is the point of this case study) integer multiples of the fundamental frequency f_1. On stringed instruments the reason is that the normal frequencies of a vibrating string are close to integer multiples of the fundamental, and wind instruments either have regular vibrating cavities (an organ tube is an example) with the same integer harmonic ratios or, as for a French horn, they are carefully (and expensively) made to sound this way.

When two or more notes are played together, their fundamentals and harmonics all interact. The tonal system of Western music has been strongly influenced by the features of this interaction; the mathematical construction we now outline goes back at least to the Pythagoreans of ancient Greece. Suppose that we have an ideal instrument for which the harmonics of each note are exactly integer multiples of the fundamental. Now suppose that we play a note, called for example A, with fundamental frequency f_1^A; we hear frequencies f_1^A, $f_2^A = 2f_1^A$, $f_3^A = 3f_1^A$ and so on. We might expect the note A', with fundamental frequency $f_1^{A'} = 2f_1^A$ equal to twice that of A, to sound good with A', because its fundamental

See Exercise 5 for the reason why a cymbal or gong sounds harsh.

When a violinist tunes up by playing the A and E strings together and eliminating beats by turning a tuning peg, the beats that are eliminated are probably those between the second harmonic of the A string and the first harmonic of the E string. See below for a discussion of beats.

coincides with the first harmonic of A, and the interval between the two, created by doubling the lower frequency, is called an *octave* even though, as yet, it exists only as an ideal. In a similar way, the note with fundamental frequency $3f_1^A$ should also produce a harmonious blend with A, and so should the note an octave below it, whose frequency is $\frac{3}{2}f_1^A$. This note is called E, and the interval corresponding to a frequency ratio $\frac{3}{2}$ is called a *fifth*. We can similarly construct a *fourth*, with frequency $\frac{4}{3}f_1^A$, called D. Its frequency ratio is $\frac{4}{3}$, and we notice that since $\frac{3}{2} \times \frac{4}{3} = 2$, the interval from E to A' is also a fourth. Following this, we have the *major and minor thirds*, with ratios $\frac{5}{4}$ and $\frac{6}{5}$ respectively. These are the most important intervals and they (or, rather, their practically realised equivalents) make up, for example, the harmonious-sounding chords you hear at the ends of pieces of music.

And, corresponding to the major and minor thirds, the minor and major sixths have ratios $\frac{8}{5}$ and $\frac{5}{3}$ respectively.

It is apparent that we could continue this process of interval construction indefinitely, until we have notes with all rational multiples of f_1^A; this might be plausible in the context of a 'continuous' instrument like a violin or the human voice but it is clearly impractical for a piano. Moreover, given that the amplitudes of the harmonics of a note decrease as we go to higher harmonics, it would be pointless because we could never hear the interactions.

In practice, therefore, the process is truncated. Western music is built around a tonal system consisting of 12 notes, separated by intervals, not necessarily exactly equal, called *semitones*, two of which make a *tone*; for the moment, we leave aside the question of how to fix frequency ratios involving semitones. Thus we now have a different definition of an interval such as a fifth, couched in terms of the number of semitones that make it up rather than a frequency ratio. Having stated that an octave consists of 12 semitones, we define a fifth to consist of 7 semitones, a fourth to consist of 5 semitones and major and minor thirds to consist of 4 and 3 semitones respectively; as already noted, a tone consists of 2 semitones. For historical reasons, only seven letters are used to denote notes (these are the 'white notes' on a piano), the remaining ones being described with the help of two operators, ♯ (pronounced 'sharp') and ♭ ('flat'), which, when placed after a note, move it up by a semitone for a sharp and down for a flat.[1] The sequence of notes can be written

..., A, A♯/B♭, B, C, C♯/D♭, D, D♯/E♭, E, F, F♯/G♭, G, G♯/A♭, A, ...'

repeated up and down the piano in octaves.

If we look at this semitone scheme more closely, we see that it is not consistent with the construction of fifths and other intervals, outlined above, as exact integer ratios. One manifestation of the inconsistency is that an octave should consist of three consecutive major thirds of four

---

[1] For musical reasons, other notations such as C♭ B or even G♭♭ F are possible, but they are irrelevant here.

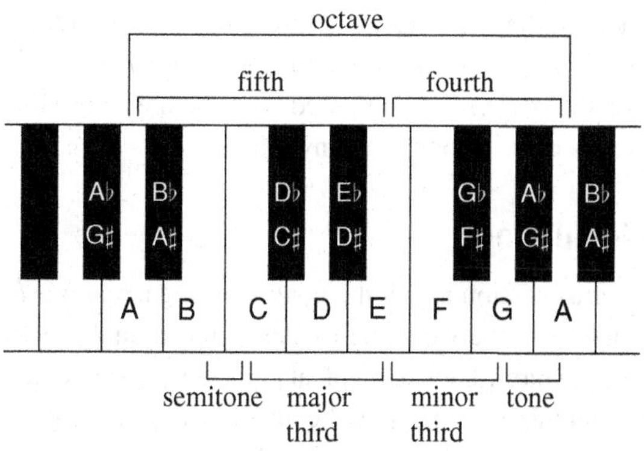

Figure 15.1 A section of a piano keyboard.

semitones each, for example A–C♯–F–A. However, in the integer-ratio scheme this gives a frequency ratio of $(\frac{5}{4})^3 = \frac{125}{64} < 2$, whereas it should give 2 exactly. Similarly the octave should be four consecutive minor thirds; but $(\frac{6}{5})^4 = \frac{1296}{625} > 2$. Another famous illustration of the inconsistency is obtained by constructing the 'circle of fifths', in which we go up by fifths, dropping down an octave as convenient:

$$A \to E \to B \to F\sharp \to C\sharp = D\flat \to A\flat$$

(A♭ this is the note 'furthest removed' from A)

$$\to E\flat \to B\flat \to F \to C \to G \to D \to A.$$

The frequency of the last A, $(\frac{3}{2})^{12}$, isn't a power of 2 as it should be. It's slightly sharp, since $531441/4096 \approx 129.746 > 2^7 = 128$.

As a consequence of this inconsistency in construction, we can never tune an instrument with exact integer-ratio harmonic frequencies in such a way that all the intervals on it are perfectly in tune. For example, if we tune the fifths to be a perfect ratio, then moving away from A in both directions, we get two different frequencies for the furthest-removed note, A♭. Going up, we get $(\frac{3}{2})^6$ and going down we get $(\frac{2}{3})^6$; the ratio of these is not a power of 2. Any other interval gives a similar result. How, then, are we to choose the fundamental frequencies of our twelve notes? The sound of two notes played together depends very strongly on their interaction. Harmonics that are close together can give unpleasant sounding beats and sound out of tune, especially on an instrument like an organ in which the volume does not decay after a note is sounded. What compromise system should we use?

This question of *temperament* caused a great deal of trouble in the past, and I don't want to go into great detail about it here; literally hundreds of solutions have been proposed (see [30] for a popular history and [24] for a more technical derivation of some popular temperaments). The currently accepted solution[2] is to insist that each interval of a semitone corresponds to the *same* frequency ratio, which, given that notes

[2] Some composers are returning to 'microtonal' music.

an octave apart are taken to have a frequency ratio of 2, must therefore be $2^{1/12} \approx 1.0595$. With this compromise, called *equal temperament*, all notes and intervals are slightly wrong compared with the natural ratios, but at least no one note is more wrong than any other.

15.2 Tuning an ideal piano

The upshot of the previous section is that the goal of tuning a piano is to obtain certain frequency ratios between the fundamental frequencies of pairs of notes. Because the harmonics of an ideal piano string are integer multiples of the fundamental, they too are to be tuned in specified ratios. Moreover, these ratios are close to, but (apart from the octave) not exactly equal to, integer ratios. For example, the equal-temperament fifth has a ratio $2^{7/12} \approx 1.4983$.

The easiest intervals to tune are the octaves. If we use a tuning fork (mechanical or electronic) to tune one note of our piano, say the A above middle C, to its standard frequency of 440 Hz, then we can tune all the other As on our instrument to frequencies of $2^{\pm k} \times 440$ Hz by eliminating beats between the fundamentals and the first harmonics of notes an octave apart. Then we can tune other notes by using intervals such as fifths, listening to the calculable (and measurable) beat rates on the appropriate harmonics; see Exercise 2.

Interlude: beats.   How would we tune a note on a piano to be the same as a standard tone? The standard way is to play them together, and listen for the beats. Suppose that they have the same amplitude a and phase (see Exercise 1 for when these are not the same) but slightly different frequencies ω and $\omega + \epsilon$, where ϵ is small. The sum of the signals is

$$a \cos \omega t + a \cos(\omega + \epsilon)t = 2a \cos(\omega + \tfrac{1}{2}\epsilon)t \cos \tfrac{1}{2}\epsilon t.$$

Not $\frac{1}{2}\epsilon$: we hear two amplitude peaks for each cycle of $\cos \frac{1}{2}\epsilon t$.

This is a modulated wave: it oscillates at the fast frequency $\omega + \tfrac{1}{2}\epsilon$, which is very close to ω, and its amplitude[3] is modulated at the slow beat frequency ϵ. So the aim in tuning is to get the beat frequency to zero (or another specified rate) by tightening or loosening the piano strings (a very skilled business).

---

[3] Irrelevant digression: how loud do n instruments of an orchestra sound compared to one on its own? Answer: $\sqrt{n}$ times as loud, because the phases of the instruments are random. The sound signal from the whole orchestra is $\sum_i a_i \cos(\omega_i t + \phi_i)$ where a_i are the individual amplitudes, ω_i the frequencies and ϕ_i the phase shifts. Even if all the a_i are the same, the ϕ_i are in practice randomly distributed and independent so the root mean square amplitude (standard deviation) of the sum is $\sqrt{n}$ times an individual amplitude (the variances of independent random variables add up). This is one reason why the concerto can succeed as an art form; the detailed workings of the ear–brain combination is another, as picking out important sounds from a background is an important survival skill; and of course, skilful writing by composers may have something to do with it too.

15.3 A real piano

Now let's look at a real piano string. The displacement of an ideal string, which has no bending stiffness, satisfies the wave equation

$$\rho A \frac{\partial^2 y}{\partial t^2} - T \frac{\partial^2 y}{\partial x^2} = 0, \qquad 0 < x < L,$$

$$y = 0 \quad \text{at} \quad x = 0, L.$$

and it's a piece of classical applied mathematics to show that the normal modes are

$$y_n = e^{-i\omega_n t} \sin \frac{n\pi x}{L}$$

where $\omega_n = n\pi c / L$ and $c^2 = T/(A\rho)$.

For later convenience, we give the dimensionless versions of these results. Scaling x with L and t with L/c and immediately dropping the primes, the equation is

$$\frac{\partial^2 y}{\partial t^2} - \frac{\partial^2 y}{\partial x^2} = 0, \qquad 0 < x < 1,$$

with

$$y = 0 \quad \text{at} \quad x = 0, 1;$$

the normal modes are now

$$y_n = e^{i\Omega_n t} \sin n\pi x$$

and the dimensionless frequencies are

$$\Omega_n = n\pi.$$

The ω_n are of course angular frequencies: the frequencies in Hz are $f_n = \omega_n / (2\pi) = nc/(2L)$, so that $1/f_n$ is the time taken for a signal travelling at the wave speed c to travel from one end of the string and back n times.

However, a real piano string does have a small bending stiffness. A combination of the string model above and the beam models we used earlier (see Exercise 3 at the end of Chapter 4) gives us the dimensionless equation

$$\rho A \frac{\partial^2 y}{\partial t^2} - T \frac{\partial^2 y}{\partial x^2} + E A k^2 \frac{\partial^4 y}{\partial x^4} = 0$$

for the string displacement. We can assess the size of the fourth-derivative term by scaling x and t as above, to get the dimensionless equation

$$\frac{\partial^2 y}{\partial t^2} - \frac{\partial^2 y}{\partial x^2} + \epsilon \frac{\partial^4 y}{\partial x^4} = 0,$$

where

$$\epsilon = \frac{E k^2}{\rho L^2 c^2} = \frac{E A k^2}{T L^2}.$$

Note the very sensitive dependence on the string thickness; this goes as the fourth power of the radius since $k \propto a$ and $A \propto a^2$.

Now for a circular string of radius a we have $k^2 = \frac{1}{2}a^2$, so if $a = 1$ mm then $k^2 = \frac{1}{2} \times 10^{-6}$ m^2. If the string is made of steel, in SI units it has $E \approx 2 \times 10^{11}$ and $\rho = 7800$. Suppose that the string is 1 m long and has a tension of 1000 N (this is quite typical: the combined force of all the strings on a grand piano is several tonnes worth). Then

How many newtons in a tonne weight?

$$\epsilon = \frac{EAk^2}{TL^2} = \frac{2 \times 10^{11} \times \pi \times 10^{-6} \times \frac{1}{2} \times 10^{-6}}{10^3 \times 1^2} \approx 3.1 \times 10^{-4},$$

which is small indeed but nevertheless has a noticeable effect, as we shall see. The frequency of this string, $c/(2L)$, is about 280 Hz, close to the D above middle C.

Now let's calculate the normal modes of a string. In order to do this we need boundary conditions, two at each end. The simplest are that $y = 0$ (obviously) and that $\partial^2 y / \partial x^2 = 0$, so-called simply supported conditions, which are probably not a bad approximation to the truth as the string passes over a 'bridge' at each end. We shortcut the process of finding normal modes, which you would usually do by looking for separable solutions $y_n(x, t) = e^{-i\Omega_n t} Y_n(x)$, by noting that with our choice of boundary conditions there are solutions

$$y_n = e^{-i\Omega_n t} \sin n\pi x$$

provided that

$$\Omega_n^2 = n^2 \pi^2 + \epsilon n^4 \pi^4.$$

So, the normal frequencies are

$$\Omega_n = n\pi \left(1 + \epsilon n^2 \pi^2\right)^{1/2}$$
$$\sim n\pi \left(1 + \tfrac{1}{2}\epsilon n^2 \pi^2 + o(\epsilon)\right).$$

The fundamental frequency of our string is thus

$$\Omega_1 \sim \pi \left(1 + \tfrac{1}{2}\epsilon \pi^2\right)$$

and so the $(n - 1)$th harmonic has frequency

$$\Omega_n \sim n\pi \left(1 + \tfrac{1}{2}\epsilon n^2 \pi^2\right)$$
$$\sim n\Omega_1 \frac{1 + \tfrac{1}{2}\epsilon n^2 \pi^2}{1 + \tfrac{1}{2}\epsilon \pi^2}$$
$$\sim n\Omega_1 \left(1 + \tfrac{1}{2}\epsilon \pi^2 \left(n^2 - 1\right)\right),$$

using the binomial expansion to simplify the fraction.

We see that the higher harmonics have slightly larger frequencies than the theoretical integer multiples of the fundamental, a property known as *inharmonicity*. So, if we tune the string A$'$, one octave above our

A, by eliminating beats between its fundamental and the first harmonic of the lower A string, the fundamental frequency of the higher string will be $2(1 + \frac{3}{2}\epsilon\pi^2)$ times that of the lower one, not the theoretical twice. This phenomenon is known as *octave stretch*; over the seven octaves of my piano, making the very crude assumption that the inharmonicities of the strings are all the same, the stretch is by a factor

$$\left(1 + \frac{3}{2}\epsilon\pi^2\right)^7 \sim 1 + \frac{21}{2}\epsilon\pi^2$$

$$\approx 1.033.$$

This may not look much, but it is more than half a semitone; in fact the inharmonicities on pianos can add up to as much as a whole semitone (the higher strings especially are very short and so have larger values of ϵ, and there are other effects due to the ends of the strings). It is not at all well known, even among pianists, that the treble strings of a piano are this much sharp of 'theoretical' values; fortunately there are no other instruments with a similar range that might accompany it. In Exercise 6 you can work out how to deduce the inharmonicity by measuring beat rates, a first step in calculating the optimal tuning for a given instrument.

15.4 Sources and further reading

This case study has described joint work in progress with Paul Duggan, who tunes my piano while I do the calculations. There is a huge amount of fascinating stuff about musical instruments in [17]. The book [16] describes a variety of applications of mathematics to music and has an article on beats and consonance.

15.5 Exercises

1 Beats.   Suppose that we combine two signals

$$a \cos(\omega t + \phi_1), \qquad a \cos(\omega t + \epsilon + \phi_2).$$

Show that the beats analysis given in Section 15.2 is unaffected.
    Now combine signals with different amplitudes,

$$a_1 \cos \omega t, \qquad a_2 \cos(\omega + \epsilon)t.$$

Show that the output consists of a constant-amplitude signal at frequency ω together with a signal that beats at frequency ϵ. To see this in detail, suppose that $a_1 > a_2$ and write the combined signal as

$$(a_1 - a_2) \cos \omega t + 2a_2 \cos \left(\omega + \tfrac{1}{2}\epsilon\right) t \cos \tfrac{1}{2}\epsilon t.$$

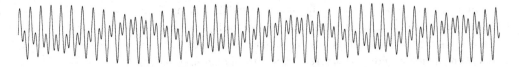

Figure 15.2 Beats generated by $\sin 30t + \sin 61t$.

Then work out the average of the squared amplitude over a 'moving window' time interval which is large compared with the period of the fast oscillation at frequency ω but small compared with the period of the slow modulation at frequency ϵ. (Making the substitution $\epsilon t = \tau$, you should get an integral something like

$$\int_{\tau}^{\tau+\delta} \left[(a_1 - a_2) \cos\left(\frac{\omega \tau'}{\epsilon}\right) + 2a_2 \cos\left(\frac{\omega}{\epsilon} + \frac{1}{2}\right) \tau' \cos \frac{1}{2}\tau' \right]^2 \, d\tau'$$

where $\epsilon \ll \delta \ll 1$. The squares of the first and second terms in the brackets average over many periods to a constant and a modulated amplitude respectively, and after a bit of diddling around the cross term is found to average to zero. Try it and see; use the Riemann–Lebesgue lemma if you want to be more rigorous.)

Now investigate the signal $\sin \omega t + \sin(2\omega + \epsilon)t$ as shown in Figure 15.2. Show that the difference between this signal and that with $\epsilon = 0$ is a modulated wave. (The beats are clearly visible in the figure and would be clearly audible.) Plot other waveforms to visualise chords consisting of two notes that are a fifth, or some other interval, apart.

For more about beats see [16], which includes a brief discussion of the way in which the ear–brain combination can perceive low frequencies even when there is very little energy in them.

The decorations at the ends of the chapters in this book show the waveforms for $\sin 12t + \sin kt$, where k is the chapter number and $0 \le t \le 4\pi$. Look through them and identify the principal intervals (an octave, a fifth etc.) discussed in the present chapter. (Some software, such as Mathematica, allows you to listen to a function as well as plot it.)

2 **Equal temperament.**   Assume for this question that the harmonics of a string are integer multiples of the fundamental. A piano tuner tunes concert A at 440 Hz and wishes to tune the E a fifth above, using equal temperament. This is to be done by counting the beat rate between the second harmonic of the A and the first harmonic of the E (in practice, it might be done with the sixth and fourth harmonics). Find the theoretical frequencies of these harmonics, deduce a formula for the required beat rate and evaluate it numerically. Repeat for the sixth-fourth pair.

3 Pianos and harpsichords. Suppose that you have a solution $y(x, t)$ of the wave equation

$$\frac{\partial^2 y}{\partial t^2} = c^2 \frac{\partial^2 y}{\partial x^2}, \qquad 0 < x < L,$$

that is periodic in time with period T. Assuming sufficient smoothness, show that

$$\int_0^T \int_0^L c^2 \left(\frac{\partial y}{\partial x}\right)^2 dx \, dt = \int_0^L \int_0^T \left(\frac{\partial y}{\partial t}\right)^2 dt \, dx$$

and interpret this statement in terms of energy. If the solution to a general initial value problem for this string is expanded in a Fourier series in x, in the form

$$y(x, t) = \sum_1^\infty \left(a_n \cos \frac{n\pi ct}{L} + b_n \sin \frac{n\pi ct}{L}\right) \sin \frac{n\pi x}{L},$$

what is the ratio of the energy in each harmonic to that in the fundamental?

A piano string is set in motion by a hammer, which imparts an instantaneous velocity V to the small segment $x_0 < x < x_0 + h$, the remainder of the string being initially at rest. Approximating this initial velocity distribution by a delta function, calculate the energy in each mode relative to the fundamental. Repeat for a harpsichord (or guitar) which is plucked by being let go from the piecewise linear static displacement you get when you displace the point x_0 normally to the string (note, to make the calculation easy, that $\partial^2 y/\partial x^2$ is a delta function at $t = 0$, so you can work out $n^2 a_n$ easily). Comment on the results. (The sound you hear is considerably modified by the soundboard and other parts of the instrument.)

4 Waves on a circular membrane. Recall from the exercises at the end of Chapter 1 that waves on a circular membrane of radius a and density ρ per unit area, stretched to tension T, satisfy

$$\frac{\partial^2 u}{\partial t^2} = c^2 \nabla^2 u,$$

where $c^2 = T/\rho$ is the wave speed.

Show that there are solutions

$$u(r, \theta, t) = e^{-i\omega t} e^{im\theta} R(r),$$

where

$$\frac{d^2 R}{dr^2} + \frac{1}{r} \frac{dR}{dr} + \left(k^2 - \frac{m^2}{r^2}\right) R = 0.$$

Putting $x = kr = \omega r/c$, reduce this to *Bessel's equation* of order m,

$$\frac{d^2 R}{dx^2} + \frac{1}{x}\frac{dR}{dx} + \left(1 - \frac{m^2}{x^2}\right)R = 0.$$

Or just look for a solution $R \sim x^\alpha$ as $x \to 0$ and find possible values of α by balancing the most singular terms.

Perform a local (Frobenius-style) analysis near $x = 0$ to show that there is only one solution that is bounded at $x = 0$; it is called $J_m(x)$. Deduce that the normal frequencies for a membrane clamped at its edges are $\omega_{m,n}$, where n labels the roots of $J_m(\omega_{m,n}a/c) = 0$; it can be shown that there are infinitely many roots of $J_m(x) = 0$ and that they are asymptotic to $(n + \frac{1}{4})\pi$ as $n \to \infty$. However, the low harmonics are far from being integer multiples of the fundamental. Sketch some nodal lines, i.e. lines where $R = 0$, for low values of m and n. Timpani (kettledrums) are much more complicated than this membrane because of the coupling with the air chamber.

5 **Cymbals and gongs.** A simple model for a cymbal or gong treats it as a circular elastic plate. It can be shown that the equation of motion for small displacements $u(\mathbf{x}, t)$ of such a plate is

$$\rho\frac{\partial^2 u}{\partial t^2} + \frac{Eh^2}{12(1-\nu)^2}\nabla^4 u = 0,$$

where ρ is the density, E is the Young's modulus and h the thickness; the parameter ν, called Poisson's ratio, is a material property whose numerical value is often about $\frac{1}{3}$. Compare this equation with that of a beam and convince yourself that it is plausible (Poisson's ratio appears because the geometry of a plate is different from that of a beam).

To save space, define

$$c_{\mathrm{L}} = \sqrt{\frac{E}{\rho(1-\nu)^2}},$$

which is the wave speed for longitudinal waves in a plate. Show that time-periodic solutions $u(\mathbf{x}, t) = e^{-i\omega t}U(\mathbf{x})$ satisfy

$$\nabla^4 U - \frac{12\omega^2}{h^2 c_{\mathrm{L}}^2}U = \nabla^4 U - k^4 U = 0.$$

Check the dimensions.

Deduce that one-dimensional waves for which $u = e^{i(kx-\omega t)}$ are dispersive, with wavenumber k and angular frequency related by

$$\omega = \frac{c_{\mathrm{L}}hk^2}{\sqrt{12}}.$$

Now consider a circular plate and look for a solution

$$U(r, \theta, t) = R(r)\cos m\theta.$$

Noting that $\nabla^4 - k^4 = (\nabla^2 - k^2)(\nabla^2 + k^2)$, show that the general bounded solution for $R(r)$ is

$$R(r) = A J_m(kr) + B I_m(kr);$$

$I_m(kr) = i^{-m} J_m(ikr)$ is sometimes called a modified Bessel function of order m (see Exercise 4).

Write down (but do not attempt to solve) the normal-frequency equation when the plate is clamped at its edges ($u = 0$ and $\partial u / \partial r = 0$). It is fairly clear that the roots are not in a harmonic progression, so the higher harmonics will clash with the fundamental. It is possible (but not recommended, on account of the heavy arithmetic) to find the normal frequencies for the more realistic case of free edges, with a similar lack of harmonicity.

This model is also useful in analysing flat-panel loudspeakers.

6 Piano tuning.   Suppose we don't know the properties of the piano strings, but we believe that the frequencies of the harmonics (in Hz) of string k are given approximately by the formula

$$f_{k,n} = n f_{k,1} \left(1 + \varepsilon_k (n^2 - 1)\right),$$

where the inharmonicity coefficient ε_k may vary from string to string. A good piano tuner can hear the beats not just between the fundamental of one string and the harmonics of another, but also between pairs of harmonics. For example, if we have strings A, E, A', where A' is an octave above A and E is in between, the beats between $f_{A,3}$ and $f_{E,2}$ can be used to tune the E relative to the A, and then the beat rate between $f_{A,6}$ and $f_{E,4}$ can be measured. Show how to take measurements between pairs of harmonics (at most two per string) to determine the inharmonicity coefficients of the three strings above. (In practice, A and E are taken in the middle of the piano, and the beat rate between $f_{A,3}$ and $f_{E,2}$ is set to be 'narrow' by about 1 Hz in order to achieve equal temperament. That is, the frequency of the higher string is lowered from the beat-free $\frac{3}{2}$ times that of the lower string until beats occur at 1 per second; see Exercise 2.)

'A rigid pulsating cylinder . . .'

16

Boundary layers

16.1 Introduction

When might we not be able to construct a regular perturbation expansion for a function in terms of a parameter $\epsilon \to 0$? Or, if we have one, where might it not be valid? One thing that might go wrong is that either the function we are approximating, or the approximation itself, may have singularities. Another is that the approximation may slowly drift away from the true solution, as we saw for the second term of the small-amplitude regular expansion for the pendulum. A third possibility is that our function oscillates very rapidly, with a period of, say, $O(\epsilon)$: we look at this case in Chapter 23. A fourth possibility is that the function changes rapidly in a very small layer, say of width $O(\epsilon)$, but is smooth elsewhere. Such a small layer is known as a *boundary layer* if attached to the boundary of the solution interval or domain, and an *interior layer* if it is internal; see Figure 16.1.

16.2 Functions with boundary layers; matching

Some functions come with built-in boundary layers. A prototype example, which crops up all over the place in applications, is

$$f(x; \epsilon) = e^{-x/\epsilon} \qquad \text{for} \quad 0 < x < 1, \quad \epsilon \to 0.$$

This function starts off with a value of 1 at $x = 0$, and becomes negligibly small, certainly smaller than any power of ϵ, by the time $x \gg O(\epsilon)$. All its effort is concentrated in a boundary layer of thickness $O(\epsilon)$ near the origin. This example is rather trivial, but it is fairly clear that if f is a bit

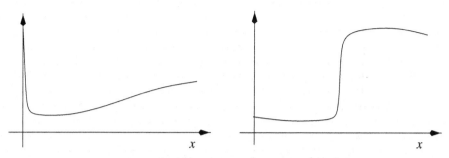

Figure 16.1 A function with a boundary layer at the origin, and one with an interior layer.

more complicated, say

$$f(x;\epsilon) = e^{-x/\epsilon}g(x) + h(x),$$

where $g(x)$ and $h(x)$ are $O(1)$ functions, then we don't need to know all the details of g and h to have a pretty good idea of what f does. When $x = O(1)$, the term $e^{-x/\epsilon}$ is so small that we can usually forget about it, and we have the *outer expansion*

$$f(x;\epsilon) \sim h(x) + \text{exponentially small correction}$$

(the exponentially small correction often goes by the catch-all name of *transcendentally small terms*). When x is small, however, we expect $g(x)$ and $h(x)$ to be close to their initial values $g(0)$ and $h(0)$, so that

$$f(x;\epsilon) \sim g(0)e^{-x/\epsilon} + h(0),$$

although here it is not quite so obvious how big the error is.

Notice that the limit as $x \to 0$ of the outer solution, $h(0)$, is not in general equal to $f(0;\epsilon)$. It is the job of the boundary layer to accommodate this discrepancy.

The real point of this discussion is not to tell us how to expand functions that we already know. It is that we can often describe a function with a boundary layer using two expansions, an outer expansion valid away from the boundary layer and an *inner expansion* valid in the boundary layer. In an application, the full function may be the solution of some horrendously difficult problem;[1] but if we can identify where the boundary layers are we may be able to formulate simpler problems for the inner and outer expansions and thereby obtain a good description of the full solution without actually having to find it as a whole.

Before we plunge into a series of examples, we should first look a little more closely at the question of how we 'join up', or *match*, the inner and outer expansions. We'll do this first assuming we know the full function, so that we are just verifying that we can do it. Later, we

---

[1] The Navier–Stokes equations spring to mind: the viscous boundary layer in high Reynolds number flow is an early and classic example of the technique in action.

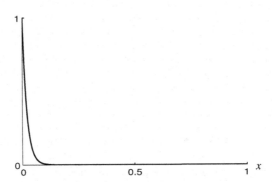

Figure 16.2 The function $e^{-x/\epsilon}$, $\epsilon = 0.02$.

will use the matching to convey information between the two regions so as to complete the solution. For example, we may have undetermined constants as the result of solving a differential equation, and if so we fix these by matching.

16.2.1 Matching

There are various ways of joining together inner and outer expansions, and it is in the nature of the subject that no way is universal; there are examples for which any method fails. However, the Van Dyke rule, which we now discuss, is as robust as any and it certainly works for all the problems treated in this book.[2]

I strongly suggest that you work through the details.

Let us return to the example we have just discussed, but with a slightly more complicated function

$$f(x;\epsilon) = e^{-x/\epsilon} g(x;\epsilon) + h(x;\epsilon),$$

where $g(x;\epsilon)$ and $h(x;\epsilon)$ have regular expansions

$$g(x;\epsilon) \sim g_0(x) + \epsilon g_1(x) + \cdots, \qquad h(x;\epsilon) \sim h_0(x) + \epsilon h_1(x) + \cdots,$$

valid in the whole domain, say the interval [0, 1]. For example, take

$$f(x;\epsilon) = e^{-x/\epsilon}(1+x) + x + e^{\epsilon x},$$

so that here

$$g(x;\epsilon) = 1+x, \qquad h(x;\epsilon) = x + e^{\epsilon x} \sim 1 + x + \epsilon x + \tfrac{1}{2}\epsilon^2 x^2 + \cdots.$$

The function $f(x;\epsilon)$ is plotted as the solid curve in Figure 16.3. Now for any fixed value of $x > 0$, the exponential term $e^{-x/\epsilon}$ tends to zero so

---

[2] A popular alternative is matching via an 'intermediate region' between the boundary layer and the outer solution; see Exercise 3. Much cruder is 'patching', in which we simply equate the values of the inner and outer expansions at a set value of (say) x: this cannot inform us about the structure of the problem but it can be a useful part of a numerical attack.

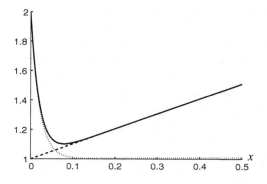

Figure 16.3 Solid line:
$e^{-x/\epsilon}(1+x) + x + e^{\epsilon x}$,
$\epsilon = 0.02$. Dashed line:
$1 + x + \epsilon x + \frac{1}{2}\epsilon^2 x^2$. Dotted
line: the inner expansion
$1 + e^{-X} + \epsilon(X e^{-X} + X) + \epsilon^2 X$ plotted on the outer scale.

fast as $\epsilon \to 0$ that we can neglect it by comparison with any power of ϵ.
The outer expansion for this example is therefore

$$f(x;\epsilon) \sim 1 + x + \epsilon x + \tfrac{1}{2}\epsilon^2 x^2 + \cdots ,$$

and it is valid provided that $x \gg O(\epsilon)$. However, it does not give a good
picture of what happens near $x = 0$; and indeed, its limit as $x \to 0$, which
is 1, is not equal to $f(0,\epsilon) = 2$. This is the dashed curve in Figure 16.3.

This is an example of the failure of the limits $x \to 0$ and $\epsilon \to 0$ to commute.

We can investigate the behaviour near the origin more closely by
rescaling x in the boundary layer, writing $x = \epsilon X$; the variable X is
often known as a *boundary layer* or *inner* variable. This gives

$$f(x;\epsilon) = F(X;\epsilon)$$
$$= g(\epsilon X)e^{-X} + h(\epsilon X).$$

Now it should be safe to construct a regular expansion of $g(\epsilon X)$ and
$h(\epsilon X)$, to give

$$F(X;\epsilon) \sim e^{-X}\left(g(0) + \epsilon X g'(0) + \cdots\right) + h(0) + \epsilon X h'(0) + \cdots .$$
$$= F_0(X) + \epsilon F_1(X) + \cdots .$$

This is the *inner expansion*. For our example, we have

$$F(X;\epsilon) = e^{-X}(1 + \epsilon X) + \epsilon X + e^{\epsilon^2 X}$$
$$\sim 1 + e^{-X} + \epsilon(X e^{-X} + X) + \epsilon^2 X + \cdots .$$

This is the dotted curve in Figure 16.3.

How should we achieve the joining of the inner and outer expansions?

Van Dyke's matching principle. Van Dyke's matching rule is a way
of achieving the joining up. It is stated as follows.

> The *m*-term inner expansion of the *n*-term outer expansion
>
> matches with
>
> the *n*-term outer expansion of the *m*-term inner expansion.

What on earth does this tell us? It says that we must do the following.

1. Construct n terms of the outer expansion in terms of the outer variable x. That is, expand up to the first n of the gauge functions (for example, powers of ϵ).
2. Rewrite this expansion in terms of the inner variable X.
3. Expand again in terms of the gauge functions for the inner expansion. (These are often, but not always, the same as the outer gauge functions.)
4. Retain the first m terms.

This constructs the first line of the matching principle above. Then, proceed the other way round:

1. Construct m terms of the inner expansion in terms of the inner variable X.
2. Rewrite this expansion in terms of the outer variable x.
3. Expand again in terms of the gauge functions for the outer expansion.
4. Retain the first n terms.

That gives the third line of Van Dyke's rule. Finally, these two expansions should *match*: that is, they should represent the same function. Notice that you simply swap the positions of 'm-term inner' and 'n-term outer' in stating the two parts of the rule, so it does express a kind of commutativity.

Let's see how this works for the example above. Recall that the outer expansion is

$$f(x;\epsilon) \sim f(x;\epsilon) \sim 1 + x + \epsilon x + \tfrac{1}{2}\epsilon^2 x^2 + \cdots$$

and the inner expansion is

$$F(X;\epsilon) \sim 1 + e^{-X} + \epsilon(Xe^{-X} + X) + \epsilon^2 X + \cdots.$$

Start, as always, with the easiest problem.

One-term outer and inner expansions, $m = n = 1$. The one-term ($n = 1$) outer expansion is

$$1 + x.$$

In inner variables, this is

$$1 + \epsilon X.$$

Expanded to one term ($m = 1$), which means that we truncate it by leaving out all smaller terms, we have

$$1.$$

Going the other way, the one-term ($n = 1$) inner expansion is

$$1 + e^{-X}.$$

In outer variables, this is

$$1 + e^{-x/\epsilon}$$

which, expanded to one term, is

$$1,$$

because the exponential is small. The two expansions do indeed match. To see this, let's do two terms in each expansion:

$1 + x + \epsilon x$	two-term outer ...
$1 + \epsilon X + \epsilon^2 X^2$	... in inner variables ...
$1 + \epsilon X$	... expanded to two terms;
$1 + e^{-X} + \epsilon(Xe^{-X} + X)$	two-term inner ...
$1 + e^{-x/\epsilon} + x(1 + e^{-x/\epsilon})$	... in outer variables ...
$1 + x$	... expanded to two terms.

Exercise: do the cases $m = 1$, $n = 2$, and $m = 2$, $n = 1$. Then do $m = n = 3$.

Again, the two sides of Van Dyke agree.

16.3 Examples from ordinary differential equations

For most of us, the first encounter with boundary layers comes via an ordinary differential equation, and there are many fascinating problems arising in this area. Boundary layers commonly occur when a small parameter in a problem multiplies the highest derivative, because then that derivative can become large without imbalancing the equation. Typically, the outer expansion, in which the higher derivative term is neglected, fails to satisfy one or more of the boundary conditions. Rescaling in the boundary layer allows us to rectify this situation (and should lead to a simpler 'inner' problem). A rather similar situation was discussed earlier, in the context of the quadratic equation $\epsilon x^2 + x - 1 = 0$; there dropping the term of highest degree lost us one of the roots.

Our first example is a first-order differential equation where, having constructed the inner and outer expansions, they match automatically. In the second example, a second-order equation, we have to go through the matching process in order to determine some of the constants in the solution, so it plays a vital role.

A first-order equation.   Consider the first-order equation

$$\epsilon \frac{dy}{dx} + y = \sin x, \qquad x > 0, \quad y(0) = 1.$$

Or a sensible guess.

It is easy enough to solve by an integrating factor:

$$y = \frac{\sin x - \epsilon \cos x}{1 + \epsilon^2} + \frac{\epsilon}{1 + \epsilon^2} e^{-x/\epsilon},$$

and this solution can be used to verify all the results of the approximate expansion. For the outer expansion, we try a regular perturbation series

$$y \sim y_0 + \epsilon y_1 + \cdots ,$$

to find that

$$\epsilon \left(\frac{dy_0}{dx} + \epsilon \frac{dy_1}{dx} + \cdots \right) + y_0 + \epsilon y_1 + \cdots = \sin x.$$

Successive terms are just read off:

$$\text{at } O(1), \qquad y_0 = \sin x,$$

$$\text{at } O(\epsilon), \qquad y_1 = -\frac{dy_0}{dx} = -\cos x$$

and so on. However, this expansion does not satisfy the initial condition, the danger signal being the ϵ multiplying the highest derivative, as a consequence of which we never have to solve a differential equation but merely need to differentiate functions that are already known.

The answer is ϵ.

Now for the inner expansion. We don't, at this stage, know how big it should be, although we can have a pretty good guess, as it will be determined by a balance between the omitted highest derivative and some other term. Still, suppose that it is a region near $x = 0$ of size $O(\delta)$, where $\delta \ll 1$. So, write

I am not going to do this procedure in any of the other examples, leaving it to you.

$$x = \delta X, \qquad y(x) = Y(X).$$

Then the rescaled (inner) problem is

$$\frac{\epsilon}{\delta} \frac{dY}{dX} + Y = \sin \epsilon X, \qquad X > 0, \quad Y(0) = 1.$$

Now, in the rescaled differential equation the term Y is a priori $O(1)$ and $\sin \epsilon X$ is $O(\epsilon)$. So the only way to get a balance is to take $\delta = \epsilon$, leaving

$$\frac{dY}{dX} + Y = \sin \epsilon X$$

$$\sim \epsilon X - \cdots .$$

Solving this by a regular expansion, the first two terms in the inner expansion are

$$Y \sim e^{-X} + \epsilon(X - 1 + e^{-X}) + \cdots .$$

Now for the matching, beginning with the one-term outer expansion. This is $\sin x$, which in inner variables is $\sin \epsilon X \sim 0 + O(\epsilon)$. So, the one-term inner of the one-term outer is zero. Now dig out the the one-term inner, which is $e^{-X} = e^{-x/\epsilon}$ in outer variables. The one-term outer is also zero, and so matching works at this order. If we take two terms, the two-term outer is $\sin x - \epsilon \cos x$, which to two terms in the inner variable is $\epsilon(X - 1)$ and matches with what is left of the two-term inner $e^{-X} + \epsilon(X - 1 + e^{-X})$ after it has been written in outer variables and expanded to two terms (so the exponentials both go). Higher-order matching can be carried out in a similar fashion.

The n-term outer of the one-term inner is zero for all n if the gauge functions are ϵ^n; why?

A second-order equation. In second-order problems, the matching can convey useful information from the outer expansion to the inner one or vice versa. Suppose for example that

$$\epsilon \frac{d^2 y}{dx^2} + \frac{dy}{dx} = \frac{1}{1 + x^2},$$

with $y(0) = 0$ and $y \to 1$ as $x \to \infty$, a two-point boundary value problem on an infinite interval. We can solve it explicitly, with some work, but that is weightlifting.

Expanding in powers of ϵ, the leading-order outer problem is

$$\frac{dy_0}{dx} = \frac{1}{1 + x^2},$$

from which

$$y_0 = \tan^{-1} x + c_0,$$

where c_0 is a constant. Now, we can satisfy the boundary condition at infinity by choosing $1 = \frac{1}{2}\pi + c_0$, so that $c_0 = 1 - \frac{1}{2}\pi$, but then y does not vanish at $x = 0$. However, rescaling $x = \epsilon X$ near $x = 0$ gives the inner problem

Check that this is the correct scaling by looking at the possible balances in the equation.

$$\frac{d^2 Y}{dX^2} + \frac{dY}{dX} = \frac{\epsilon}{1 + \epsilon^2 X^2},$$

with $Y(0) = 0$. The leading-order solution is

$$Y_0(X) = C_0(1 - e^{-X})$$

where C_0 is still unknown. However, matching with the outer solution is achieved at this order if $C_0 = y_0(0) = c_0 = 1 - \frac{1}{2}\pi$: the matching is essential to specify the inner solution fully. Higher-order terms in the expansion will throw up further undetermined constants, which will be determined by higher-order matching.

An important practical point to note is that the inner problem is effectively solved on an infinite domain. Suppose that the outer problem has a

This translates into replacing $\frac{1}{2}\pi$ by $\frac{1}{4}\pi$ above.

boundary condition specified at $x = 1$, say $y(1) = 1$. In inner variables, this translates into a condition at the far end $X = 1/\epsilon$. However, this equals infinity for all practical purposes, and the condition at $X = 1/\epsilon$ is replaced by a matching condition at infinity (in the inner region).

We have just scratched the surface of the many fascinating examples that have been devised and investigated for ordinary differential equations alone. For more details, see books such as [27, 33, 36, 49]. We now return to an earlier case study, before moving on to look at some partial differential equations.

16.4 Case study: cable laying

Recall that in our case study of laying an undersea cable (see Section 4.3), we wrote down a model in which the angle θ between the cable and the horizontal satisfies

$$\epsilon \frac{d^2\theta}{ds^2} - F^* \sin\theta + (F_0 + s)\cos\theta = 0,$$

where F_0 is an unknown constant (equal to the dimensionless vertical force on the sea bed at the point where the cable touches down), F^* is a known dimensionless constant and ϵ is a small dimensionless constant measuring the relative importance of the cable rigidity and the cable weight. The boundary conditions for the problem are

$$\theta = 0, \quad \frac{d\theta}{ds} = 0 \quad \text{at} \quad s = 0,$$

and θ is prescribed at $s = \lambda$.

This problem is ideally suited to a boundary layer expansion, with a small parameter multiplying the highest derivative. The leading-order outer solution $\theta_0(s)$ satisfies

$$\tan\theta_0 = \frac{s + F_0}{F*},$$

but it does not satisfy the conditions at $s = 0$. Before investing too much energy in it, let us look at the possibility of a boundary layer near $s = 0$. Clearly θ is small in such a layer, and a little playing around, starting with the obvious guess that the boundary layer is where $s = O(\epsilon^{1/2})$, suggests the scalings

$$s = \epsilon^{1/2}\xi, \qquad \theta = \epsilon^{1/2}\phi, \qquad F_0 = \epsilon^{1/2} f_0,$$

following which the leading-order term in a regular expansion for ϕ satisfies

$$\frac{d^2\phi_0}{d\xi^2} - F^*\phi_0 + s + f_0 = 0.$$

Because there can be no exponentially growing term, the two boundary conditions at $\xi = 0$ tell us both ϕ_0 and f_0:

$$\phi_0(\xi) = \frac{\xi}{F^*} - \frac{1}{(F^*)^{3/2}}\left(1 - e^{-\xi(F^*)^{1/2}}\right), \qquad f_0 = -\frac{1}{(F^*)^{1/2}}. \quad (16.1)$$

It is easy to see that this matches with the outer solution since, substituting for F_0 and writing $s = \epsilon^{1/2}\xi$ in our expression for θ_0, we see that the inner limit of the outer solution is

$$\epsilon^{1/2}\frac{\xi + f_0}{F^*},$$

which is just the same as the outer limit of the inner solution, obtained by neglecting the exponential term in (16.1).

We have also learned that F_0 is small, so away from the boundary layer the outer solution satisfies

$$\tan\theta_0 = \frac{s}{F^*}.$$

See the exercises at the end of the chapter for a demonstration that the solution of this equation is a catenary, as we might expect if the bending stiffness of the cable is negligible.

16.5 Examples for partial differential equations

Although ordinary differential equations lead to many interesting boundary layers, the technique has its greatest impact when applied to partial differential equations, simply because they are so much more difficult. The original boundary layer was Prandtl's analysis of the high-Reynold's-number flow of a viscous fluid past a flat plate, and fluid mechanics remains a prolific source of these problems. However, we use simpler examples from heat flow and potential theory to illustrate the ideas involved.

16.5.1 Large-Peclet-number advection–diffusion past an infinite flat plate

Suppose that a liquid flows with velocity $(U, 0)$ past a flat plate along the positive x axis, that the temperature at infinity is zero and that the plate is heated to a temperature $T(x)$. Finding the heat transfer from the plate to the fluid is a prototype problem for many practical situations. Recall from Chapter 3 that the relevant dimensionless model for the temperature $u(x, y)$ is

$$\text{Pe}\frac{\partial u}{\partial x} = \frac{\partial^2 u}{\partial x^2} + \frac{\partial^2 u}{\partial y^2},$$

with the boundary condition

$$u(x, 0) = T(x), \qquad y = 0, \quad x > 0,$$

and the condition that $u \to 0$ at infinity. By symmetry, we need solve only for $y > 0$.

Suppose that the Peclet number is large (advection-dominated heat transfer), so that we can write

Note the preference for writing a large parameter (Pe) in terms of a small one. The square of ϵ is taken for convenience, to avoid the square root of $1/$Pe.

$$\text{Pe} = \frac{1}{\epsilon^2},$$

where $0 < \epsilon \ll 1$. Thus

$$\frac{\partial u}{\partial x} = \epsilon^2 \left(\frac{\partial^2 u}{\partial x^2} + \frac{\partial^2 u}{\partial y^2} \right),$$

with

$$u(x, 0) = T(x), \qquad y = 0, \quad x > 0, \qquad u \to 0 \quad \text{at infinity.}$$

What is the structure of the problem?

First we try a regular expansion in powers of ϵ, $u \sim u_0 + \epsilon u_1 + \cdots$. The lowest-order equation is

$$\frac{\partial u_0}{\partial x} = 0,$$

which, with the fact that u vanishes at upstream infinity, means that $u_0 = 0$. This in turn means that there can be no 'source' term for u_1, and it too vanishes identically, as do all the higher-order terms in the expansion. As in the ordinary-differential-equation examples, the regular perturbation expansion fails to satisfy both boundary conditions (at infinity and on the x-axis). The danger signal is, as before, that the small parameter ϵ^2 multiplies the highest derivatives in the advection–diffusion equation.

We rectify this situation with a boundary layer near the plate; it is known as a *thermal boundary layer*. The scaling we need is $y = \epsilon Y$, $u(x, y) = U(x, Y)$, leading directly to

$$\frac{\partial U}{\partial x} = \epsilon^2 \frac{\partial^2 U}{\partial x^2} + \frac{\partial^2 U}{\partial Y^2}.$$

The first term in a regular perturbation, $U_0(x, Y)$, satisfies

$$\frac{\partial U_0}{\partial x} = \frac{\partial^2 U_0}{\partial Y^2}, \qquad U_0(x, 0) = T(x), \qquad U_0(x, Y) \to 0 \text{ as } Y \to \infty$$

(the last condition is a simple matching condition with the outer solution: again, note how the matching replaces a boundary condition far away from the boundary in the inner region). This is a *parabolic* equation for U_0, in which x, which measures distance along the plate, plays the role of time and Y is the space variable; with the initial condition $U(0, Y) = 0$, also obtained by matching back to the outer expansion, its solution can be written down as an integral using Duhamel's principle. When $T(x) = 1$ there is a similarity solution $U_0(x, Y) = F(Y/\sqrt{2x})$, where

$$\frac{\mathrm{d}^2 F}{\mathrm{d}z^2} + z \frac{\mathrm{d}F}{\mathrm{d}z} = 0, \qquad z = \frac{Y}{\sqrt{2x}},$$

so that

$$U_0(x, Y) = \int_{Y/\sqrt{2x}}^{\infty} e^{-s^2/2} \, \mathrm{d}s.$$

This formula (and the Duhamel solution for non-constant plate temperature) shows that the thermal boundary layer grows as the square root of distance along the plate.

Further aspects of this problem, including a direct comparison with the exact solution when $T(x) = 1$, are dealt with in Exercise 7. A more complicated example, large-Peclet-number flow past a cylinder, is the subject of Exercise 8; it explains how a balance between conduction and advection leads to the amplification of conduction known as wind-chill.

16.5.2 Traffic flow with small anticipation

In Chapter 8 we looked at the simple model

$$\frac{\partial \rho}{\partial t} + \frac{\partial (\rho U(\rho))}{\partial x} = 0 \tag{16.2}$$

for the density $\rho(x, t)$ of traffic travelling along a road. We saw that shocks, described by curves $x = S(t)$, can form and that their speed is given by the Rankine–Hugoniot relation

$$\frac{\mathrm{d}S}{\mathrm{d}t} = \frac{[\rho U(\rho)]_-^+}{[\rho]_-^+}.$$

We also suggested that if drivers anticipate the traffic density, rather than simply responding to its local value, the model

$$\frac{\partial \rho}{\partial t} + \frac{\partial (\rho U(\rho))}{\partial x} = \epsilon \frac{\partial}{\partial x} \left(\rho \frac{\partial \rho}{\partial x} \right) \tag{16.3}$$

might be appropriate. In Exercise 5 on p. 113 you showed that travelling waves of this equation moving with speed V, in which ρ changes from ρ_- at $\xi = x - Vt = -\infty$ to ρ_+ at $\xi = \infty$, also lead to the Rankine–Hugoniot condition in the form

$$V = \frac{[\rho U(\rho)]_{-\infty}^{\infty}}{[\rho]_{-\infty}^{\infty}}.$$

We can now tie these results together using matched expansions for (16.3) rather than Rankine–Hugoniot for (16.2).

The idea is to treat the shock as an interior layer – a boundary layer that is not fixed onto a boundary – for equation (16.3). A regular expansion of the solution to (16.3) away from the shock (the outer expansion) simply leads to (16.2) at leading order. There will be a shock at an as yet unknown location $x = S(t)$.[3] Near this location, introduce the inner variable

The scaling has to be determined: try $x = S(t) + \delta X$ and show that the sensible choice is $\delta = \epsilon$.

$$x = S(t) + \epsilon X, \quad \text{so that} \quad \frac{\partial}{\partial x} \leftrightarrow \frac{1}{\epsilon}\frac{\partial}{\partial X}, \quad \frac{\partial}{\partial t} \leftrightarrow \frac{\partial}{\partial t} - \frac{1}{\epsilon}\frac{\mathrm{d}S}{\mathrm{d}t}\frac{\partial}{\partial X}.$$

With $\rho(x, t) = R(X, t)$, the inner problem is

$$\epsilon\frac{\partial R}{\partial t} - \frac{\mathrm{d}S}{\mathrm{d}t}\frac{\partial R}{\partial X} + \frac{\partial(RU(R))}{\partial X} = \frac{\partial}{\partial X}\left(R\frac{\partial R}{\partial X}\right).$$

Once again, in the spirit of matched expansions, we solve this equation for $-\infty < X < \infty$, with matching conditions at $X = \pm\infty$. When we construct the leading-order term R_0 in a regular expansion, the time derivative does not feature. That is, time only appears as a parameter, through $\mathrm{d}S/\mathrm{d}t$; the solution is 'slowly varying' in t on the inner scale, although on the outer ($O(1)$) scale t is fully involved.

The final piece of the jigsaw is the matching procedure: we need to impose $R_0 \to \rho_\pm$ as $X \to \pm\infty$, where $\rho_\pm$ are the limiting (leading-order) outer values of ρ on either side of the shock. Putting the whole lot together, this gives precisely the Rankine–Hugoniot condition for $\mathrm{d}S/\mathrm{d}t$: we see that we can interpret the inner layer as a smoothed-out shock.

16.5.3 A thin elliptical conductor in a uniform electric field

Sometimes it is the geometry of the solution domain, rather than the differential equation itself, that leads to a matched-expansion problem. As a simple example, suppose that, in two dimensions, a perfect conductor in the shape of the ellipse

The factor 2 is for later convenience.

$$x^2 + \frac{y^2}{2\epsilon^2} = 1$$

is placed in a uniform electric field $(E, 0)$, for which the potential (without the ellipse) is $\phi(x, y) = -Ex$. We want to calculate the electric potential when the ellipse is present.

---

[3] Technically, we should expand $S(t)$ in terms of ϵ, but we are not going to calculate to the order of accuracy that would warrant this step.

Figure 16.4 Inner regions for a thin ellipse in a uniform electric field.

The problem to solve is $\nabla^2 \phi = 0$ outside the ellipse, with $\phi = 0$ on the ellipse and $\phi \sim -Ex + O(1)$ at infinity. As it happens, it can be solved explicitly by conformal maps, but suppose that we did not know this. What can we do when $\epsilon \ll 1$?

What should we expect? When $\epsilon = 0$, the ellipse is a conducting plate from $(-1, 0)$ to $(1, 0)$. In this case, we should see the field lines bent towards it, because it is a short circuit for the field, and in particular the field should be very high (indeed, singular) at the ends. In short, the plate acts as a lightning conductor does, collecting the field at one end and ejecting it at the other.[4] When the ellipse is thin but not a plate, we expect to see something similar.

First, construct a regular expansion valid away from the ellipse and, in particular, away from its tips $x = \pm 1$. The leading-order problem in this expansion is to solve $\nabla^2 \phi_0$ away from the slit from $(-1, 0)$ to $(1, 0)$, with $\phi_0 = 0$ on this slit and $\phi_0 \to -Ex$ at infinity. The solution is found by standard complex-variable methods to be

> Note that the condition $\phi = 0$ on the ellipse has been linearised onto the x-axis, just as in water-wave problems.

$$\phi_0(x, y) = -E \, \Re \big(z^2 - 1 \big)^{1/2},$$

where $z = x + iy$ and the branch cut for the square root is taken along the slit, so that $\big(z^2 - 1 \big)^{1/2} \sim z$ at infinity. This is a splendid approximation (it is the exact potential for a zero-thickness conductor) but it is singular at $x = \pm 1$. For example, as $z \to 1$, $\phi \sim -E\sqrt{2} \, \Re (z - 1)^{1/2}$ and hence $\mathbf{E} = -\nabla \phi$ is infinite.

To investigate further, look near the right-hand end by setting $x = 1 + \epsilon^2 X$, $y = \epsilon^2 Y$; see Figure 16.4. In these inner variables, the tip of the ellipse becomes, approximately, the parabola $Y^2 = -4X$ (the factor 4 is the reason for the factor 2 in the equation of the ellipse). The inner problem for $\Phi(X, Y) = \phi(x, y)$ is to solve Laplace's equation outside this curve, with $\Phi = 0$ on the approximate parabola and, from matching, $\Phi \sim -E\sqrt{2} \, \Re Z^{1/2} + o(1)$ at infinity, where $Z = X + iY$. This latter is a matching condition with the inner limit of the outer solution ϕ_0. There are various ways to solve this problem; the use of parabolic coordinates, as introduced for heat transfer from a flat plate, is one and conformal

> You should derive them by putting $x = 1 + \delta X$, $y = \delta Y$ and looking for balances in the equation of the ellipse. Note that x and y are scaled in the same way, so that Laplace's equation is not altered. There is no reason for it to be changed; only a long-thin geometry, or some other external reason for differential scaling, would have that effect.

---

[4] Aeroplanes have small spikes in strategic places to help lightning on its way after it has struck the fuselage. Passengers inside are protected by the Faraday cage effect of the metal skin of the aircraft.

mapping is another. They all lead to the solution

$$\Phi_0(X, Y) = -E\sqrt{2}\left(\Re(Z+1)^{1/2} - 1\right)$$

and we see that the apparent singularity is indeed resolved by the inner region, since the branch-point for $(Z+1)^{1/2}$ is safely inside the conductor.

Notice that this procedure would work for any conductor whose shape is $y^2 = \epsilon^2 f(x)$, provided that the ends are approximately parabolic. We only see the effect of the details of the shape at $O(\epsilon)$ in the outer expansion. Expanding to this order also brings in eigensolutions with $(z^2 - 1)^{-1/2}$ singularities at $z = \pm 1$. Although ostensibly worrying, they simply tell us that the outer expansion breaks down near the tips, the symptom being that when $|z^2 - 1| = O(\epsilon)$ the second term in the outer expansion, which is $O(\epsilon(z^2 - 1)^{-1/2})$, is the same size as the first term, $-E(z^2 - 1)^{1/2}$. There is no contradiction, and the constants that multiply the eigensolutions can be determined by matching with the inner region.

16.6 Exercises

1 **A simple expansion near a singularity.**   Consider the function

$$f(x; \epsilon) = \frac{1}{x + \epsilon}$$

as $\epsilon \to 0$. If $x = O(1)$, expand by the binomial theorem to show that

$$f(x; \epsilon) \sim \frac{1}{x} - \frac{\epsilon}{x^2} + \cdots .$$

Of course, the series in powers of ϵ/x does not converge if $|x| < \epsilon$ (what is the correct series representation in this case?), but let us pretend we don't know this.

Clearly this expansion is invalid near $x = 0$, as the first term is singular and the second term is larger than the first. Rescale $x = \epsilon X$ to find a valid approximation for small x. (This technique is useful for integrals of the form $\int_0^1 g(x)/(x + \epsilon)\, dx$.)

2 **Expanding a function.**   Find inner and outer expansions, correct to $O(\epsilon^2)$, for the function

$$f(x; \epsilon) = \frac{e^{-x/\epsilon}}{x} + \frac{\sin \epsilon x}{x} - \epsilon \coth \epsilon x.$$

3 **Matching by intermediate regions.**   The idea behind this matching principle is to choose a range of values of the independent variable(s) that is large compared with the boundary layer but small compared with the outer region. For example, in the problem described in subsection 16.2.1, the intermediate region might be $x = O(\epsilon^{1/2})$. Then both inner and outer expansions are written in terms of an intermediate variable $x = \epsilon^{1/2}\xi$, re-expanded as asymptotic series in this

new variable, and compared; they should be the same. Carry out this procedure for the example of subsection 16.2.1,

$$f(x;\epsilon) = e^{-x/\epsilon}(1+x) + x + e^{\epsilon x},$$

for which the outer expansion is

$$f(x;\epsilon) \sim 1 + x + \epsilon x + \tfrac{1}{2}\epsilon^2 x^2 + \cdots$$

and the inner expansion is

$$F(X;\epsilon) \sim 1 + e^{-X} + \epsilon(Xe^{-X} + X) + \epsilon^2 X + \cdots.$$

Set $x = \epsilon^{1/2}\xi$, $X = \epsilon^{-1/2}\xi$, and show that the intermediate expansion of both functions is

$$1 + \epsilon^{1/2}\xi + \epsilon^{3/2}\xi + \cdots.$$

4 **A two-point boundary value problem.** Use matched asymptotic expansions to find an approximate solution to the two-point boundary value problem

$$\epsilon\frac{d^2 y}{dx^2} + x\frac{dy}{dx} + y = 0, \quad 1 < x < 2,$$

$$y(1) = 0, \quad y(2) = 1, \quad 0 < \epsilon \ll 1.$$

How can you tell that there is a boundary layer at $x = 1$ but not at $x = 2$? What happens if ϵ is small and *negative*?

5 **An artificial example.** Find an approximate solution to

$$\epsilon\frac{d^2 u}{dx^2} + \frac{du}{dx} = \frac{u + u^3}{1 + 3u^2}, \quad u(0) = 0, \quad u(1) = 1.$$

First find the outer solution: which boundary condition will it satisfy and why? Then find the solution in the boundary layer near $x = 0$ and carry out the matching. (The right-hand side of this example is selected so that it (a) gives an easy solution to the outer problem and (b) is uniformly Lipschitz in u, and there is no question of blow-up. I very much doubt that the full equation can be solved explicitly, but the approximation tells you all about the structure of the solution.)

You may need to convince yourself by drawing a graph that the equation $u + u^3 = a$ has a unique real root for each a.

6 **Cable laying with small bending stiffness.** In Section 16.4, we derived the equation

$$\tan\theta_0 = \frac{s}{F^*}$$

for the leading order gradient of a cable with small bending stiffness. Remembering that $\tan\theta_0 = dy/dx = y'$ and that

$$\frac{ds}{dx} = \left(1 + (y')^2\right)^{1/2},$$

show that the solution consistent with $y(0) = 0$ and $y'(0) = 0$ (because $\theta_0(0) = 0$) is

$$y = F^* \left(\cosh(x/F^*) - 1 \right).$$

Deduce that the ship's dimensionless position is given by

$$x^* = F^* \cosh^{-1}(1 + 1/F^*)$$

and that the tensioner angle θ^* and dimensionless thrust F^* are related by

$$\tan^2 \theta^* = \frac{1 + 2F^*}{(F^*)^2}.$$

7 Large-Peclet-number flow past a flat plate heated to a constant temperature. Suppose that $u(x, y)$ satisfies

$$\frac{\partial u}{\partial x} = \epsilon^2 \left(\frac{\partial^2 u}{\partial x^2} + \frac{\partial^2 u}{\partial y^2} \right),$$

with

$$u(x, 0) = 1, \quad y = 0, x > 0, \quad u \to 0 \quad \text{at infinity.}$$

What is the associated conformal map? Show that it maps the right-hand half plane $\xi > 0$ onto the flow domain. Because it is a conformal change of variables, the Laplacian transforms nicely: how?

Introducing parabolic coordinates ξ, η satisfying

$$x + iy = -(\xi + i\eta)^2, \quad \xi > 0, \quad -\infty < \eta < \infty,$$

show that the curves ξ = constant are parabolae wrapped around the positive x-axis, that the curves η = constant are parabolae wrapped around the negative x-axis and that the two families of curves are orthogonal. Show that the problem becomes

$$-\xi \frac{\partial u}{\partial \xi} + \eta \frac{\partial u}{\partial \eta} = \epsilon^2 \left(\frac{\partial^2 u}{\partial \xi^2} + \frac{\partial^2 u}{\partial \eta^2} \right), \quad \xi > 0,$$

with $u(0, \eta) = 1$ and $u \to 0$ at infinity. Further, show that the solution takes the form $u(\xi, \eta) = f(\xi)$ and find f. When ϵ is small, eliminate η to show that $\xi \sim \epsilon y/2\sqrt{x}$ and hence confirm the correctness of the thermal boundary layer solution of subsection 16.5.1.

Now return to the problem of subsection 16.5.1, with $u = T(x)$ on $y = 0, x > 0$, where $T(x)$ is smooth and $T(0) \neq 0$. Show that, in addition to the thermal boundary layer described in the text, there is a small region centred at the tip of the plate $(0, 0)$ in which both x and y are $O(\epsilon)$ and for which the leading-order problem is a version of the first part of this question.

8 Wind-chill. Consider the large-Peclet-number version of the advection–diffusion problem of subsection 3.1.1, steady heat transfer from a circular cylinder in a potential flow with velocity **u**, with $\mathrm{Pe} = 1/\epsilon^2$, $\epsilon \ll 1$. In plane polar coordinates r, θ, we have $\mathbf{u} = \nabla(\cos\theta(r - 1/r))$, and the scaled temperature $T(r, \theta)$ satisfies

$$\cos\theta\left(1 - \frac{1}{r^2}\right)\frac{\partial T}{\partial r} - \frac{\sin\theta}{r}\left(1 + \frac{1}{r^2}\right)\frac{\partial T}{\partial \theta}$$

$$= \epsilon^2\left(\frac{\partial^2 T}{\partial r^2} + \frac{1}{r}\frac{\partial T}{\partial r} + \frac{1}{r^2}\frac{\partial^2 T}{\partial \theta^2}\right)$$

with $T = 1$ on $r = 1$ and $T \to 0$ at infinity.

    Show that the only solution for the first term in a regular expansion in the outer region (away from the cylinder) that is consistent with the condition at infinity is $T_0 = 0$. Deduce that there must be a boundary layer on the cylinder. Define an inner variable R by $r = 1 + \delta R$, where δ is small. Show that a consistent balance can be achieved in the partial differential equation if $\delta = \epsilon$.

    With this scaling, write $u(r, \theta) = U(R, \theta)$ in the boundary layer, and show that the first term $U_0(R, \theta)$ of the inner expansion satisfies

$$2R\cos\theta\,\frac{\partial U_0}{\partial R} - 2\sin\theta\,\frac{\partial U_0}{\partial \theta} = \frac{\partial^2 U_0}{\partial R^2},$$

with $U_0(0, \theta) = 1$ and $U_0 \to 0$ as $R \to \infty$.

Note that there is no $1/r$ in front of $\partial U_0/\partial\theta$: r is nearly constant in the boundary layer.

    Show that there is a similarity solution in the form

$$U_0(R, \theta) = F(Rg(\theta))$$

and, by finding the differential equation that it satisfies, that

$$g(\theta) = \frac{\sin\theta}{\sqrt{1 + \cos\theta}}.$$

Show also that $F(z)$ satisfies $F'' + zF' = 0$, and find it.

    Notes: (i) The differential equation for $g(\theta)$ is first-order and its solution contains an arbitrary constant, so that $1 + \cos\theta$ is replaced by $c + \cos\theta$. The choice $c = 1$ is motivated by matching with an $O(\epsilon) \times O(\epsilon)$ region around the upstream stagnation point in which the full problem must be solved (but in a simplified geometry): it says that as we approach this point, the solution depends on $Y/\sqrt{2X}$ in coordinates $x = 1 + \epsilon X$, $y = \epsilon Y$ centred there.

    (ii) There is also a small region near the downstream stagnation point in which the full problem must be solved. This is succeeded by a thin wake of hot liquid that carries the heat away from the cylinder.

    (iii) The total heat transfer from the cylinder is, in dimensionless terms, $O(1) \times O(1/\epsilon)$ (length $\times$ heat flux). This is large: wind-chill!

9 Potential outside an ellipse. Show that the conformal map $z = \zeta + 1/\zeta$ maps circles $|\zeta| = r, r \geq 1$, into ellipses in the z-plane, that $|\zeta| = 1$ maps to the slit from -1 to 1 along the real axis and that $\zeta \sim z$ at infinity. Show also that the real part of $W(\zeta) = -E(\zeta - 1/\zeta)$

Compare with Milne-Thomson's circle theorem for potential flow.

where $\zeta = \xi + i\eta$ is a harmonic function that vanishes on $|\zeta| = 1$ and tends to $-E\xi$ at infinity. Hence find the exact solution to the problem of a conducting ellipse in a uniform electric field. Also, find the circle that maps onto the thin ellipse of the text, and hence confirm the accuracy of the asymptotic approximation given there.

10 Logarithms.   Logarithms pose considerable difficulties in matching. The following example shows why. Let

$$f(x;\epsilon) = 1 + \frac{\log x}{\log \epsilon}, \qquad x > 0.$$

Show that its one-term outer expansion for $x = O(1)$ is $f \sim 1$.

Write $x = \epsilon X$ to focus on the boundary layer near $x = 0$. Show that the one-term inner expansion for $F(X;\epsilon) = f(x;\epsilon)$ is $F \sim 2$. Deduce that Van Dyke's matching rule fails when matching the one-term inner and outer expansions.

Show that the rule works if, when matching, we take terms involving $\log \epsilon$ together with those of $O(1)$.

(The practical cure for this failure is exactly as in the exercise, to take $\log \epsilon$ terms together with $O(1)$ terms when matching. After all, what is $\log 10^5$?)

'Instead of considering 10 as large, let's consider 10 as small.'

17

Case study: the thermistor (2)

17.1 Strongly temperature-dependent conductivity

The thermistor problem was introduced in Chapter 5. In that chapter, we wrote down a one-dimensional model for the heat and current flow in such a device, in which the dimensionless temperature $u(z, t)$, measured from room temperature, and potential $\phi(z, t)$ satisfy

$$\frac{\partial}{\partial z}\left(\sigma(u)\frac{\partial\phi}{\partial z}\right) = 0, \qquad \frac{\partial u}{\partial t} - \frac{\partial^2 u}{\partial z^2} = \gamma\sigma(u)\left|\frac{\partial\phi}{\partial z}\right|^2$$

for $0 < z < 1$. When a constant voltage is applied across the thermistor, the boundary conditions are

$$\phi = 0 \quad \text{on} \quad z = 0, \qquad \phi = 1 \quad \text{on} \quad z = 1,$$

$$\frac{\partial u}{\partial n} + \beta u = 0 \quad \text{on} \quad z = 0, 1.$$

Here β and γ are dimensionless parameters representing the heat transfer coefficient (this may be small) and the heating rate for a cold thermistor (which is large).

In this short chapter we introduce the crucial new feature that the electrical conductivity $\sigma(u)$ varies dramatically as u increases from 0 to 1. It is reasonable to write $\sigma(u)$ in the form

$$\sigma(u) = e^{-f(u)/\epsilon},$$

where the choice $f(0) = 0$ ensures that $\sigma(0) = 1$, which is the dimensionless value of $\sigma(u)$ at room temperature, where $u = 0$. The parameter

Figure 17.1 Conductivity of a thermistor, on a log scale. The units are dimensional but the graph indicates the shape of the dimensionless form of $\sigma(u)$.

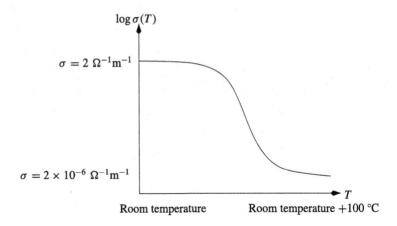

ϵ is small, a value of 10^{-1} being typical. In this way, the very rapid decrease of $\sigma(u)$ with u is translated into an $O(1)$ change in $f(u)$, and the parameter ϵ quantifies the magnitude of the change. This kind of rewriting of a rapidly varying function is common in combustion theory, where it goes by the name of large-activation-energy asymptotics. The crucial feature of this formulation is that if $f(u)$ (and hence u) changes by $O(\epsilon)$, then $\sigma(u)$ changes by $O(1)$. If $f(u)$ changes by more than $O(\epsilon)$ then the change in $\sigma(u)$ is very large. This is exactly the same behaviour as we saw in boundary layers, and we can hope that analogously we only have to consider the behaviour of $\sigma(u)$ in small regions. As we see below, this hope is realised.

Let us assume for the moment that the heat transfer coefficient β is $O(1)$, and let us look at the steady-state operation of the device, assuming that the applied voltage is sufficient to drive the temperature well above $u = 0$. Thus $u(z, t)$ and $\phi(z, t)$ are functions of z alone, which we still call $u(z)$ and $\phi(z)$. The equation for $\phi(z)$ can be integrated once to give

$$\frac{d\phi}{dz} = \frac{I}{\sigma(u(z))} = I e^{f(u(z))/\epsilon},$$

where the constant I is the current through the device. A second integration gives

$$\phi(z) = I \int_0^z e^{f(u(s))/\epsilon} \, ds,$$

and the boundary condition at $z = 1$ fixes the constant I, so that

$$\phi(z) = \frac{\int_0^z e^{f(u(s))/\epsilon} \, ds}{\int_0^1 e^{f(u(s))/\epsilon} \, ds}.$$

Then the equation for $u(z)$ is

$$\frac{d^2u}{dz^2} + \frac{\gamma e^{f(u)/\epsilon}}{\left(\int_0^1 e^{f(u(s))/\epsilon}\,ds\right)^2} = 0, \quad 0 < z < 1,$$

$$\pm\frac{du}{dz} + \beta u = 0 \quad \text{on} \quad z = 0, 1. \tag{17.1}$$

with the interesting feature that it is nonlocal.

Let us pause to think what the structure of the solution may be. If there is a hot region, this will be near the centre of the device, and $\sigma(u)$ will be small in it. This means that the term $e^{f(u)/\epsilon}\ (= 1/\sigma(u))$ is large. Now if the hot region is of size $O(1)$, so that most of the thermistor runs hot, the integral in the denominator of the last term in (17.1) is both large and comparable to $e^{f(u)/\epsilon}$; because it is squared, the result is that d^2u/dz^2 is small and u is nearly uniform. However, the boundary conditions for u only allow this to happen if u itself is small, which contradicts the assumption that u is significantly above zero.[1] We conclude that the hot region must be small, and a little experimentation shows that it should have extent $O(\epsilon)$. Furthermore, the temperature u needs to change only by $O(\epsilon)$ in this region, as larger changes mean that the numerator in (17.1) is exponentially negligible. As a consequence, all the resistance is concentrated in a thin hot layer.

Let us, then, write $u(\frac{1}{2}) = u^*$, say, for the maximum achieved value of u, and

The reason for dividing w by $f'(u^*)$ appears very shortly . . .

$$z = \tfrac{1}{2} + \epsilon\xi, \qquad u = u^* + \frac{\epsilon w(\xi)}{f'(u^*)},$$

so that we have an inner expansion in a region of size $O(\epsilon)$ near the middle of the thermistor. The first task is to find u^*, at least to an accuracy of $O(1)$. Let us write $\lambda^* = e^{f(u^*)/\epsilon}$, the maximum value of $\sigma(u)$ (which is large). Then we write

. . . as we then don't have to carry the constant $f'(u^*)$ in the exponents for the rest of the calculation.

$$e^{f(u)/\epsilon} = e^{(f(u^*)+\epsilon w+\cdots)/\epsilon}$$

$$\sim \lambda^* e^w,$$

so that (17.1) becomes

Some details to check; note especially the limits on the integral.

$$\frac{d^2w}{d\xi^2} + \frac{\gamma f'(u^*)}{\epsilon\lambda^*} \frac{e^w}{\left(\int_{-\infty}^{\infty} e^{w(\xi')}d\xi'\right)^2} = 0.$$

We conclude from this equation that we should choose u^* to be the

We could use any other $O(1)$ constant, but 1 is simplest.

---

[1] The case in which β is small is different and is dealt with in Exercise 2.

root of

$$\frac{\gamma f'(u^*)}{\epsilon \lambda^*} = 1,$$

that is,

$$\epsilon e^{-f(u^*)/\epsilon} \frac{f'(u^*)}{\epsilon} = \gamma$$

or, rearranging,

$$\frac{d\sigma}{du} = -\frac{1}{\gamma}.$$

This simple equation tells us the approximate operating point of the thermistor.

It remains to find the inner and outer solutions and to match them. With our choice of u^* the inner equation is

$$\frac{d^2 w}{d\xi^2} + c^2 e^w = 0,$$

where $c \int_{-\infty}^{\infty} e^{w(\xi')} \, d\xi' = 1$. By symmetry $dw/d\xi = 0$ at $\xi = 0$, and we expect linear behaviour as $\xi \to \pm\infty$ in order for matching with the outer expansion, which is the solution of $d^2 u/dz^2 = 0$ (as explained above, the exponential term is negligible in the outer region). The details of the solution are requested in Exercise 1.

We could go much further with this problem; in particular, we could look at the effect of adding an external resistance and we could look at the unsteady approach to the steady solution described above. These are both interesting problems, and we refer the reader to the paper [20] for more details.

17.2 Exercises

1 The inner equation.   Consider the equation

$$\frac{d^2 w}{d\xi^2} + c^2 e^w = 0,$$

where $c \int_{-\infty}^{\infty} e^{w(\xi')} \, d\xi' = 1$. Take $w = w_0$ (which gives the $O(\epsilon)$ correction to u^*) and $dw/d\xi = 0$ at $\xi = 0$. Explain why $w \le w_0$. Integrate once to show that

$$\frac{1}{2} \left(\frac{dw}{d\xi} \right)^2 + c^2 \left(e^w - e^{w_0} \right) = 0.$$

Writing

$$d\xi' = \frac{dw}{dw/d\xi'}$$

in the integral that defines c, find a relation between w_0 and c. Integrate once more to find a relation between w and ξ. Hence show that $w \sim \mp K\xi$ as $\xi \to \pm\infty$ and find the constant K.

Finally calculate the outer solution for $\frac{1}{2} < z < 1$ with $du/dz + \beta u = 0$ on $z = 1$, and match it with the inner solution (assume that $\beta = O(1)$).

2 **The case of small β.** Suppose that β and ϵ are approximately the same size, and write $\beta = \epsilon b$. Show that the steady solution is of the form

$$u(z) = u^* + \frac{\epsilon w(z)}{f'(u^*)}$$

where u^* is as before, but now the approximation holds for all z, i.e. there is no thin hot layer. Find the equation and boundary conditions satisfied by w.

What is the corresponding system of partial differential equations for ϕ and w (again, $u = u^* + \epsilon w/f'(u^*)$) for a two-dimensional model?

'We've just got to number the integers right.'

18

'Lubrication theory' analysis in long thin domains

18.1 'Lubrication theory' approximations: slender geometries

We now turn to a class of approximations that derives its name from the classical theory of lubricated bearings in machinery, associated with Reynolds (the end of the nineteenth century was a great time to be a hydrodynamicist: there were indeed giants on the earth in those days). The distinguishing feature of problems to which the lubrication approximation can be applied is that the physical domain is 'long and thin' in at least one direction, like a plate or rod. One might think of a lubrication solution as being 'all boundary layer'; moreover, the geometry tells us where the boundary layer is. The key technical step is to scale the co-ordinate(s) in the 'thin' direction differently from the rest and thereby hope to formulate a simpler problem by exploiting the smallness of the *slenderness parameter*

$$\epsilon = \frac{\text{typical thickness}}{\text{typical length}}.$$

Indeed, the full problem is usually very hard, if not impossible, to solve either explicitly or numerically, and even if we could solve it we would not necessarily gain understanding. As so often, it is usually very difficult even to prove that the lubrication approximation converges to the full solution in the appropriate limit (one variety of 'rigorous asymptotics').

This chapter is longer than most in the book. You can find excellent descriptions of most of the earlier material in other standard texts but, although lubrication expansions are common in practice and in research papers, they don't feature prominently in textbooks. We'll see applications to sheets and jets of fluids, as well as the original Reynolds problem, but we'll start with some simple problems in heat flow.

18.2 Heat flow in a bar of variable cross-section

We start with a very simple example: heat flow in a bar of variable cross-section with insulated sides. Only in very rare cases can this problem be solved exactly, and a geometry of this kind does not lend itself very readily to a simple numerical discretisation. However, we can find a very good approximation to the solution with relatively little effort.

Consider steady heat flow in the domain $0 < x < L$, $-h(x) < y < h(x)$, where

$$\epsilon = \frac{H_0}{L} \ll 1,$$

in which H_0 is a 'typical size' for the bar thickness $h(x)$; this means that we can write

$$h(x) = H_0 H(x/L)$$

for some $O(1)$ function H. Let us also impose a temperature drop from $T = T_i$ at the inlet $x = 0$ to $T = 0$ at $x = L$ and assume perfectly insulated sides. The temperature T satisfies

$$\frac{\partial^2 T}{\partial x^2} + \frac{\partial^2 T}{\partial y^2} = 0, \qquad 0 < x < L, \quad -h(x) < y < h(x),$$

with

$$T(0, y) = T_i, \qquad T(1, y) = 0,$$

and

$$\mathbf{n} \cdot \nabla T = \frac{\partial T}{\partial n} = 0 \qquad \text{on} \quad y = \pm h(x).$$

Let us first see what the answer is, by a physical argument. Then we'll derive it more mathematically. We argue as follows.

1 The heat flux is approximately unidirectional, along the bar, because no heat is lost through the sides. Thus, $T(x, y)$ is approximately independent of y (that is, it is approximately equal to its average across the bar), and we write $T(x, y) \approx T_0(x)$.

2 The heat flux $Q(x)$ across any line $x = \text{constant}$ is exactly equal to

$$\int_{-h(x)}^{h(x)} -k\frac{\partial T}{\partial x} \, \mathrm{d}y.$$

Because $\partial T/\partial y \approx 0$, we have

$$Q(x) \approx -2hk\frac{\partial T}{\partial x}, \quad \text{that is} \quad Q(x) \approx -2hk\frac{dT_0}{dx}.$$

3 Heat is conserved, so $dQ/dx = 0$, that is

$$\frac{d}{dx}\left(h(x)\frac{dT_0}{dx}\right) \approx 0,$$

and the solution of this ordinary differential equation, with $T_0(0) = T_i$, $T_0(L) = 0$, gives the leading-order behaviour of $T(x, y)$.

There is nothing at all wrong with this excellent argument. However, we would like to be able to do a bit better. We would like to know how big the error is, when the approximation is valid and, most of all, how to attack more complicated long–thin problems where the physical argument is less clear cut. This is what lubrication theory, in its general sense, does.

The crux of the lubrication approach is to exploit the slenderness by scaling x and y differently. We write

$$x = LX, \qquad y = H_0 Y;$$

that is, *we scale each variable with its own natural length scale*. This is the distinctive feature of the lubrication approach. Making the trivial scaling of T with T_i and dropping the prime on the scaled variable, we

Recall that $h(x) = H_0 H(X)$.   find that

$$\epsilon^2\frac{\partial^2 T}{\partial X^2} + \frac{\partial^2 T}{\partial Y^2} = 0, \qquad 0 < X < 1, \quad -H(X) < Y < H(X),$$

with

$$T = 1 \quad \text{on} \quad X = 0, \qquad T = 0 \quad \text{on} \quad X = 1.$$

The conditions on $y = \pm h(x)$ take a little more work to scale. We have

$$\mathbf{n} = \frac{\left(-h'(x) \pm 1\right)}{\left(1 + (h'(x))^2\right)^{1/2}} \qquad \text{for } y = \pm h(x) \text{ respectively,}$$

and so $\mathbf{n} \cdot \nabla T = 0$, namely $\pm \partial T/\partial y - h'(x)\partial T/\partial x = 0$, becomes

$$\pm\frac{\partial T}{\partial Y} - \epsilon^2 H'(X)\frac{\partial T}{\partial X} = 0 \quad \text{on} \quad Y = \pm H(X).$$

Notice that in the new variables the solution domain is $O(1) \times O(1)$, but the small parameter ϵ has been moved into the field equation and

It is relatively easy to see that the   boundary conditions.
expansion is only in powers of ϵ^2.   Let us try writing
But you may want to write it all
out to get a feel for how it works.

$$T(X, Y) \sim T_0(X, Y) + \epsilon^2 T_1(X, Y) + \cdots.$$

Then we find that

$$\frac{\partial^2 T_0}{\partial Y^2} = 0,$$

which, together with the leading-order approximate boundary condition

$$\frac{\partial T_0}{\partial Y} = 0 \quad \text{on} \quad Y = \pm H(X),$$

means that

$$T_0 = T_0(X),$$

a function of X *that is as yet unknown*. This is all the information we get from the leading-order equations and boundary conditions.

In order to find T_0, we have to look at the problem for T_1. This is

$$\frac{\partial^2 T_1}{\partial Y^2} = -\frac{\partial^2 T_0}{\partial X^2},$$

whose symmetric solution is

$$T_1(X, Y) = -\frac{Y^2}{2}\frac{\partial^2 T_0}{\partial X^2} + \text{an arbitrary function of } X.$$

The $O(\epsilon^2)$ terms in the boundary conditions are

$$\pm\frac{\partial T_1}{\partial Y} - H'(X)\frac{d^2 T_0}{dX^2} = 0 \quad \text{on} \quad Y = \pm H(X),$$

and putting these together we find that

$$-H(X)\frac{d^2 T_0}{dX^2} - H'(X)\frac{dT_0}{dX} = 0,$$

which is the same as

$$\frac{d}{dX}\left(H(X)\frac{dT_0}{dX}\right) = 0,$$

in confirmation of our intuitive argument. We have found out more, though: we now know that the error is $O(\epsilon^2)$ (and we could calculate it if we felt strong enough). We also know that the expansion will not work if any of the terms that we have assumed are $O(1)$ are in fact large. In particular, it is not guaranteed to work if $H'(X)$ is large.

Note the following features of the analysis, which are very common in this and other approximation schemes.

- The full problem (which here is an elliptic partial differential equation, Laplace's equation) has a unique solution.
- The leading-order approximate problem (here the ordinary differential equation $\partial^2 T_0/\partial Y^2 = 0$) does not have a unique solution.

Aide-arithmétique:

$$\frac{\partial^2 T_0}{\partial Y^2} + \epsilon^2\frac{\partial^2 T_0}{\partial X^2}$$

$$+\epsilon^2\frac{\partial^2 T_1}{\partial Y^2} + \cdots = 0.$$

$$\pm\frac{\partial T_0}{\partial Y} + \epsilon^2\frac{\partial T_1}{\partial Y}$$

$$-\epsilon^2 H'\frac{\partial T_0}{\partial X} + \cdots = 0.$$

Notice that the arbitrary function of X disappears; it is only found at $O(\epsilon^4)$.

- We eliminate the nonuniqueness by going to higher order, $O(\epsilon^2)$, in the expansion, and find a solvability condition that resolves the indeterminacy in the lowest-order solution. This condition is essentially the Fredholm Alternative theorem. In Exercise 7 you can see the the process of introducing indeterminacy at one order in an expansion, then resolving it at the next, for the very simple linear algebra problem

$$\begin{pmatrix} 1+\epsilon & \epsilon \\ 1-\epsilon & 2\epsilon \end{pmatrix} \begin{pmatrix} x \\ y \end{pmatrix} = \begin{pmatrix} 1 \\ 1 \end{pmatrix}.$$

18.3 Heat flow in a long thin domain with cooling

Let us look more briefly at a variation on this problem. Consider steady heat flow in the rectangular domain $0 < x < L$, $-H_0 < y < H_0$, where

$$\frac{H_0}{L} = \epsilon \ll 1,$$

again with a temperature drop from $T = T_i$ at $x = 0$ to $T = 0$ at $x = L$, but now with Newton cooling at the sides, with a background temperature of 0 and heat transfer coefficient Γ. The temperature $T(x, y)$ satisfies

$$\frac{\partial^2 T}{\partial x^2} + \frac{\partial^2 T}{\partial y^2} = 0,$$

with

$$T(0, y) = T_i, \qquad T(L, y) = 0,$$

and

$$\pm k \frac{\partial T}{\partial y} + \Gamma T = 0 \quad \text{on} \quad y = \pm H_0.$$

We can solve this problem by an eigenfunction expansion (see Exercise 2). But in a more complicated problem this might not be feasible, so let's see what the lubrication approach has to say.

We first find the answer by an elementary physical argument. *If* the heat flux is mostly in the x-direction, which is not quite so obvious as before, because now heat is lost through the sides, we can still work with the average of the temperature across the bar. And *if* the heat loss is proportional to this average temperature, a straightforward 'box' argument then shows that

$$\text{gradient of heat flux} = \text{rate of cooling},$$

or, again writing $T_0(x)$ for the approximate temperature,

$$-k \frac{d^2 T_0}{dx^2} \approx \Gamma T_0,$$

an ordinary differential equation for the approximate temperature, to be solved with $T_0 = T_i$ at $x = 0$ and $T_0 = 0$ at $x = L$. However, there is more prima facie doubt about this argument: for example, it requires heat to flow out of the bar, so that $\partial T / \partial y$ cannot vanish, while maintaining that it is all right to work with the averaged value of T. Is this consistent? We know that it works when $\Gamma = 0$, the insulated case treated above, and it would be nice to know the other values of the heat transfer coefficient for which this approximation is valid and also to know the approximate temperature profile within the material (so we can verify that it is indeed nearly one dimensional).

As above, we write

$$x = LX, \qquad y = H_0 Y$$

and scale $T(x, y)$ with T_i, to find that

$$\epsilon^2 \frac{\partial^2 T}{\partial X^2} + \frac{\partial^2 T}{\partial Y^2} = 0, \qquad 0 < X < 1, \quad -1 < Y < 1,$$

with

$$T(0, Y) = 1, \qquad T(1, Y) = 0,$$

and

$$\frac{\partial T}{\partial Y} \pm \gamma T = 0 \quad \text{on} \quad Y = \pm 1;$$

here $\gamma = \epsilon L \Gamma / k$ is the dimensionless heat transfer coefficient, also called a Biot number (see p. 41).

As the analysis below confirms, the most interesting case is when $\gamma = O(\epsilon^2)$, say $\gamma = \epsilon^2 \alpha^2$ where $\alpha^2 = O(1)$ and the square is used for later convenience, so we will proceed on this basis. If on the one hand $\gamma \ll O(\epsilon^2)$ then there is no heat loss through the sides $y = \pm H_0$ to leading order: that is, almost all the heat is conducted linearly from $X = 0$ to $X = 1$. (The small correction can be calculated by a regular perturbation expansion.) On the other hand, if $\gamma \gg O(\epsilon^2)$ then almost all the heat is lost in a small region near $X = 0$ (see Exercise 4).

As above, we expand in the form

$$T(X, Y) \sim T_0(X, Y) + \epsilon^2 T_1(X, Y) + \cdots .$$

Substituting, we have

$$\epsilon^2 \left(\frac{\partial^2 T_0}{\partial X^2} + \epsilon^2 \frac{\partial^2 T_1}{\partial X^2} + \cdots \right) + \left(\frac{\partial^2 T_0}{\partial Y^2} + \epsilon^2 \frac{\partial^2 T_1}{\partial Y^2} + \cdots \right) = 0$$

with

$$\frac{\partial T_0}{\partial Y} + \epsilon^2 \frac{\partial T_1}{\partial Y} + \cdots + \epsilon^2 \alpha^2 \left(T_0 + \epsilon^2 T_1 + \cdots \right) = 0$$

on $Y = 1$. As before,

$$T_0 = T_0(X)$$

is as yet unknown. So, we move on to the problem for T_1, which is

$$\frac{\partial^2 T_1}{\partial Y^2} = -\frac{d^2 T_0}{dX^2} \tag{18.1}$$

with

$$\frac{\partial T_1}{\partial Y} \pm \alpha^2 T_0(X) = 0 \quad \text{on} \quad Y = \pm 1. \tag{18.2}$$

The solution of (18.1) is

$$T_1(X, Y) = -\frac{Y^2}{2}\frac{d^2 T_0}{dX^2} + \text{an arbitrary function of } X,$$

and then from the boundary condition (18.2) we can find an equation for $T_0(X)$,

$$-\frac{d^2 T_0}{dX^2} + \alpha^2 T_0 = 0.$$

After undoing the scalings, this is exactly what we derived by a physical argument earlier. Incorporating the boundary conditions at $X = 0$ and $X = 1$, we have

$$T_0(X) = \frac{\sinh \alpha (1 - X)}{\sinh \alpha},$$

and of course we can construct higher-order corrections if we want.

18.4 Advection–diffusion in a long thin domain

Let's look at an extension of our previous examples, to include advection along the domain. Suppose that the material of our domain $0 < x < L$, $-H_0 < y < H_0$ is moving with speed U in the x-direction (think of modelling the heat lost by hot water flowing through a radiator). The model for the steady temperature field is

$$\rho c U \frac{\partial T}{\partial x} = k \nabla^2 T,$$

and let's take the boundary conditions

$$T = T_{\mathrm{i}} \quad \text{on} \quad x = 0,$$

modelling a specified inlet temperature,

$$T = 0 \quad \text{on} \quad y = \pm H_0,$$

modelling excellent heat transfer to the surroundings, and

$$\frac{\partial T}{\partial x} = 0 \quad \text{at} \quad x = L.$$

This last condition is not in fact an insulating boundary condition (remember that the heat flux is $\rho c T (U, 0) - k \nabla T$) but, rather, a rough guess at a plausible outflow condition; it's always hard to know what to prescribe on an outflow boundary of this kind. But in any case the message of our analysis below is that it doesn't much matter what we do at this downstream end. We can even impose the condition $T = 0$, which is physically more or less impossible to realise, and the solution upstream won't be enormously affected (this case is treated in Exercise 4).

As in the previous examples, we scale x with L, y with $H_0 = \epsilon L$ and T with T_i to get

$$\frac{\rho c U H_0^2}{kL} \frac{\partial T}{\partial X} = \frac{\partial^2 T}{\partial Y^2} + \epsilon^2 \frac{\partial^2 T}{\partial X^2},$$

with boundary conditions

$$T(0, Y) = 1, \qquad T(X, \pm 1) = 0, \qquad T(1, Y) = 0.$$

The dimensionless number

$$\mathrm{Pe} = \frac{\rho c U H_0^2}{kL}$$

is a Peclet number, measuring the relative effects of advection in the x-direction and conduction in the y-direction. We assume that it is $O(1)$ and, just for clarity, that it is equal to 1. As in the previous example, this is the only balance for which interesting action happens over all the length of our domain. Put another way, this is the balance for which the system can effectively transfer heat from the interior to the exterior.

We could of course again use an eigenfunction expansion (see Exercise 5). But again this is messy. Instead, write

$$T(X, Y) \sim T_0(X, Y) + \epsilon^2 T_1(X, Y) + \cdots,$$

and it soon emerges that the leading-order problem is

$$\frac{\partial T_0}{\partial X} = \frac{\partial^2 T_0}{\partial Y^2}, \qquad 0 < X < 1, \qquad (18.3)$$

with

$$T_0 = 0 \quad \text{on} \quad Y = \pm 1.$$

We now have to choose whether to impose $T_0 = 1$ at $X = 0$ or $\partial T_0 / \partial X = 0$ at $X = 1$. We can't have both, as (18.3) is a parabolic equation with X as the 'timelike' direction. This gives us the clue: the equation is forward parabolic from $X = 0$ and backward parabolic from

$X = 1$, and only the former gives a well-posed problem. So, we take $T_0 = 1$ at $X = 0$.

The solution with $T_0(0, Y) = 1$ is found by the standard separation-of-variables method in the form

$$T_0(X, Y) = \sum_{n=0}^{\infty} \frac{2(-1)^n}{n + \frac{1}{2}} \cos\left((n + \tfrac{1}{2})\pi Y\right) e^{-(n+1/2)^2 \pi^2 X},$$

and of course it does not satisfy the condition at $X = 1$. We deal with this by introducing a boundary layer there. We want to rescale $X - 1$ so as to bring back the neglected term $\partial^2 T_0 / \partial X^2$. So, we write $X - 1 = \delta\xi$, where $\xi < 0$ and δ is still to be found, to give

$$\frac{1}{\delta} \frac{\partial T_0}{\partial \xi} = \frac{\epsilon^2}{\delta^2} \frac{\partial^2 T_0}{\partial \xi^2} + \frac{\partial^2 T_0}{\partial Y^2}.$$

The only plausible choice is to balance the terms involving δ and ϵ, which gives

$$\delta = \epsilon^2$$

(so this boundary layer is very small and might not be easy to resolve numerically); then, writing T_b for the temperature in the boundary layer, we have

$$\frac{\partial T_b}{\partial \xi} = \frac{\partial^2 T_b}{\partial \xi^2} + \epsilon^2 \frac{\partial^2 T_b}{\partial Y^2}, \qquad \xi < 0,$$

with

$$T_b(\xi, \pm 1) = 0, \qquad \frac{\partial T_b}{\partial \xi}(0, Y) = 0.$$

The leading-order term in a regular expansion in powers of ϵ^2 is easily found to be

$$T_{b0}(\xi, Y) = A(Y),$$

where all we know about the arbitrary function A is that $A(\pm 1) = 0$. So how do we find it?

We have not yet exploited the information coming into our boundary layer from the main flow. That is, we have to match with the 'outer' solution. At leading order, this is easy. We use the Van Dyke rule with one term in the inner and outer expansions. This tells us that

$$A(Y) = T_0(1, Y) = \sum_{n=0}^{\infty} \frac{2(-1)^n e^{-(n+1/2)^2 \pi^2}}{n + \frac{1}{2}} \cos\left((n + \tfrac{1}{2})\pi Y\right).$$

This is almost trivial, but perhaps counter-intuitively it shows that the matching at leading order is between the values of the temperature and not its gradient, even though the boundary condition we have imposed

The condition at $\xi = 0$ rules out the exponential solution of the differential equation.

is on the latter. We have to go to higher orders to see this matching too; this is requested in Exercise 6. Here we just note that it is very plausible that the $O(\epsilon^2)$ term in the inner expansion, $\epsilon^2 T_b(\xi, Y)$, can match with an $O(1)$ outer temperature gradient because $\epsilon^2 \partial / \partial \xi = \partial / \partial X$.

Notice again that where the full problem is elliptic, requiring boundary conditions all round the domain, the approximate problem is parabolic and we cannot impose a condition at $x = L$. The deficit is made up by the boundary layer, which allows the outer solution to accommodate to whatever we want at $x = L$. Again, the approximate analysis tells us a lot about the structure of the problem, in both qualitative and quantitative terms.

18.5 Exercises

1 Heat flow in a bar of variable cross-section. In this exercise we find an exact solution for steady heat flow in a long thin bar of variable cross-section, exploiting the fact that Laplace's equation is invariant under conformal maps.

Consider the long thin rectangle $1 < \xi < e, -\epsilon < \eta < \epsilon, \epsilon \ll 1$. Show that its image under the conformal map $x + iy = \log(\xi + i\eta)$, with the branch cut out of the way on the negative real axis, is very close to the region between the curves $y = \pm \epsilon e^{-x}, 0 < x < 1$. Show that the solution for steady heat flow in the rectangle having $T = 0$ at $\xi = 1, T = 1$ at $\xi = e$ and insulated sides is $T = (\xi - 1)/(e - 1)$. Writing this in terms of x and y, verify that this exact solution is consistent with the approximate solution derived in Section 18.2. Use other conformal maps to construct similar examples.

> How big is the difference between the two regions?

2 Heat flow in a long thin domain. Consider an eigenfunction expansion solution to the problem

$$\frac{\partial^2 T}{\partial x^2} + \frac{\partial^2 T}{\partial y^2} = 0, \qquad 0 < x < L, \quad -H < y < H,$$

with

$$T(0, y) = 1, \qquad T(L, y) = 0,$$

and

$$\pm k \frac{\partial T}{\partial y} + \Gamma T = 0 \quad \text{on} \quad y = \pm H.$$

Separate the variables to find eigenfunctions of the form

$$T_n(x, y) = \cos \alpha_n y \sinh \alpha_n (L - x)$$

and show that the homogeneous boundary conditions on $x = L$,

$y = \pm H$ are all satisfied provided that

$$\alpha_n \tan \alpha_n H = \frac{\Gamma}{k}$$

(note that α_n has dimensions 1/length). Verify that the eigenfunctions are orthogonal in y, and hence use the condition on $x = 0$ to calculate the coefficients in the expansion

$$T(x, y) = \sum_n a_n T_n(x, y).$$

Verify that as $H/L \to 0$ this solution is accurately approximated by the 'lubrication' model described in the text; consider all cases for the size of Γ.

3 **More heat flow in a long thin domain.**   Suppose that the domain of the previous exercise is

$$0 < x < L, \quad -H_0 \left(1 + f(x/L)\right) < y < H_0 \left(1 + f(x/L)\right),$$

where $f(0) = 0$, $f(1) = 1$ and f is smooth. Suppose that the heat transfer boundary condition is

$$k\mathbf{n} \cdot \nabla T + \Gamma T = 0$$

on the lateral boundaries, where $\mathbf{n}$ is the unit normal. Would you be able to write down an eigenfunction expansion now? Show that the dimensionless model is

$$\epsilon^2 \frac{\partial^2 T}{\partial X^2} + \frac{\partial^2 T}{\partial Y^2} = 0, \quad 0 < X < 1, \quad -1 - f(X) < Y < 1 + f(X),$$

with

$$T = 1 \quad \text{on} \quad X = 0, \qquad T = 0 \quad \text{on} \quad X = 1$$

and

$$\frac{\partial T}{\partial Y} \mp \epsilon f'(X)\frac{\partial T}{\partial X} \pm \epsilon^2 \alpha^2 \left(1 + \epsilon^2 f'^2(X)\right)^{1/2} T = 0$$

on $Y = \pm (1 + f(X))$ (α is as defined in the text on p. 245). Deduce that there is now a term of $O(\epsilon)$ in the expansion for T and find the ordinary differential equation for T_0.

4 **Still more heat flow in a long thin domain.**   Consider the model above but for a rectangular domain, namely

$$\epsilon^2 \frac{\partial^2 T}{\partial X^2} + \frac{\partial^2 T}{\partial Y^2} = 0, \qquad 0 < X < 1, \quad -1 < Y < 1,$$

with

$$T = 1 \quad \text{on} \quad X = 0, \qquad T = 0 \quad \text{on} \quad X = 1$$

and

$$\frac{\partial T}{\partial Y} \pm \epsilon^2 \alpha^2 T = 0 \quad \text{on} \quad Y = \pm 1.$$

Suppose that $\alpha^2 = 1/\delta$, where $\epsilon^2 \ll \delta \ll 1$, so that the heat transfer coefficient is larger than in the example in the text. Show that scaling X with ϵ via $X = \epsilon \xi$ leads to the problem

$$\frac{\partial^2 T}{\partial \xi^2} + \frac{\partial^2 T}{\partial Y^2} = 0, \qquad 0 < \xi < 1/\epsilon, \quad -1 < Y < 1,$$

with boundary conditions

$$T = 1 \quad \text{on} \quad X = 0, \qquad T = 0 \quad \text{on} \quad X = 1/\epsilon$$

and

$$T \pm \delta \frac{\partial T}{\partial Y} = 0 \quad \text{on} \quad Y = \pm 1.$$

Show that, for $O(1)$ values of ξ, the leading-order term in a regular expansion in powers of δ is the solution of

$$\frac{\partial^2 T_0}{\partial \xi^2} + \frac{\partial^2 T_0}{\partial Y^2} = 0, \qquad 0 < \xi < \infty, \quad -1 < Y < 1,$$

with boundary conditions

$$T_0 = 1 \quad \text{on} \quad X = 0, \qquad T_0 \to 0 \quad \text{as} \quad X \to \infty$$

and

$$T_0 = 0 \quad \text{on} \quad Y = \pm 1.$$

Solve this problem by conformal mapping, an eigenfunction expansion or a Fourier sine transform in ξ. Verify that the solution decays exponentially as $\xi \to \infty$, thereby justifying the replacement of the condition at $\xi = 1/\epsilon$ by one at $\xi = \infty$.

Show further that $\partial T/\partial Y$ becomes very large as $\xi \to 0$ on $Y = \pm 1$. Deduce that the expansion is not valid near the two corners $(0, \pm 1)$. Consider an inner expansion near $(0, -1)$: show that in coordinates

$$\xi = \delta \tilde{\xi}, \qquad Y = -1 + \delta \tilde{Y},$$

and, with $T_0(\xi, Y) \sim \tilde{T}_0(\tilde{\xi}, \tilde{Y}) + \cdots$, the inner problem is, to leading order,

$$\frac{\partial^2 \tilde{T}_0}{\partial \tilde{\xi}^2} + \frac{\partial^2 \tilde{T}_0}{\partial \tilde{Y}^2} = 0, \qquad 0 < \tilde{\xi}, \tilde{Y} < \infty,$$

with

$$\tilde{T}_0 = 1 \quad \text{on} \quad \tilde{\xi} = 0, \qquad \frac{\partial \tilde{T}_0}{\partial \tilde{Y}} - \tilde{T}_0 = 0 \quad \text{on} \quad \tilde{Y} = 0$$

and the matching condition

$$\tilde{T}_0 \to \frac{2\theta}{\pi} \quad \text{as} \quad \tilde{\xi}^2 + \tilde{Y}^2 \to \infty,$$

where θ is the local polar angle. This problem is not easy to solve; the Mellin transform may be best.

Repeat the calculation (with minor variations) when $\alpha = O(1)$.

The point of this exercise is that when the heat transfer coefficient is large enough, all the action takes place near the end $x = 0$ of the rod, and we can lose the geometrical complications associated with its finite length (and, indeed, irregular shape).

5 Heat flow with advection in a long thin domain.   Suppose that $T(x, y)$ satisfies

$$\rho c U \frac{\partial T}{\partial x} = k \left(\frac{\partial^2 T}{\partial x^2} + \frac{\partial^2 T}{\partial y^2} \right) + Q, \quad 0 < x < L, \quad -H < y < H,$$

where Q is a constant, with the boundary conditions

$$T = 0 \quad \text{on} \quad x = 0 \quad \text{and} \quad y = 0, h, \quad \frac{\partial T}{\partial x} = 0 \quad \text{on} \quad x = L.$$

What is being modelled here? Now suppose that $H/L = \epsilon \ll 1$. Make the equation dimensionless by scaling x with L and y with H, and suppose that the Peclet number $\rho c U H^2/(kL)$ turns out to be equal to 1. What is the appropriate scale for T?

You should have arrived at the dimensionless equation

$$\frac{\partial T}{\partial X} = \epsilon^2 \frac{\partial^2 T}{\partial X^2} + \frac{\partial^2 T}{\partial Y^2} + 1, \quad 0 < X < 1, \quad 0 < Y < 1,$$

with the boundary conditions

$$T = 0 \quad \text{on} \quad X = 0 \quad \text{and} \quad Y = 0, 1, \quad \frac{\partial T}{\partial X} = 0 \quad \text{on} \quad X = 1.$$

Is this equation elliptic, parabolic or hyperbolic? Briefly indicate how you would find a separation-of-variables solution in the form

$$T(X, Y) = \sum a_n e^{\lambda_n X} \sin n\pi Y,$$

where

$$\epsilon^2 \lambda_n^2 - \lambda_n - n^2 = 0.$$

For each $O(1)$ value of n, find expressions for the positive and negative roots of this equation as $\epsilon \to 0$. Find the leading-order terms in an approximate solution to the original problem, and explain why the positive roots of the eigenvalue equation for λ_n correspond to the boundary layer contribution to the approximate solution.

6 And still more on heat conduction in a long thin domain. Find terms up to $O(\epsilon^2)$ in the outer and inner expansions of the solution of

$$\frac{\partial T}{\partial x} = \epsilon^2 \frac{\partial^2 T}{\partial X^2} + \frac{\partial^2 T}{\partial Y^2}, \qquad 0 < X < 1,$$

with boundary conditions

$$T(0, Y) = 1, \qquad T(X, \pm 1) = 0, \qquad T(1, Y) = 0, \qquad \frac{\partial T}{\partial X}(1, Y) = 0$$

(you will have to go to $O(\epsilon^4)$ in the outer solution; save ink by writing b_n for the Fourier coefficients). Now carry out the matching to second order using Van Dyke's matching principle,

two-term inner expansion of two-term outer expansion

= two-term outer expansion of two-term inner expansion,

i.e. write the outer expansion in terms of the inner variable given by $\xi = (X - 1)/\epsilon^2$. Expand the result, keeping terms of $O(\epsilon^2)$. Repeat this procedure for the inner expansion (notice how terms involving e^ξ are neglected in this expansion, being exponentially small). Compare the two expansions to identify all the unknown functions in the inner expansion.

Repeat the whole problem for the (physically unrealistic) condition $T = 0$ at $X = 1$ (if you are feeling tired by now, just do the $O(1)$ terms).

Can you see physically why we may have to do something different if we try to impose a zero-heat-flux condition at $x = L$?

7 Singular expansion for a linear algebra problem. Consider the problem

$$\begin{pmatrix} 1 + \epsilon & \epsilon \\ 1 - \epsilon & 2\epsilon \end{pmatrix} \begin{pmatrix} x \\ y \end{pmatrix} = \begin{pmatrix} p \\ q \end{pmatrix},$$

where $0 < \epsilon \ll 1$ and p, q are given. Draw the two lines whose point of intersection is the solution of these equations (a) when $p = q = 1$, (b) when $p = 1, q = 2$. What happens as $\epsilon \to 0$? Calculate the exact solution and verify that it is consistent with your graphical analysis and that it is large when ϵ is small unless $p = q + O(\epsilon)$.

Write the problem as

$$\mathbf{Ax} = \mathbf{b}.$$

Recall the Fredholm Alternative theorem for the linear equations $\mathbf{Ax} = \mathbf{b}$ (p. 132).

- If $\mathbf{A}$ is invertible (in particular if none of its eigenvalues vanishes, so that the homogeneous problem $\mathbf{Ax} = \mathbf{0}$ has only the trivial solution $\mathbf{x} = \mathbf{0}$), the solution is unique.

- If, however, $\mathbf{A}$ is not invertible then there is a vector $\mathbf{v}$ such that $\mathbf{A}\mathbf{v} = \mathbf{0}$. That is, 0 is an eigenvalue of $\mathbf{A}$ and $\mathbf{v}$ is the corresponding (right) eigenvector. Let us assume for clarity that 0 is a simple eigenvalue. Then there is also a nontrivial (left eigen)vector $\mathbf{w}$ such that $\mathbf{A}^{\mathsf{T}}\mathbf{w} = \mathbf{0}$ and
  - if $\mathbf{w}^{\mathsf{T}}\mathbf{b} \neq 0$, then there is no solution to the original problem;
  - if, however, $\mathbf{w}^{\mathsf{T}}\mathbf{b} = 0$ then there is a solution but it is not unique: the difference between any two solutions is a multiple of $\mathbf{v}$.

Now find an asymptotic expansion for the solution of $\mathbf{A}\mathbf{x} = \mathbf{b}$. Write

$$\mathbf{A} = \mathbf{A}_0 + \epsilon\mathbf{A}_1, \qquad \mathbf{x} \sim \mathbf{x}_0 + \epsilon\mathbf{x}_1 + \cdots.$$

Show that $\mathbf{A}_0$ has 0 for an eigenvalue and calculate the corresponding right and left eigenvectors $\mathbf{v}_0$ and $\mathbf{w}_0$. Deduce that the leading-order problem $\mathbf{A}_0\mathbf{x}_0 = \mathbf{p}$ has a solution only if $p - q = O(\epsilon)$.

From now on, take $p = q = 1$. Write down the general solution of $\mathbf{A}_0\mathbf{x}_0 = \mathbf{p}$ in the form 'particular solution + complementary solution', where the latter is a multiple α of $\mathbf{v}_0$ that is not determined at this order in the expansion. Write down the problem from the $O(\epsilon)$ terms in the expansion, and use the Fredholm Alternative to show that it has a solution only if $\alpha = 2$. Noting that $\mathbf{x}_0$ is now uniquely determined, verify that it agrees with the small-ϵ expansion of the exact solution.

The analogy with linear ordinary differential equations suggested by this terminology is exact: see Exercise 6 on p. 132.

'... in a long thin spherical region.'

19

Case study: continuous casting of steel

19.1 Continuous casting of steel

An important technique in steel manufacture is the production of a seam-less bar, or 'strand', by continuous casting. Molten steel is poured into a large 'tundish' from which it emerges through a mould slot in the bot-tom. It is cooled by water pipes in the sides of the mould and, once it has emerged, by water sprays and jets (see Figure 19.1). When the steel emerges it has a thin solid skin, which becomes thicker as the steel moves down. Nevertheless, the liquid steel extends far down the strand.

It is important to be able to control the location of the solid–liquid boundary, for safety reasons (the molten steel must not be allowed to spill out) and in order to get the metallurgy right. The former involves a very complicated situation near to the mould, which we will not attempt to model here. Instead, we will write down a simple two-dimensional model for the latter (we shall not tackle the problem of controlling the heat fluxes to achieve a desired solidification rate).

This kind of model can also be used for other solidification processes such as a Bridgeman crystal grower, in which a continuous strand of silicon is solidified very slowly (in order to minimise defects) by being passed along a conveyor belt and cooled from above and below.

Before proceeding, we remind ourselves of the Stefan model for the solidification of a pure material. It is an experimental observation that a fixed amount of energy per unit mass is required to melt a pure[1] solid without changing its temperature, and the same amount of energy

---

[1] In this case study, we are going to ignore the complication that steel is an alloy.

Figure 19.1 The continuous
casting of steel.

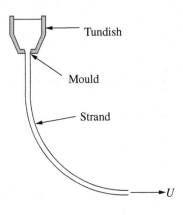

Tundish

Mould

Strand

U

Figure 19.2 Derivation of the
Stefan condition.

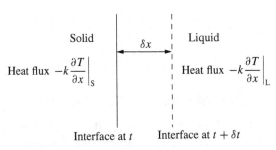

must be removed to solidify it. This heat is supplied or removed by
the difference between the heat fluxes into and out of the solid–liquid
interface. In one space dimension, we can carry out a 'box' argument
for the configuration of Figure 19.2, in which the solid is to the left of
the interface.

If the interface moves a distance δx in time δt then the latent heat
absorbed (for melting, $\delta x < 0$) or released (for solidification, $\delta x > 0$)
by that amount of material in changing phase is

$$\rho \lambda \, \delta x.$$

This must be balanced by the difference in heat fluxes over time δt,

$$\left[-k \frac{\partial T}{\partial x} \right]_{\mathrm{S}}^{\mathrm{L}} \delta t.$$

Hence we derive the Stefan condition for the speed of the interface,

$$\rho \lambda \frac{dx}{dt} = \left[-k \frac{\partial T}{\partial x} \right]_{\mathrm{S}}^{\mathrm{L}}.$$

The right-hand side of this condition is the net rate at which heat is
supplied to the interface, while the left-hand side is the rate at which it
is used up or produced as the interface moves. In more dimensions, this

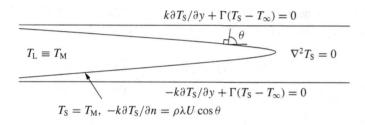

Figure 19.3 Model for continuous casting.

argument is simply generalised, so that for an interface with unit normal **n** from solid to liquid, and normal velocity V_n in that direction,

$$\rho \lambda V_n = [-k\mathbf{n} \cdot \nabla T]_\text{S}^\text{L}.$$

If this brings the Rankine–Hugoniot condition to mind (it should), see Exercise 1 for more details.

We are now in a position to write down a model for steady heat flow in the strand of steel. We will straighten the strand out, modelling it by a rectangle, thus assuming that the effects of curvature are small, as can be verified later. We write U for the speed of the strand and immediately note that the Peclet number is very small, so that the temperature approximately satisfies Laplace's equation. We make the further simplification that the liquid steel is exactly at its melting temperature, so we only have to find the temperature in the solid (a *one-phase problem*). We'll also take Newton cooling with a background temperature of T_∞ as a crude model for the effect of the water cooling.

It is important to notice that the liquid–solid interface is unknown: we have to find it as part of the solution. Let's write it as $y = \pm f(x)$ (see Figure 19.3). The solid temperature $T_\text{S}(x, y)$ satisfies

$$\frac{\partial^2 T_\text{S}}{\partial x^2} + \frac{\partial^2 T_\text{S}}{\partial y^2} = 0$$

in the solid region, with

$$\pm k \frac{\partial T_\text{S}}{\partial y} + \Gamma\,(T_\text{S} - T_\infty) = 0$$

on the edges $y = \pm H$.

On the liquid–solid interface $y = \pm f(x)$, we have

$$T_\text{S} = T_\text{M},$$

the melting or solidification temperature, and the Stefan condition in the form

$$-k \frac{\partial T_\text{S}}{\partial n} = \rho \lambda U \cos\theta,$$

where θ is the angle between the normal to the interface and the x-axis, so that the normal velocity of the interface is $U \cos \theta$. This condition can also be written

Some details for checking.

$$-k\left(\frac{\partial T_S}{\partial y} - \frac{\mathrm{d}f}{\mathrm{d}x}\frac{\partial T_S}{\partial x}\right) = -\rho\lambda U \frac{\mathrm{d}f}{\mathrm{d}x}.$$

For large x, we impose $T_S \to T_\infty$, and we won't be too specific about the inlet conditions at this stage.

Now let's make the problem dimensionless. Obviously we'll scale y and f with H, but the length scale L for x is less obvious. We could of course use the length of the strand but a better idea is to *derive* the length scale from the balance between latent heat release and cooling. This also has the merit of telling us directly when our approximation is valid and how long the molten region is expected to be. So we write $x = LX$ and $y = HY$, where L is yet to be found but as usual $\epsilon = H/L \ll 1$, and we write $y = \pm f(x)$ as $Y = \pm F(X)$. We also need a scale for the temperature; this is built into the problem as

$$T_S = T_M + (T_M - T_\infty)\, T(X, Y).$$

So, dropping the primes, we have the dimensionless model

$$\frac{\partial^2 T}{\partial Y^2} + \epsilon^2 \frac{\partial^2 T}{\partial X^2} = 0$$

in the solid, with the interface conditions

$$T = 0, \qquad \frac{\partial T}{\partial Y} + \epsilon \frac{\mathrm{d}F}{\mathrm{d}X}\frac{\partial T}{\partial X} = \epsilon\tilde{\lambda}\frac{\mathrm{d}F}{\mathrm{d}X} \qquad (19.1)$$

on $Y = \pm F(X)$, where

$$\tilde{\lambda} = \frac{\rho\lambda U}{k(T_M - T_\infty)/H}$$

is a dimensionless number which is written in this way to show that it measures the balance between latent heat release from an interface moving with speed U and conduction due to a temperature difference $T_M - T_\infty$ across a distance of $O(H)$. The factor ϵ on the right-hand side of (19.1) arises because the interface only has a very small normal velocity.

Lastly the scaled cooling conditions are

$$\frac{\partial T}{\partial Y} \pm \alpha(T + 1) = 0 \quad \text{on} \quad Y = \pm 1, \qquad (19.2)$$

with $\alpha = \Gamma H (T_M - T_\infty)$. Bearing in mind the previous examples we need the cooling rate to be small, so that $\alpha \ll 1$, and we also need it to balance the rate of latent heat loss. We therefore determine L by making the choice

$$\epsilon = \alpha,$$

and check later that it is consistent.

Let's concentrate on the part of the strand where the liquid has not all solidified, and expand $T(X, Y)$ in the form

$$T(X, Y) \sim T_0(X, Y) + \epsilon T_1(X, Y) + \cdots.$$

By symmetry, we need only focus on the top half of the strand. We have easily

$$T_0 = A_0(X) + B_0(X)Y,$$

There are details here that should be worked through.

where from the cooling condition (19.2) at lowest order $B(X) = 0$, and then, from the melting temperature condition, $A(X) = 0$ as well. So

$$T(X, Y) \sim \epsilon T_1(X, Y) + \cdots,$$

telling us that the temperature is everywhere within $O(\epsilon)$ of the melting temperature. Continuing, we have

$$T_1(X, Y) = C_1(X)(Y - F(X)),$$

Automatically incorporating the melting temperature, much more economical than grinding out $T_1 = A_1 Y + B_1$.

and now the '1' in the cooling condition (19.2) comes in to give

$$C_1(X) = -1.$$

Lastly we use the hitherto unexploited latent heat condition (19.1) to find that

$$\frac{\mathrm{d}F}{\mathrm{d}X} = -\frac{1}{\tilde{\lambda}},$$

so that, if the interface starts off from $Y = 1$ at $X = 0$,

$$F(X) = 1 - X/\tilde{\lambda}.$$

We have thus predicted the length of the liquid region ($L\tilde{\lambda}$) and the shape of the interface to lowest order (linear).

Clearly this analysis is not valid near the tip of the strand, where the upper and lower free surfaces meet, as the heat flow is obviously two-dimensional there. In fact one can carry out a more detailed analysis, involving at least six regions of solid, as well as the liquid (see Figure 19.4). Region 1 is an inlet region, from which all we need to know is a starting value for the interface. We have just analysed region 2, which matches into region 3, centred on the tip of the liquid region. This is essentially the problem of a half-line at temperature 0 with temperature -1 on $Y = \pm 1$. Region 4 is necessary to resolve the singularity at the end of the half-line, and the solution in this region shows that the tip of the liquid region is parabolic. Returning to region 3, it matches into

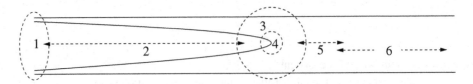

Figure 19.4 Regions for the continuous-casting problem.

region 5, an intermediate region of length $O(H/\epsilon^{1/2})$, that enables the transition into region 6, in which we finally have an eigenfunction expansion decaying exponentially as $x \to \infty$. Further details are given in the exercises, and the problem is described in the paper [11].

19.2 Exercises

1 The Stefan condition and Rankine–Hugoniot. Think about the internal energy in a material that can change phase from solid to liquid and convince yourself that it can be written

$$H(T) = \begin{cases} \rho c T, & T < 0, \\ \rho c T + L, & T > 0. \end{cases}$$

Express the conservation of energy as

$$\frac{\partial H}{\partial t} + \frac{\partial}{\partial x}\left(-k\frac{\partial T}{\partial x}\right) = 0,$$

and then deduce the Stefan condition as a Rankine–Hugoniot condition for this conservation law.

2 The other regions in the continuous-casting problem. This exercise brings together boundary layer and lubrication theory techniques. It is probably the hardest in the book.

Look at Figures 19.5 and 19.4, which show the dimensionless model for continuous casting and the regions in it. We have analysed region 2 in the text, where we showed that the tip of the molten region is approximately at $X = \tilde{\lambda}$ and that the temperature in region 2 is

$$T^{(2)} \sim \epsilon \left(1 - X/\tilde{\lambda} - |Y|\right) + o(\epsilon), \qquad 1 - X/\tilde{\lambda} < |Y| < 1.$$

Start with region 6, for which $X - \tilde{\lambda} = O(1)$, far down the strand, and show that in this region the solution has the form

$$T^{(6)}(X, Y) \sim -1 + \sum_{n=1}^{\infty} a_n \cos k_n Y e^{-k_n(X-\tilde{\lambda})/\epsilon},$$

where the k_n are the roots of $k_n \tan k_n = \epsilon$. The coefficients a_n are determined by matching back towards the tip region. Show that the k_n are all $O(1)$ except for k_1, which is $O(\epsilon^{1/2})$.

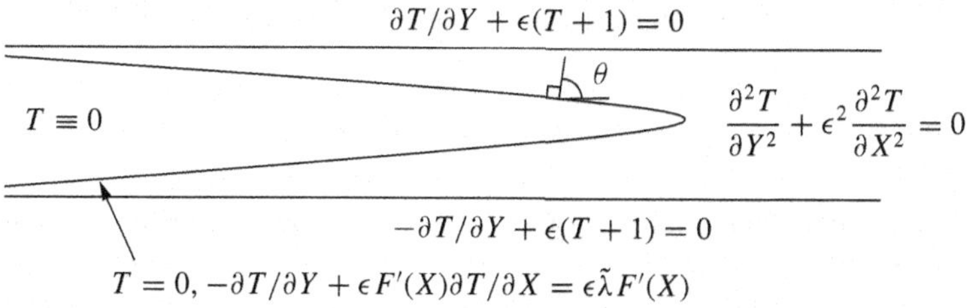

$$\partial T/\partial Y + \epsilon(T+1) = 0$$

Figure 19.5 Model for continuous casting.

$T \equiv 0$

$$\frac{\partial^2 T}{\partial Y^2} + \epsilon^2 \frac{\partial^2 T}{\partial X^2} = 0$$

$$-\partial T/\partial Y + \epsilon(T+1) = 0$$

$$T = 0, \quad -\partial T/\partial Y + \epsilon F'(X)\partial T/\partial X = \epsilon\tilde{\lambda}F'(X)$$

Because there is one small eigenvalue k_1, we cannot match directly from region 6 to region 3 (you could try this, to see why it does not work). Instead, we need to interpose the intermediate region 5, in which $X - \tilde{\lambda} = \epsilon^{1/2}X_5$, and we have

$$\frac{\partial^2 T^{(5)}}{\partial Y^2} + \epsilon \frac{\partial^2 T^{(5)}}{\partial X_5^2} = 0.$$

Show that the leading-order term in an expansion $T^{(5)} \sim T_0^{(5)} + \cdots$ is independent of Y. Calculate $T_1^{(5)}$ and use the boundary conditions at $Y = \pm 1$ to show that the solution that matches with region 6 is

$$T_0^{(5)} = -1 + a_1 e^{X_5}.$$

Note that we have to match with region 3 in which $T^{(3)}$ is small, hence $T_0^{(5)}$ tends to zero as $X_5 \to 0$. Deduce that $a_1 = 1$.

Now consider region 3, in which $X_3 = \epsilon(X - \tilde{\lambda})$. Show that the free-surface conditions can be linearised onto $Y = 0, X_3 < 0$, and that $T^{(3)}$ satisfies the problem

In dimensional terms, this is the $O(H_0) \times O(H_0)$ region around the tip of the molten region.

$$\frac{\partial^2 T^{(3)}}{\partial X_3^2} + \frac{\partial^2 T^{(3)}}{\partial Y^2} = 0$$

in the strip $-1 < Y < 1, -\infty < X_3 < \infty$, with the negative X_3-axis removed; the boundary conditions are

Draw a picture.

$$\pm\frac{\partial T^{(3)}}{\partial Y} + \epsilon = 0 \quad \text{on} \quad Y = \pm 1$$

and

$$T^{(3)} = 0 \quad \text{on} \quad Y = 0, \quad X_3 < 0.$$

Use the Van Dyke matching rule in the form 'two-term inner of one-term outer matches with one-term outer of two-term inner' to show that, for large values of X_3,

$$T^{(3)} \sim \epsilon^{1/2}X_3 + \tfrac{1}{2}\epsilon X_3^2 + \cdots.$$

Deduce that the expansion for $T^{(3)}$ proceeds in powers of $\epsilon^{1/2}$,

$$T^{(3)} \sim \epsilon^{1/2}T_1^{(3)} + \epsilon T_2^{(3)} + \cdots,$$

and confirm that matching with regions 2 and 4 is accomplished if

$$T_1^{(3)} \sim X_3, \qquad T_2^{(3)} \sim \tfrac{1}{2}X_3^2, \qquad \text{as} \quad X_3 \to \infty,$$

$$T_1^{(3)} \to 0, \qquad T_2^{(3)} \sim -|Y|, \qquad \text{as} \quad X_3 \to -\infty.$$

Write down the problems for $T_1^{(3)}$ and $T_2^{(3)}$. Show that the hodograph variable

$$\frac{\partial T_1^{(3)}}{\partial X_3} - i\frac{\partial T_1^{(3)}}{\partial Y}$$

is analytic and that either its real or its imaginary part is known on all the boundary of region 3. Draw the hodograph plane. Find the conformal mapping from the hodograph plane to the physical plane to show that

$$\frac{\partial T_1^{(3)}}{\partial X_3} - i\frac{\partial T_1^{(3)}}{\partial Y} = \left(1 - e^{-\pi Z_3}\right)^{-1/2},$$

where $Z_3 = X_3 + iY$, and find $T_2^{(3)}$ (note that its Z_3-derivative satisfies the same problem as $T_1^{(3)}$).

Lastly note that this solution is clearly not valid near the tip $Z_3 = 0$, where the gradient of $T_1^{(3)}$ is infinite. Show that a rescaling via $Z_3 = \epsilon^{1/2} Z_4$ leads to the full free-boundary problem in an infinite region (you will need to reinstate the full Stefan condition). Use parabolic coordinates or the mapping $Z_4 = (\zeta_4 + c)^2$ to show that there is an explicit solution in which the free boundary is $Y_4^2 = -\pi X_4/4$ (the *Ivantsov parabola*, which is a famous exact solution of zero-specific-heat solidification).

'They can curve as much as they like and still be linear.'

20

Lubrication theory for fluids

20.1 Thin fluid layers: classical lubrication theory

In this chapter, we describe the lubrication-theory analysis of a variety of thin fluid flows. Having done the heat conduction problems of Chapter 18, we shouldn't have too much trouble with the original (eponymous) lubrication theory model of flow of a viscous fluid in a thin bearing bounded by rigid surfaces. Then we'll generalise the approach to find equations for thin viscous sheets with free surfaces.

The simplest configuration is that of a *slider bearing*, in which one rigid surface slides over another as in Figure 20.1. These bearings are common in machinery ranging from the head floating over the hard disc of a computer[1] to enormous pumps and other engines. When the bearing is wrapped round into a circle, so that a rotating shaft can be supported, it is known as a *journal bearing*. Knees and other joints are examples of natural-grown bearings.

We'll look at two-dimensional flows only. Let us call the upper surface $y = H_0 H(x/L)$, where L is the length of the bearing and H_0 a representative value for the separation; as usual, $\epsilon = H_0/L \ll 1$. Let us also take axes in a frame in which the upper surface is stationary and the lower surface $y = 0$ moves to the right with velocity $(U, 0)$. The idea behind this bearing is to choose the shape $H(x/L)$ of the upper surface so that fluid dragged into the bearing (remember the no-slip condition on the lower surface) generates a high load-bearing pressure as it is forced through the converging part of the gap.

---

[1] Nowadays the gap between the head and the disc is so small that it is not clear that ordinary continuum models can safely be used for the fluid.

Figure 20.1 A slider bearing.

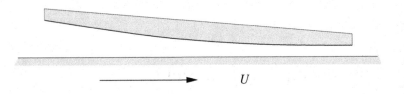

$$U$$

We'll start from the Navier–Stokes equations for the velocity $\mathbf{u} = (u, v)$ and the pressure p:

$$\rho\left(\frac{\partial \mathbf{u}}{\partial t} + \mathbf{u}\cdot\nabla\mathbf{u}\right) = -\nabla p + \mu\nabla^2\mathbf{u}, \qquad \nabla\cdot\mathbf{u} = 0,$$

with the no-slip boundary conditions

$$(u, v) = (U, 0) \quad \text{on} \quad y = 0,$$
$$(u, v) = (0, 0) \quad \text{on} \quad y = H_0 H(x/L).$$

Following our long–thin analysis above, and in contrast to our scaling when we last used these equations, we'll scale x and y differentially, x with L and y with H_0. When we come to the velocity $\mathbf{u} = (u, v)$, we have to scale its two components differentially as well, or we will not conserve mass. In view of the imposed motion of the lower plate and the no-slip condition, we want to scale u with U and, since (in unscaled variables)

$$\frac{\partial u}{\partial x} + \frac{\partial v}{\partial y} = 0,$$

we need to scale v with ϵU. In the absence of any forced unsteady motion of the upper surface, the natural time scale is then L/U. Lastly, we need a scale P for the pressure p. In the absence of any obvious 'exogenous' scale, we'll work this out from the equations.

As in the analysis of Chapter 18, we use X and Y for the scaled co-ordinates, but we will 'drop the primes' and stick with lower-case letters for the dependent variables. It's a nasty hybrid notation, but capitals are so much harder to read, and we'll also be using dimensional equations later in the chapter so we want to be able to distinguish them at a glance.

The X-component of the scaled momentum equation is

$$\frac{\rho U^2}{L}\left(\frac{\partial u}{\partial T} + u\frac{\partial u}{\partial X} + v\frac{\partial u}{\partial Y}\right) = -\frac{P}{L}\frac{\partial p}{\partial X} + \frac{\mu U}{H_0^2}\left(\epsilon^2\frac{\partial^2 u}{\partial X^2} + \frac{\partial^2 u}{\partial Y^2}\right).$$

If viscous shear forces are to do their job in generating pressure,[2] we have to choose P so as to balance terms on the right-hand side of this

---

[2] At the other end of the viscosity range, one can make a model of an inviscid surf-skimmer held up over a thin layer of water by inertial forces only.

equation. Thus we choose $P = \mu U L / H_0^2$. We now have the back-of-the-envelope estimate $LP = \mu U L^3 / H_0^2$ for the load per unit distance in the z-direction that this bearing can support.

This scaling for p leaves one dimensionless parameter in the problem,

$$\mathrm{Re}' = \epsilon^2 \frac{UL}{\nu} = \epsilon^2 \mathrm{Re},$$

known as the *reduced Reynolds number*. Our analysis, which leads to the lubrication theory model, is valid when Re' is small, as we assume henceforth. This entails in particular that all the inertial terms, some of which are nonlinear, are neglected.

Crossing off lots of small terms, our leading-order model is then

$$\frac{\partial^2 u}{\partial Y^2} = \frac{\partial p}{\partial X}, \qquad \frac{\partial p}{\partial Y} = 0, \qquad \frac{\partial u}{\partial X} + \frac{\partial v}{\partial Y} = 0$$

for $0 < Y < H(X)$, with

$$u = 1, v = 0 \quad \text{on} \quad Y = 0, \qquad u = v = 0 \quad \text{on} \quad Y = H(X).$$

It is straightforward to integrate these equations, first noting that $p = p(X)$, then finding that

$$u = 1 - \frac{Y}{H(X)} - \tfrac{1}{2} Y (H(X) - Y) \frac{\mathrm{d}p}{\mathrm{d}X}$$

and lastly using the continuity equation integrated with respect to Y,

$$\frac{\mathrm{d}}{\mathrm{d}X} \int_0^{H(X)} u(X, Y) \, \mathrm{d}Y = 0,$$

to find *Reynolds' equation*

$$\frac{\mathrm{d}}{\mathrm{d}X} \left(H^3 \frac{\mathrm{d}p}{\mathrm{d}X} \right) = 6 \frac{\mathrm{d}H}{\mathrm{d}X}$$

for the pressure. Given $H(X)$, we can solve this with ambient-pressure conditions at each end and then calculate the load our bearing can support.

20.2 Thin viscous fluid sheets on solid substrates

For our next application of the lubrication theory approach, we'll derive approximate equations for the evolution of thin sheets or fibres of a viscous fluid. These problems are a little more difficult because the fluid has one or more *free surfaces*, whose locations have to be determined as part of the solution of the problem. We start with the case of a thin layer spreading out on a horizontal surface, a situation that arises in applications ranging in lateral scale from microns (layers of

---

Compare with the scaling $\mu U / L$ we used in deriving the slow-flow equations earlier: here we have $(1/\epsilon^2)\mu U / L$, indicating the effectiveness of the long thin geometry in generating high pressures.

Exercise ...

The flow is a combination of a Couette shear (the first two terms) and a Poiseuille flow with pressure gradient $\mathrm{d}p/\mathrm{d}X$, so with hindsight we could have written this down.

Figure 20.2 (a) Viscous layer spreading under gravity. (b) Velocity profile for Poiseuille flow.

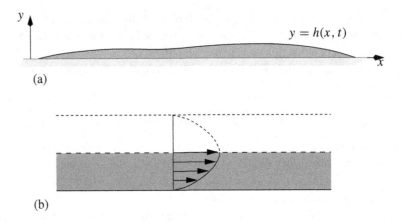

(a)

(b)

conductor applied in liquid form to a printed circuit board before being baked solid) through millimetres (paint on a wall, honey spilled on a table) to kilometres (magma flow from a volcano). We will also describe briefly the corresponding model for flow on a vertical surface before, in Section 20.3, looking at free sheets such as the glass sheets you would use when making a bottle by blowing, or a window by the float-glass process. Lastly we look more briefly at the manufacture of fibres of, for example, glass (optical fibres) or polymer (artificial fabrics).

20.2.1 Viscous fluid spreading horizontally under gravity: intuitive argument

Imagine that you spill a puddle of honey on a table. How does it spread out? Assume that the depth is much smaller than the spread, and for now take the two-dimensional situation shown in Figure 20.2(a). Here is a physical argument, in four steps (all variables are dimensional).

1. The flow is slow, so we use the Stokes equations (uncontroversial).
2. The flow is driven by hydrostatic pressure and resisted by viscous shear forces (uncontroversial). The pressure is approximately hydrostatic, because vertical velocities are small enough that the viscous contribution to the forces in that direction is small (not so obvious; believe it for now). Thus

$$p(x, y, t) = \rho g \left(h(x, t) - y \right).$$

3. The horizontal velocity u is much greater than the vertical velocity v and the free surface is almost horizontal. Moreover, on the free surface the shear stress, which is approximately $\mu \, \partial u / \partial y$, vanishes (uncontroversial, although we might want to check this later). We can

also regard these two statements as constituting a symmetry condition and thus, locally, the flow looks like the bottom half of a flow between two parallel plates separated by $2h$ under a pressure gradient $\partial p / \partial x$ (see Figure 20.2(b)). The velocity profile is therefore parabolic,

$$u(x, y, t) = -\frac{1}{2\mu} y \, (2h(x, t) - y) \frac{\partial p}{\partial x},$$

and the horizontal flux is

$$
\begin{aligned}
Q(x, t) &= \int_0^{h(x,t)} u(x, y, t) \, dy \\
&= -\frac{1}{3\mu} h^3 \frac{\partial p}{\partial x} \\
&= -\frac{\rho g}{3\mu} h^3 \frac{\partial h}{\partial x}.
\end{aligned}
$$

4. Mass conservation (uncontroversial) in the form

$$\frac{\partial h}{\partial t} + \frac{\partial Q}{\partial x} = 0$$

gives us a nonlinear diffusion equation for $h(x, t)$:

$$\frac{\partial h}{\partial t} = \frac{\rho g}{3\mu} \frac{\partial}{\partial x} \left(h^3 \frac{\partial h}{\partial x} \right). \tag{20.1}$$

Note that this dimensional equation tells us the timescale for the spreading out of the layer. If x is scaled with L and h with a representative initial value H_0, then the timescale emerges immediately as $\mu L^2 / (\rho g H_0^3)$. This looks reasonable: stickier fluids (larger μ) spread out more slowly, as do thin layers or fluids in regions of low g.

If, instead of gravity, surface tension at the interface drives the motion (as would be appropriate for thin layers of paint or conductor on a printed circuit board), a very similar argument (see the exercises) shows that we get the *fourth-order* nonlinear diffusion equation

$$\frac{\partial h}{\partial t} + \frac{\gamma}{3\mu} \frac{\partial}{\partial x} \left(h^3 \frac{\partial^3 h}{\partial x^3} \right) = 0. \tag{20.2}$$

Not surprisingly, these equations with their evident structure have attracted a lot of theoretical analysis; natural questions to ask include 'if we start with a solution that is positive, does it remain so?' (yes for (20.1) and (20.2), but if h^3 in (20.2) had been h the answer would have been no) or 'if we have a dry patch where $h = 0$, what conditions apply at its edges?' (conserving mass is not too hard, but the extra condition for the fourth-order equation (20.2) is rather more problematic). Suggestions for further reading are given at the end of the chapter.

20.2.2 Viscous fluid spreading under gravity: systematic argument

You may be convinced by the derivation just given (I think I am). However, there are situations where a more precise approach is essential, so let's warm up for that by rederiving equation (20.1) by a systematic asymptotic approach.

Let's start with the slow-flow equations

$$\nabla^2 \mathbf{u} = \nabla p - \rho \mathbf{g}, \qquad \nabla \cdot \mathbf{u} = 0, \tag{20.3}$$

for the velocity $\mathbf{u} = (u, v)$, for which we have the no-slip condition

$$u = v = 0 \quad \text{on} \quad y = 0.$$

The big new feature in this problem is the free surface $y = h(x, t)$. It is unknown – we have to find it as part of the solution – and the boundary conditions applied on it are more complicated. The kinematic condition

$$v = \frac{\partial h}{\partial t} + u \frac{\partial h}{\partial x} \quad \text{on} \quad y = h(x, t)$$

is easy enough,[3] and the other conditions, which say that no stresses act at the free surface, are written

$$\sigma_{ij} n_j = 0,$$

where

$$\mathbf{n} = (n_j) = \left(-\frac{\partial h}{\partial x}, 1 \right) \bigg/ \left(1 + \left(\frac{\partial h}{\partial x} \right)^2 \right)^{1/2}$$

is the unit normal to the surface and

$$\sigma_{ij} = -p \delta_{ij} + \mu \left(\frac{\partial u_i}{\partial x_j} + \frac{\partial u_j}{\partial x_i} \right)$$

is the stress tensor for a Newtonian viscous fluid. We recall from Chapter 1 that the components of σ are

$$(\sigma_{ij}) = \begin{pmatrix} -p + 2\mu \dfrac{\partial u}{\partial x} & \mu \left(\dfrac{\partial u}{\partial y} + \dfrac{\partial v}{\partial x} \right) \\[2ex] \mu \left(\dfrac{\partial u}{\partial y} + \dfrac{\partial v}{\partial x} \right) & -p + 2\mu \dfrac{\partial v}{\partial y} \end{pmatrix}.$$

---

[3] Either think of this as

$$\frac{D}{Dt}(y - h(x, t)) = 0$$

to express the fact that a particle in the surface remains there, or show that the normal to any curve $f(x, y, t) = 0$ is $\mathbf{n} = \nabla f / |\nabla f|$ and the normal velocity of a point on the curve is $-(\partial f / \partial t) / |\nabla f|$, then equate the latter to $\mathbf{n} \cdot \mathbf{u}$.

Thus, the two components of the zero-stress condition

$$\sigma_{11}n_1 + \sigma_{12}n_2 = 0, \qquad \sigma_{12}n_1 + \sigma_{22}n_2 = 0$$

are

$$-\frac{\partial h}{\partial x}\left(-p + 2\mu\frac{\partial u}{\partial x}\right) + \mu\left(\frac{\partial u}{\partial y} + \frac{\partial v}{\partial x}\right) = 0, \qquad (20.4)$$

$$-\mu\left(\frac{\partial u}{\partial y} + \frac{\partial v}{\partial x}\right)\frac{\partial h}{\partial x} - p + 2\mu\frac{\partial v}{\partial y} = 0. \qquad (20.5)$$

Scale using $x = LX$, $y = H_0 Y = \epsilon L Y$ as usual. Since we expect the flow to be driven by hydrostatic pressure, scale p with $\rho g H_0$. Then the horizontal component of the momentum, on balancing $\mu \, \partial^2 u/\partial y^2$ with $\partial p/\partial x$, tells us the scale for u, namely $U_0 = \rho g H_0^3/(\mu L)$, and the scale for v must be ϵU_0 so that mass conservation is not violated (it never is). Lastly our timescale is L/U_0 (or $H_0/(\epsilon U_0)$). We get the equations

> Again we employ a hybrid notation in which independent variables are capitalised and dependent ones are lower case.

$$\frac{\partial^2 u}{\partial Y^2} + \epsilon^2 \frac{\partial^2 u}{\partial X^2} = \frac{\partial p}{\partial X}, \qquad (20.6)$$

$$\epsilon^2 \frac{\partial^2 v}{\partial Y^2} + \epsilon^4 \frac{\partial^2 v}{\partial X^2} = \frac{\partial p}{\partial Y} - 1, \qquad (20.7)$$

$$\frac{\partial u}{\partial X} + \frac{\partial v}{\partial Y} = 0, \qquad (20.8)$$

(the -1 in (20.7) is due to gravity), the no-slip condition

$$u = v = 0 \quad \text{on} \quad Y = 0 \qquad (20.9)$$

and lastly the kinematic condition

$$v = \frac{\partial h}{\partial T} + u\frac{\partial h}{\partial X} \qquad (20.10)$$

and the stress-free conditions

> Note that unlike, say, water waves, our different scalings for u and v mean that we keep the term $u\,\partial h/\partial X$. This is sometimes called a long-wave approximation.

$$-\frac{\partial h}{\partial X}\left(-p + 2\epsilon^2\frac{\partial u}{\partial X}\right) + \frac{\partial u}{\partial Y} + \epsilon^2\frac{\partial v}{\partial X} = 0, \qquad (20.11)$$

$$-\epsilon^2\left(\frac{\partial u}{\partial Y} + \epsilon^2\frac{\partial v}{\partial X}\right)\frac{\partial h}{\partial X} - p + 2\epsilon^2\frac{\partial v}{\partial Y} = 0, \qquad (20.12)$$

both on $Y = h(X, T)$.

We've done the hard work. Now we expand u, v, p in regular expansions[4]

$$u \sim u_0 + \epsilon^2 u_1 + \cdots, \quad \text{etc.}$$

---

[4] Strictly speaking, we should expand h as well, but as in practice there is no real benefit in going beyond the leading-order terms we won't bother, sticking with $h(X, T)$.

and then we decide in which order to solve the equations. Clearly from (20.7) the pressure is hydrostatic to leading order and from (20.12) it vanishes on $Y = h$, so we put a tick against those two equations, as we won't use them again at this order, and write down

$$p_0 = h(X, T) - Y.$$

Now we find u_0 using (20.6), with (20.9) and (20.11) for boundary conditions:

$$u_0(X, Y, T) = -\frac{1}{2} \frac{\partial h}{\partial X} Y(2h - Y).$$

> A quicker way to do this is to note that the leading-order flux $Q_0(X, T)$ is
>
> $$\int_0^{h(X,T)} u_0(X, Y, T) \, dY,$$
>
> and, to leading order, mass conservation is
>
> $$\frac{\partial h}{\partial T} + \frac{\partial Q_0}{\partial X} = 0.$$

Next integrate (20.11) with respect to Y and use the other part of (20.9) to find v_0; then substitution into (20.10) gives, as promised,

$$\frac{\partial h}{\partial T} = \frac{1}{3} \frac{\partial}{\partial X} \left(h^3 \frac{\partial h}{\partial X} \right).$$

Undoing the nondimensionalisation leads immediately to (20.1).

The situation is rather different if the fluid is on a vertical surface, as we now see.

20.2.3 A viscous fluid layer on a vertical wall

> Think of paint.

Suppose that the layer of fluid is on a vertical wall (or an inclined plane that is not almost horizontal). In this case gravity acts along the film, with the result that it balances the shear forces directly rather than being transmitted through the pressure. The intuitive argument to derive the equation of motion is as follows.

1. The flow is approximately unidirectional with velocity $u(x, y, t)$ in the x-direction, down the wall (y is measured out from the wall).
2. The pressure is everywhere very small (because the stress is zero on the free surface $y = h(x, t)$). Instead, the body force ρg in the x-momentum equation drives the flow.
3. Remembering that we have no slip at $y = 0$, that is $u = 0$, and no stress at $y = h(x, t)$, that is approximately $\partial u / \partial y = 0$, the flow is the same as half a Poiseuille flow between $y = 0$ and $y = 2h$, driven by a pressure gradient ρg.
4. The flux is therefore (using a standard calculation) $\rho g h^3 / 3\mu$, and conservation of mass as above gives

$$\frac{\partial h}{\partial t} + \frac{\rho g}{\mu} h^2 \frac{\partial h}{\partial x} = 0.$$

A systematic derivation of this equation by scaling techniques is asked for in Exercise 6. Notice that the new equation is *first order*, not second order as for nearly horizontal flow.

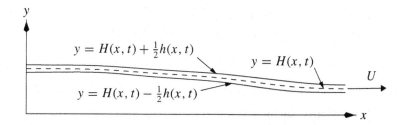

Figure 20.3 Drawing out a thin sheet of viscous fluid.

20.3 Thin fluid sheets and fibres

For our last example in this series of models for thin layers of a viscous fluid, we consider the evolution of a long thin viscous sheet stretched with characteristic speed U from its ends $x = 0$, L (see Figure 20.3). Now we have not just one free surface but two, which adds some complexity to the analysis, as we shall see. This configuration is not very easy to realise, nor is it common in practice, although pizza makers come close (not that dough is anything close to a Newtonian fluid). The best example is probably the float-glass process in which a layer of glass, which may be some hundreds of metres long and tens of metres wide but only a few millimetres thick, is floated on a bath of much less viscous molten tin. As it travels from one end to the other, it should reach a state of absolutely smooth pellucid perfection, so it is of vital importance to glass manufacturers to be able to control this process and eliminate waves and wrinkles. The corresponding axisymmetric situation of a thread or fibre of fluid, which has only one free surface but still no fixed surfaces, is very common. Examples are the manufacture of optical fibres from glass and artificial fabric fibres from polymers, both of which involve the solidification of a liquid thread (so does making candy floss).[5]

There is a simple physical argument that leads to the correct answer (more or less). Much of it is familiar.

1. The sheet is nearly flat and the velocity is approximately unidirectional, in the x-direction (along the sheet).
2. The surfaces are stress free, so the x-velocity does not vary significantly across the sheet: it has the form $u(x, t)$.
3. The stretching is resisted by viscous stresses (stress is force per unit area), which, for a Newtonian fluid, are proportional to the velocity gradient $\partial u / \partial x$. Thus the total force (per unit length perpendicular to

This will be the σ_{11} stress component, the force per unit area in the x-direction across a plane whose normal is in the same direction.

---

[5] We should really include the temperature-dependent viscosity in the model for both these fibres, and non-Newtonian fluid effects for the polymer. We'll keep things simple though.

the axes in Figure 20.3) is proportional to $h\,\partial u/\partial x$, where $h(x,t)$ is the thickness of the sheet. As there are no external forces, this must be constant along the sheet:

$$\frac{\partial}{\partial x}\left(h\frac{\partial u}{\partial x}\right) = 0. \qquad (20.13)$$

4. The second equation for h and u is mass conservation,

$$\frac{\partial h}{\partial t} + \frac{\partial (hu)}{\partial x} = 0,$$

and this completes the model. An identical argument holds for a thin fibre if $h(x,t)$ is replaced by the cross-sectional area $A(x,t)$.

As ever, this analysis raises as many questions as it answers. In particular, it says nothing about the constant of proportionality in the relationship between the resistive stress and the velocity gradient. Clearly this constant is some sort of viscosity, and indeed it has a name, the *Trouton viscosity*, but how is it related to the usual dynamic viscosity μ? We don't need to know this if no forces (such as air drag) act at the surfaces of our sheet, because we can cancel it from both sides of the momentum balance equation (20.13), but it is crucial otherwise. In any case, if we integrate (20.13) we find

$$h\frac{\partial u}{\partial x} \propto T(t),$$

Or, replacing h by A, our fibre.

where $T(t)$ is the tension applied to our sheet, so we need the constant if we are to calculate the total tension needed to stretch the sheet. Only a more detailed analysis can tell us.

20.3.1 The viscous sheet equations by a systematic argument

Let us call the surfaces of the sheet $y = \bar{h}(x,t) \pm \frac{1}{2}h(x,t)$, so that the centreline of the sheet is at $y = \bar{h}(x,t)$: we don't know a priori that it is symmetrical. The dimensional equations that we must solve are very similar to those of the previous section, but with gravity removed and the no-slip condition replaced by zero-stress conditions on both surfaces. We have the slow-flow equations

$$\nabla^2 \mathbf{u} = \nabla p, \qquad \nabla \cdot \mathbf{u} = 0$$

for the velocity $\mathbf{u} = (u, v)$, with the kinematic and dynamic (zero-stress) conditions

$$v = \frac{\partial}{\partial t}\left(\bar{h} \pm \tfrac{1}{2}h\right) + u\frac{\partial}{\partial x}\left(\bar{h} \pm \tfrac{1}{2}h\right), \qquad \sigma_{ij}n_j = 0,$$

on $y = \bar{h}(x,t) \pm \frac{1}{2}h(x,t)$.

The scaling of x with L, y, $\bar{h}$ and h with a typical thickness H_0 and of u, v with U, ϵU is much as before. It's not so easy to see a pressure scale here, so let's use the standard slow-flow scale $\mu U/L$ as a first guess and let the equations tell us how (if at all) this should be corrected. The scaled equations are pretty much a cut-and-paste job too:

This pressure scale is $O(\epsilon^2)$ smaller than the slider-bearing scale because there are no solid surfaces to generate high pressures.

$$\frac{\partial^2 u}{\partial Y^2} + \epsilon^2 \frac{\partial^2 u}{\partial X^2} = \epsilon^2 \frac{\partial p}{\partial X}, \qquad (20.14)$$

$$\frac{\partial^2 v}{\partial Y^2} + \epsilon^2 \frac{\partial^2 v}{\partial X^2} = \frac{\partial p}{\partial Y}, \qquad (20.15)$$

$$\frac{\partial u}{\partial X} + \frac{\partial v}{\partial Y} = 0, \qquad (20.16)$$

The ϵ's crop up now in different places because of our different pressure scale, so you may want to work through the details, for which the margin is too small.

with the kinematic condition

$$v = \frac{\partial}{\partial T}\left(\bar{h} \pm \tfrac{1}{2}h\right) + u\frac{\partial}{\partial X}\left(\bar{h} \pm \tfrac{1}{2}h\right) \qquad (20.17)$$

on $y = \bar{h} \pm \tfrac{1}{2}h$. The stress-free conditions become

$$-\epsilon^2\left(-p + 2\frac{\partial u}{\partial X}\right)\frac{\partial}{\partial X}\left(\bar{h} \pm \tfrac{1}{2}h\right) + \frac{\partial u}{\partial Y} + \epsilon^2\frac{\partial v}{\partial X} = 0, \qquad (20.18)$$

$$-\left(\frac{\partial u}{\partial Y} + \epsilon^2\frac{\partial v}{\partial X}\right)\frac{\partial}{\partial X}\left(\bar{h} \pm \tfrac{1}{2}h\right) - p + 2\frac{\partial v}{\partial Y} = 0. \qquad (20.19)$$

Now we expand in the form

$$u \sim u_0 + \epsilon^2 u_1 + \cdots, \qquad v \sim v_0 + \epsilon^2 v_1 + \cdots,$$
$$p \sim p_0 + \epsilon^2 p_1 + \cdots, \qquad \bar{h} \sim \bar{h}_0 + \epsilon^2 \bar{h}_1 + \cdots,$$
$$h \sim h_0 + \epsilon^2 h_1 + \cdots,$$

take a deep breath and insert.[6]

At $O(1)$, equation (20.14) tells us that $\partial^2 u_0/\partial Y^2 = 0$, so that

$$u_0 = u_0(X, T),$$

and, to leading order, the flow is unidirectional (extensional) as promised. Looking through our equations, we see that (20.18) is also satisfied at this order, and so we tick it off and move on to the continuity equation (20.16), which tells us that

$$v_0 = -Y\frac{\partial u_0}{\partial X} + V_0(X, T),$$

---

[6] Warning: in this problem we are going to go to $O(\epsilon^2)$. If we were going to be consistently accurate, at each stage we would have to remember to expand the location where the free surface conditions are applied about the leading-order location $Y = \bar{h}_0 \pm \tfrac{1}{2}h_0$ (as in subsection 13.5.1). Fortunately, we don't need to do that here.

where V_0 is found from the kinematic condition (20.17) as

$$V_0(X, T) = \frac{\partial}{\partial T} \left(\bar{h}_0 \pm \tfrac{1}{2} h_0 \right) + \frac{\partial}{\partial X} \left(u_0 \left(\bar{h}_0 \pm \tfrac{1}{2} h_0 \right) \right).$$

This is true for both $+$ and $-$ signs and so, subtracting, we find

$$\frac{\partial h_0}{\partial T} + \frac{\partial (u_0 h_0)}{\partial X} = 0,$$

which is the conservation of mass to leading order. Lastly we see from (20.15) that $\partial p_0 / \partial Y = 0$ and so, from (20.19) and the expression we have just found for v_0,

$$p_0(X, T) = -2 \frac{\partial u_0}{\partial X}.$$

Let us pause and take stock. On the one hand we have shown that, to leading order, the flow is extensional and that mass is conserved, but this is only one equation for u_0, $\bar{h}_0$ and h_0. On the other hand, we have also shown that, in scaled terms,

$$(\sigma_{11})_0 = -p_0 + 2 \frac{\partial u_0}{\partial X} = 4 \frac{\partial u_0}{\partial X},$$

and so we expect the leading-order tension to be

$$(h \sigma_{11})_0 = 4 h_0 \frac{\partial u_0}{\partial X}.$$

Thus, we anticipate that

$$\frac{\partial}{\partial X} \left(4 h_0 \frac{\partial u_0}{\partial X} \right) = 0,$$

giving a second relation between h_0 and u_0 but not $\bar{h}_0$. In dimensional terms, this says that

$$\sigma_{11} \sim 4 \mu \frac{\partial u}{\partial x}$$

and so the Trouton viscosity for a sheet is 4μ, a result we could never have guessed. (For a slender fibre it is an even less likely 3μ.)

This is encouraging, so let us press on to the $O(\epsilon^2)$ equations. We solve them in the same order as the $O(1)$ equations, first (20.14) with (20.18), then (20.16) with (20.17) and lastly (20.15) with (20.19). We have reached the stage at which the arithmetical details become unedifying and are best dealt with in private; here is a sketch.

From (20.14)

Hic opus, hic labor est.

$$\frac{\partial^2 u_1}{\partial Y^2} = -3 \frac{\partial^2 u_0}{\partial X^2},$$

which, with (20.18) on $Y = \bar{h}_0(X, T) \pm \frac{1}{2}h_0(X, T)$, gives

$$\frac{\partial u_1}{\partial Y}\bigg|_{Y=\bar{h}_0-\frac{1}{2}h_0}^{Y=\bar{h}_0+\frac{1}{2}h_0} = -3Y\frac{\partial^2 u_0}{\partial X^2}\bigg|_{Y=\bar{h}_0-\frac{1}{2}h_0}^{Y=\bar{h}_0+\frac{1}{2}h_0}$$
$$= \text{a lot of terms involving } u_0, \bar{h}_0, h_0.$$

After simplification, we do indeed find that

$$\frac{\partial}{\partial X}\left(4h_0\frac{\partial u_0}{\partial X}\right) = 0. \tag{20.20}$$

The other equations are integrated in a similar way and lead, eventually, to the equation

$$\frac{\partial}{\partial X}\left(4h_0\frac{\partial u_0}{\partial X}\frac{\partial \bar{h}_0}{\partial X}\right) = 0.$$

Bearing in mind the equation (20.20) just found, this shows that

$$\frac{\partial^2 \bar{h}_0}{\partial X^2} = 0,$$

and so the sheet is, to lowest order, straight (the same applies to a fibre). This should not be taken to mean that all viscous sheets and fibres are straight, but rather that if they are being stretched on the timescale L/U of our analysis they must be straight. If the ends of a curved sheet are pushed together, another model must be used, as it also must when the sheet is being stretched so rapidly that the slow-flow assumption does not hold.

I cannot leave this topic without pointing out that the nonlinear equations we have derived,

$$\frac{\partial h_0}{\partial T} + \frac{\partial(u_0 h_0)}{\partial X} = 0, \qquad \frac{\partial}{\partial X}\left(4h_0\frac{\partial u_0}{\partial X}\right) = 0,$$

can be reduced to a linear system. You can find out about this by doing Exercise 13.

20.4 Further reading

There is much more on the derivation of models for thin sheets and fibres in the papers [28], [12]. For nonlinear diffusion equations, especially of higher order, see the book [50].

20.5 Exercises

1 **Tilting pad bearings.** Calculate the pressure in a slider bearing of (dimensionless) thickness $H(X) = 1 + \alpha X$. Write down an integral for the load.

Now suppose that the upper surface of the bearing is pivoted freely at a point X_0 ($0 < X_0 < 1$). Write down a moment condition for the bearing to be in equilibrium, and deduce a relation between the load and the angle α of the upper 'tilting pad'. (Don't try to simplify the integrals without a symbolic manipulator such as Maple.) This kind of bearing is self-adjusting: the pad tilts to accommodate whatever load is imposed.

2 **Two-dimensional bearings.**   Extend the analysis of the slider bearing to a rectangular upper surface above a flat lower surface, to derive a two-dimensional version of Reynolds' equation.

3 **Squeeze films.**   Suppose that, in addition to the moving lower plate as in Figure 20.1, a slider bearing has an upper surface that is time varying with characteristic frequency ω, as a result of, for example, the imposition of a periodic load, so that the gap is $H(X, T)$ in dimensionless variables, in which t is scaled with $1/\omega$. Show that mass conservation is

$$\frac{\partial}{\partial X} \int_0^{H(X)} u(X, Y)\, \mathrm{d}Y + \sigma \frac{\partial H}{\partial T} = 0,$$

where $\sigma = \omega L/U$ is a dimensionless parameter known as the *bearing number*, and that Reynolds' equation becomes

$$\frac{\partial}{\partial X}\left(H^3 \frac{\partial p}{\partial X}\right) = 6\frac{\partial H}{\partial X} + 12\sigma \frac{\partial H}{\partial T}.$$

Now suppose that $U = 0$, but that the upper surface is oscillated up and down with frequency ω and amplitude a, thus forming a *squeeze film*. Scaling t with $1/\omega$ and v with $a\omega$, what are the appropriate scales for u and p? Show that the appropriate version of Reynolds' equation is

$$\frac{\partial}{\partial X}\left(H^3 \frac{\partial p}{\partial X}\right) = 12\frac{\partial H}{\partial T}.$$

Show that by oscillating a suitably shaped upper surface in a direction normal to the lower surface it is possible to generate a non-zero pressure (averaged over one cycle of oscillation) even if $U = 0$. This effect is used to move silicon chips around semiconductor plants on oscillating tracks with saw-tooth-shaped surfaces (asymmetry in the surfaces generates a longitudinal pressure gradient, which induces motion in that direction).

Show that if a constant load is applied in a direction normal to two flat parallel plates initially a distance H_0 apart, they take an infinite time to make contact. (In practice, no surface is absolutely flat, and small *asperities* in the surfaces make contact well before $t = \infty$. It is almost impossible to pull apart two optically flat surfaces that have been put together, and it can be surprisingly hard to lift a sheet of paper away from a smooth surface. The trick is of course to slide the optically flat surfaces and to lift the paper from the edge.)

4 Surface-tension-driven thin horizontal film. Consider the evolution of a thin nearly flat horizontal fluid layer under the action of surface tension. Assume that the effect of surface tension is to give a jump in the normal stress across the fluid surface of

$$\gamma \times \text{curvature},$$

where γ is the surface tension coefficient. Show that the curvature of a nearly flat interface $y = h(x, t)$ is approximately $\partial^2 h / \partial x^2$ and hence that the pressure in the flow is

$$p(x, y, t) \sim \gamma \frac{\partial^2 h}{\partial x^2}.$$

Deduce that the thickness satisfies

$$\frac{\partial h}{\partial t} + \frac{\gamma}{3\mu} \frac{\partial}{\partial x} \left(h^3 \frac{\partial^3 h}{\partial x^3} \right) = 0.$$

What is the timescale for the flow? What is the equation when we also consider variations in the third ('into the paper') direction?

Repeat the systematic asymptotic derivation in the latter case, imposing the stress conditions

$$\sigma_{ij} n_i n_j = \gamma \kappa, \quad \sigma_{ij} n_i t_j = 0,$$

where κ is the curvature with the appropriate sign and $\mathbf{t} = (t_i)$ is the unit tangent to the free surface.

What are the dimensions of γ?

If s measures arclength along the curve $y = f(x)$ and ψ is the angle between the tangent and the x-axis then the curvature is $\kappa = d\psi/ds$, which is equal to $f''/(1 + (f')^2)^{3/2}$.

5 Similarity solution for thin fluid layer. Show that the equation

$$\frac{\partial h}{\partial t} = \frac{1}{3} \frac{\partial}{\partial x} \left(h^3 \frac{\partial h}{\partial x} \right)$$

(the dimensionless version of (20.1)) has similarity solutions of the form

$$h(x, t) = t^{-\alpha} f(x/t^{\alpha})$$

and find α and the ordinary differential equation satisfied by $f(\xi)$, where $\xi = x/t^{\alpha}$. Show further that this equation has solutions of the form

$$f(\xi) = \begin{cases} A(c^2 - \xi^2)^{\beta}, & |\xi| < c, \\ 0, & |\xi| > c, \end{cases}$$

and find the constants A, c, β. (Although this solution, which has compact support, does not have continuous derivatives at $\xi = \pm c$, it represents the evolution of a blob of fluid whose extent is always finite since the mass flux at $x = \pm ct^{\alpha}$ is bounded. This property is generated by the nonlinearity and in particular the fact that the 'diffusion coefficient' (the h^3 multiplying $\partial^2 h/\partial x^2$) vanishes at $h = 0$. A linear diffusion equation, or one whose diffusion coefficient is bounded away from zero, could never produce such a solution. Notice also that this solution tends to $\delta(x)$ as $t \to 0$. Although the thin-film assumption is not valid if h is a delta function, one can still think of this solution as representing the large-time asymptotic behaviour of any initial blob with compact support.)

6 Viscous liquid on an inclined plane.   Give a careful asymptotic derivation of the equation

Note that this is dimensional.

$$\frac{\partial h}{\partial t} + \frac{g \sin \theta}{\nu} h^2 \frac{\partial h}{\partial x} = 0$$

for the spreading of a thin viscous fluid sheet down a plane inclined at an angle θ to the horizontal. What model is appropriate if $\theta = O(\epsilon)$, where ϵ is the slenderness parameter of the sheet?

Harder: If the flow is over the surface $z = f(x, y)$, show that the generalisation is

$$\frac{\partial h}{\partial t} - \frac{g}{3\nu} F \nabla \cdot \left(h^3 F \nabla f \right) = 0,$$

where $F(x, y) = (1 + |\nabla f(x, y)|^2)^{-1/2}$ and $h(x, y, t)$ is the layer thickness measured *normally* to the surface.

7 Liquid paint flow.   A thin layer of viscous paint flows down a vertical wall. Taking x to be measured downwards, write the paint thickness as $y = h(x, t)$, and work through the following alternative derivation of the equation for h. Because the layer is thin, its velocity may be taken to be approximately $u(x, y, t)$ in the x-direction. Gravity is resisted by the viscosity of the paint, resulting in a shearing force that we assume to be approximately equal to $\mu \, \partial u/\partial y$, where μ is the viscosity of the fluid and $\partial u/\partial y$ is the velocity gradient. Use a force balance for a small fluid element to show that $\partial^2 u/\partial y^2 = -\rho g/\mu$. Using the boundary conditions $u = 0$ on $y = 0$ (no slip) and $\partial u/\partial y = 0$ on $y = h(x, t)$ (no shear at the free surface), deduce that

$$u = \frac{\rho g}{2\mu} y(2h - y).$$

Show that mass conservation requires that

$$\frac{\partial h}{\partial t} + \frac{\partial}{\partial x} \int_0^h u \, dy = 0,$$

that is,

$$\frac{\partial h}{\partial t} + \frac{\rho g}{\mu} h^2 \frac{\partial h}{\partial x} = 0.$$

Assume a lengthscale L for variations in h, and a typical thickness H. Make the equation dimensionless and write it in conservation form

$$\frac{\partial h}{\partial t} + \frac{\partial(\frac{1}{3}h^3)}{\partial x} = 0.$$

You may want to refer back to Exercise 12 at the end of Chapter 3 to remind yourself about the dimensional analysis of this problem.

Write down the characteristic equations and draw the characteristic projections in the xt-plane (a) when $h(x, 0) = h_0(x)$ is an increasing function of x, and (b) when it is a decreasing function of x. Interpret the results. Which flows quicker, a thick layer or a thin one?

Antonio, Bruno and Carlo are cooking. They have a very small amount of olive oil (a nice Newtonian fluid, unlike many liquids in the kitchen) in the bottom of one of those square bottles. Carlo suggests turning the bottle upside down and holding it vertically while Bruno says that it should be held at an angle to the horizontal with one of its flat sides facing downwards. Assuming that the oil flows down the sides of the bottle rather than simply falling through the air to the neck, who will wait longer, and why? Antonio, who has done the first part of this exercise, has a simple twist on Bruno's method that improves it significantly: what, and why?

OK, a cylinder of square cross-section ...

Returning to the equation for $h(x, t)$ on a vertical wall, show that there is a similarity solution of the form $h(x, t) = t^{-\alpha} f(x/t^\alpha)$ and find α (put this form into the equation and show that it works.) Show that the total mass of liquid in this solution is constant, and note that $h(x, 0) = \delta(x)$. Find f assuming that $h(0, t) = 0$ and solve only for $x > 0$. Show that if h is equal to this similarity solution for $0 < x < S(t)$ and is zero elsewhere, so that there is a discontinuity shock at $x = S(t)$, then the Rankine–Hugoniot shock condition

$$\frac{dS}{dt} = \frac{\left[\frac{1}{3}h^3\right]}{[h]}$$

is satisfied provided that $S = At^{1/3}$ for constant A. How should this solution be interpreted?

Suppose now that the film thickness is nearly constant, and look for small perturbations by finding solutions of the dimensionless equation in the form $h = 1 + \epsilon e^{i(kx+\omega t)}$ where $\epsilon \ll 1$ (like doing water waves). What is the relation between k and ω? In which direction do these waves travel? (Note that here there is only one direction of travel; water waves have two.) What is their dimensional speed? Show that the only smooth travelling-wave solutions (i.e. solutions of the form $h(x, t) = g(x - Ut)$ for constant U) to the full equation are $h = $ constant. However, the linearised solution you have just found looks like a travelling wave. How do you reconcile these facts?

8 Linear stability of thin films on horizontal surfaces. Investigate the linear stability of thin films under gravity or surface tension, by writing

$$h(x, t) \sim h_0 + \epsilon e^{ikx} e^{\lambda t}$$

in the dimensional equations and finding λ in terms of k. Note how the sign in front of the space derivatives changes when we go from second order to fourth order and relate this to the linear stability result.

9 Marangoni effects in a thin layer. Some flows are driven by variations in the surface tension coefficient γ. This may be due to temperature variations, or because there is a surfactant chemical in the fluid, or because some other effect such as evaporation of a solvent changes γ. The net effect is to induce a tangential (shear) stress at the interface, which acts to drag the fluid from regions of low surface tension to those where it is high. Foams are a particularly important practical example; the thin fluid sheets that form the bubble faces are stabilised by surfactants that counteract the tendency of the fluid to drain into the lower pressure regions where fluid sheets meet (known as Plateau borders; the pressure is lower there because of the curvature of the surface, as a sketch will show).

Suppose that we have a thin fluid layer as above and that the surface tension coefficient $\gamma(x)$ varies in a known way by an $O(1)$ amount (i.e. $\Delta\gamma/\gamma = O(1)$) over a horizontal distance L. Assuming that the Marangoni force translates into the (dimensional) boundary condition

$$\mu \frac{\partial u}{\partial y}\bigg|_{y=h(x,t)} = \frac{d\gamma}{dx},$$

explain why the flow is locally equivalent to Couette flow with a linear velocity profile and derive the equation

$$\frac{\partial h}{\partial t} + \frac{1}{2\mu} \frac{\partial}{\partial x}\left(h^2 \frac{d\gamma}{dx}\right) = 0.$$

What is the timescale of the motion? How small would the surface tension variation with x have to be for the normal force (surface tension $\times$ curvature) to be significant? What is the equation for h in this case?

10 Tides. Consider water waves in a basin $0 < x < L$, $-H < y < 0$. The velocity potential $\phi(x, y, t)$ and surface elevation $\eta(x, t)$ for

Paint drying, that riveting example.

small-amplitude waves satisfy

$$\frac{\partial^2 \phi}{\partial x^2} + \frac{\partial^2 \phi}{\partial y^2} = 0, \qquad 0 < x < L, \quad -H < y < 0,$$

with

$$\frac{\partial \phi}{\partial y} = 0 \text{ on } y = -H, \qquad \frac{\partial \phi}{\partial y} = \frac{\partial \eta}{\partial t}, \qquad \frac{\partial \phi}{\partial t} + g\eta = 0 \text{ on } y = 0,$$

with suitable boundary conditions on $x = 0, L$. Make the problem nondimensional, scaling x with L and y with H, and, using the timescale $\sqrt{L^2/(gH)}$, show that in the dimensionless version of the problem $\phi(x, y, t)$ satisfies the *elliptic* equation

$$\frac{\partial^2 \phi}{\partial y^2} + \epsilon^2 \frac{\partial^2 \phi}{\partial x^2} = 0 \quad \text{in } -1 < y < 0,$$

with

$$\frac{\partial \phi}{\partial y}(x, -1, t) = 0, \qquad \frac{\partial \phi}{\partial y}(x, 0, t) + \epsilon^2 \frac{\partial^2 \phi}{\partial t^2}(x, 0, t) = 0.$$

Show that an expansion in which

$$\phi \sim \phi_0(x, y, t) + \epsilon^2 \phi_1(x, y, t) + \cdots$$

satisfies the equation and the boundary condition up to terms of $O(\epsilon^2)$ if $\phi_0 = \phi_0(x, t)$, where the $O(\epsilon^2)$ equation shows that ϕ_0 satisfies the *hyperbolic* equation

$$\frac{\partial^2 \phi_0}{\partial x^2} - \frac{\partial^2 \phi_0}{\partial t^2} = 0.$$

What (in dimensional terms) is the wave speed?

This example, which is a *very* simple model for tidal flows on earth (it does not even have the daily periodicity built in, nor the rotation of the earth!), shows that the solution of an elliptic equation can sometimes be consistently approximated by that of a hyperbolic equation. The development of mathematical models for tide prediction preoccupied many famous minds – Newton and Laplace among them – and is described in [7]. One approach was to expand the water depth (as a function of time) as a series of harmonic terms, to reflect the periodic influence of the sun, moon etc. The summation of such a series by hand was a tedious and error-prone business, which was greatly facilitated by the invention by Lord Kelvin of a mechanical analogue based on pulleys. These machines were used until well into the twentieth century, and one of them can be seen in Liverpool Museum.

11 Shallow-water equations. There is no need to consider only sticky fluids in thin layers. In this question we derive the famous shallow water model for inviscid flow, starting with an intuitive derivation.

(a) Write down the two-dimensional Euler equations

$$\rho\left(\frac{\partial u}{\partial t} + u\frac{\partial u}{\partial x} + v\frac{\partial u}{\partial y}\right) = -\frac{\partial p}{\partial x},$$

$$\rho\left(\frac{\partial v}{\partial t} + u\frac{\partial v}{\partial x} + v\frac{\partial v}{\partial y}\right) = -\frac{\partial p}{\partial y} - \rho g,$$

$$\frac{\partial u}{\partial x} + \frac{\partial v}{\partial y} = 0$$

for the unsteady flow of an inviscid liquid under gravity.

(b) Assume that there is a base at $y = 0$ and a free surface at $y = h(x, t)$, that the layer is long and thin, and that the flow is fast enough that the velocity is approximately unidirectional and independent of depth and hence of the form $(u(x, t), 0)$.

(c) Assume further that the pressure is approximately hydrostatic; show that

$$p(x, y, t) = \rho g\left(h(x, t) - y\right).$$

(d) Write down mass conservation.

(e) Put these assumptions into the Euler equations to derive

$$\frac{\partial u}{\partial t} + u\frac{\partial u}{\partial x} + g\frac{\partial h}{\partial x} = 0, \qquad \frac{\partial h}{\partial t} + \frac{\partial(uh)}{\partial x} = 0.$$

Linearise the system about the constant solution $u = 0$, $h = h_0$ and find the speed of propagation of small disturbances; compare with the previous exercise.

This hyperbolic system can describe all sorts of phenomena such as the Severn Bore or its less well-known cousin the Trent Aegir (and a host of other bores around the world); they appear as shocks in the solutions. See [46] for lots more about the shallow-water equations and their properties.

Now derive these equations by a lubrication scaling of the Euler equations, to justify the (very reasonable) assumptions made above. Scaling x with L, y with $\epsilon L = H_0$, t with L/U and p with $\epsilon \rho g L$, you should get at lowest order in ϵ

$$\frac{\partial u_0}{\partial T} + u_0\frac{\partial u_0}{\partial X} + v_0\frac{\partial u_0}{\partial Y} = -\frac{1}{F^2}\frac{\partial p_0}{\partial X},$$

$$\frac{\partial p_0}{\partial Y} = -1, \qquad \frac{\partial u_0}{\partial X} + \frac{\partial v_0}{\partial Y} = 0,$$

together with the kinematic and dynamic free surface conditions on $Y = h(X, T)$. Here $F^2 = U^2/(g H_0)$ is a dimensionless parameter called the *Froude* (rhymes with crowd) number, measuring the inertia–gravity balance. Notice that the X-momentum equation is not quite the same as the first of the shallow-water equations derived above, because of the term $v_0 \partial u_0/\partial Y$. Now make the additional assumption that $\partial u_0/\partial Y = 0$ at the inlet or beginning of the flow. Calculate p_0 and deduce that $\partial u_0/\partial Y = 0$ throughout the flow. Write down the condition for irrotationality at leading order and compare with the assumption that $\partial u_0/\partial Y = 0$; relate this to Kelvin's theorem in fully two-dimensional flow. Hence derive the shallow-water equations.

Harder: Derive these equations starting from potential flow. Why do you not now have to assume irrotationality explicitly?

12 Boussinesq flow in a porous medium.   Suppose that water flows in a porous rock (an *aquifer*) under the action of gravity. The French sewage engineer Darcy established the law

$$\mathbf{u} = -K \nabla(p + \rho g y),$$

Consistency: why a minus sign?

giving the fluid velocity $\mathbf{u}$ as proportional to the gradient of the pressure (after subtracting off the hydrostatic head); here y is vertically upwards and K is called the *mobility*. Assuming that water is incompressible, show that this model is equivalent to potential flow with potential $\Phi = -K(p + \rho g y)$.

A thin layer of water (called a water mound in the trade) lies above a horizontal impermeable base at $y = 0$. Intuitively or by scaling, or using both approaches, derive the nonlinear diffusion equation

$$\frac{\partial h}{\partial t} = K \rho g \frac{\partial}{\partial x}\left(h \frac{\partial h}{\partial x}\right).$$

Find the similarity solution corresponding to an initial distribution that is a delta function.

13 Linearising the viscous sheet equations.   Take the viscous sheet equations in the form

$$\frac{\partial h}{\partial t} + \frac{\partial (uh)}{\partial x} = 0, \qquad \frac{\partial}{\partial x}\left(4h \frac{\partial u}{\partial x}\right) = 0$$

and integrate to get

$$h \frac{\partial u}{\partial x} = f(t)$$

for some $f(t)$ that is proportional to the tension. Define

$$\tau(t) = \int_0^t f(s)\, ds$$

and set $u(x, t) = f(t)v(x, \tau)$. Show that

$$h\frac{\partial v}{\partial x} = 1, \qquad \frac{Dh}{D\tau} = -1, \qquad \text{where} \qquad \frac{D}{D\tau} = \frac{\partial}{\partial \tau} + v\frac{\partial}{\partial x}.$$

Differentiate $h = h(x, \tau)$ implicitly with respect to x and τ to show that $1 = h_x x_h$, $0 = h_x x_\tau + h_\tau$, then show that $v = x_\tau - x_h$ and differentiate again with respect to x to get v_x.

Change the independent variables to h, τ (a *partial hodograph transformation*) to show that $x(h, \tau)$ satisfies

$$\frac{\partial^2 x}{\partial h \partial \tau} - \frac{\partial^2 x}{\partial h^2} = \frac{1}{h}\frac{\partial x}{\partial h}.$$

By solving this linear equation for $\partial x/\partial h$, deduce that

$$\frac{\partial x}{\partial h} = \frac{1}{hF(h + \tau)}$$

for arbitrary F, and hence that

$$\frac{\partial h}{\partial x} = hF(h + \tau).$$

Notice that F can now be determined from the initial data for h, so the whole system can be solved explicitly.

'We approximate an irrational point by a sequence of increasingly irrational rational points.'

21

Case study: turning of eggs during incubation

21.1 Incubating eggs

This case study is based on a problem that was brought by Bristol Zoo to the 2003 UK Study Group meeting, held in Bristol.[1] The problem dealt with the artificial incubation of eggs of the African penguin, whose numbers are low enough that zoo breeding programmes are important. In such programs it is vital that as many eggs as possible survive to hatch. Early literature on incubation [10] was often directed to the rearing of pheasants, destined to be shot (at) by the English moneyed class (from the royal family down) and their guests.[2] More recent literature has been driven by conservation issues but, as in many areas of biology or zoology, opportunities for mathematicians abound.

Rather than preservation, as game-keeping is sometimes called.

The specific question is this. In the wild, hen penguins lay one egg. In the nest, they turn this egg by an appreciable fraction of a whole turn at intervals of about 20 minutes. What advantage does this confer? Is it to equilibrate the temperature within the egg or for some other purpose? The turning is especially important in the first few days of incubation, and we focus on this period.

---

[1] See www.maths-in-industry.org for the reports on the problems presented at that meeting; they ranged from artificial spider-silk manufacture via sweaty feet and eggs to the design of a transport system for Cardiff.

[2] 'Up goes a guinea, bang goes sixpence, down comes half a crown.' For younger readers, a guinea was 21 shillings, now £1.05, a sixpence equalled $2\frac{1}{2}$ current pence and half a crown was two shillings and sixpence.

Figure 21.1 Schematic diagram
of an egg.

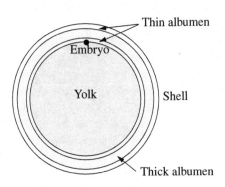

21.2 Modelling

We must first understand the structure of an egg. In presenting the model,
we are going to omit a large number of apparently inessential details of
egg composition (see [14] for more details). The minimum we need
to consider in a new egg is the yolk, the white (called *albumen*) and
the embryo, which develops into the chick, as shown schematically in
Figure 21.1. The yolk is contained in a membrane, is more or less spherical and is slightly less dense than the albumen. It floats in the albumen, which has a layered structure, and we describe this in more detail
below. Lastly, the embryo is contained in a small capsule attached to
the surface of the yolk, and it is less dense than either the yolk or the
albumen.

Heat flow

Let us first see whether turning is a good way to maintain an even temperature in the egg, counteracting the temperature gradient from a warm
hen to a cold nest floor. To do this, we calculate the thermal diffusion time
R^2/κ (see p. 18) for an egg of radius R. For penguin eggs, R is about 2 cm
and the diffusivity κ is similar to that of water, about $1.4 \times 10^{-7} \text{ m}^2 \text{ s}^{-1}$.
This gives a thermal diffusion time $t_d \approx 3000$ s, comparable to the turn-
ing interval. So it is plausible that turning maintains a roughly even
temperature, although, given the comparability of the two time scales,
the interior temperature must still vary by an $O(1)$ fraction of the top-to-
bottom temperature difference; had the turning interval been much less
than the diffusion time, the interior temperature would have been much
more uniform.

However, it is observed that eggs in incubators, which have a uniform
temperature, also need to be turned in order to hatch successfully. There
must be some other effect of turning.

Fluid flow in the egg

Another possible effect of turning is to induce a fluid flow in the egg. Because the yolk is less dense than the albumen, it will float to the top; also, the embryo, which is even lighter, will make the yolk rotate until it is also at the top. The net effect is to bring the embryo rather close to the shell. It is possible that turning counteracts these effects, and to see whether this is plausible we need to work out the timescales on which they operate.

The first point to make is that, like many biological liquids and gels, the albumen is not a Newtonian viscous fluid. A reasonable model treats it as a material that behaves like an elastic solid when subject to rapid shears or stresses, but on slower time scales behaves as a viscous fluid. This is a particular case of what is called a *viscoelastic fluid*, the modelling of which is a rather complex business. For our purposes, we just assume that the effect of the elastic behaviour is to ensure that if the egg is turned reasonable quickly then it rotates as a solid body. Subsequently, the relaxation is a slow viscous flow driven by buoyancy forces due to density differences within the egg.

Crazy putty is an example. You can roll it into a ball and bounce it off the floor but, put into its pot, over a long time it spreads out like a viscous liquid.

The next point is that the deformable part of the albumen forms a relatively thin layer surrounding the yolk. The typical thickness of the albumen is up to 25% of the egg radius, and it is in three layers, as shown in Figure 21.1. There is a thin layer of relatively low-viscosity albumen surrounding the yolk, then a thicker layer of relatively high-viscosity albumen and finally another layer of relatively low-viscosity albumen next to the shell (the viscosity ratio is more than 10 to 1, and the yolk is more viscous still). The distance across each less viscous layer is about 5% of the the radius of the egg, and the more viscous layer is three times as far across. In view of the result of Exercise 1, it is reasonable to treat the yolk and high-viscosity albumen as a rigid body, and just to look at the fluid flow in the outer low-viscosity layer.[3]

We can assume (or, if dubious, check after the fact) that lubrication theory gives a good model for the flow in the outermost layer. There are two forces driving this flow: the buoyancy of the yolk, which makes it rise, and the buoyancy of the embryo, whose principal effect is to make the yolk rotate so as to bring the embryo to the top. The densities of the embryo, yolk and albumen are very close to that of water, ρ_w, and differ by about 0.5%. Thus the density differences are of size $\Delta \rho = 0.005 \rho_w$. The buoyant force from a spherical yolk of radius R_y is

$$F_b = \tfrac{4}{3} \pi r_y^3 g \Delta \rho.$$

---

[3] A more sophisticated treatment would allow flow in the inner low-viscosity layer as well.

This drives the flow of the albumen, with viscosity μ_a, in its gap of thickness, say, h_a; as we have seen, h_a is much smaller than the length of the gap, which is comparable with r_y. This is a squeeze-film flow and, adapting the result of Exercise 3 at the end of Chapter 20, we can show that the typical pressure generated in a flow of this kind is of size $\mu_a V_y r_y^2 / h_a^3$, where V_y is the typical size for the vertical velocity of the yolk (the velocity normal to the layer). Multiplying by r_y^2 to get a typical size for the pressure force, we can equate the result with the buoyancy force to see that

Numerical constants such as 4 or π are dropped because we are only doing an order of magnitude argument.

$$r_y^3 g \Delta\rho \qquad \text{balances} \qquad \frac{\mu_a V_y r_y^4}{h_a^3},$$

and so the order of magnitude estimate for V_y is $h_a^3 g \Delta\rho /(\mu_a r_y)$. Putting in the values $h_a \approx 10^{-3}$ m, $\mu_a \approx 4 \times 10^{-3}$ kg m^{-1} s^{-1} and $\Delta\rho \approx 5$ kg m^{-3}, we find that $V \approx 6 \times 10^{-4}$ m s^{-1}. That is, the yolk takes a few seconds to float a millimetre or two towards the top of the shell (of course, as the albumen layer between the yolk and the shell gets thinner and thinner, the yolk slows down markedly). This is much shorter than the turning interval, and no amount of adjustment for the second layer of thin albumen helps. We should also ask whether flow in the more viscous albumen layer, which is two or three times as wide as the less viscous layers, will take place on the inter-turn timescale. As h_a is three times bigger and μ_a is 12 times larger than for the less viscous layers, the typical vertical velocity is $3^3/12$ times that for the thin layers, and again the timescale of the motion is less than we need if flotation is to provide a reason for the egg-turning. The details of the flow are described in Exercise 2.

Flotation is not the mechanism we require. Can rotation help?

Rotation of the yolk

Let us now see how the yolk rotates back to a position in which the embryo is at the top. In this motion, the off-centre buoyancy force of the embryo exerts a torque on the yolk, the magnitude of which is $r_y g \Delta m$, where Δm is the mass deficit of the embryo and is of $O(r_e^3 \Delta\rho)$, r_e being a typical size for the embryo. This is balanced by a shear force $\mu_a r_y \omega_y / h_a$ per unit area, where ω_y is the angular speed of the yolk, and we multiply this by r_y^2 to find an order of magnitude for the viscous force, and again by r_y to find the resistive torque. All this leads to the estimate

The inertia of the yolk is neglected. What is the small parameter that justifies this?

Check the dimensions!

$$\frac{1}{\omega_y} \approx O\left(\frac{\mu_a r_y^3}{h_a g \Delta m}\right),$$

which, with an embryo size of 1 mm and other parameter values as above,

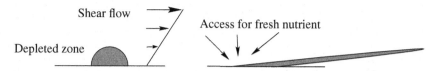

Figure 21.2 Shear-enhanced diffusion.

gives a timescale of hundreds to thousands of seconds, which is exactly comparable with the turning interval.

But why should rotation be significant? One reason may be that, after turning, the embryo rises to the top by rotation and, because the yolk has already floated to the top on a faster timescale, it then finds itself uncomfortably close to the shell. More turning returns it to a position at the side, and the timescale of rotation determines that of turning. Another possibility is that rotation enhances the supply of nutrients to the embryo, and we now explore this further.

Diffusion

The albumen is not just a passive fluid. It contains nutrients and some antiseptic agents that sustain and protect the embryo. But, in the early stages before the embryo has developed a vascular system, these agents can only get to the embryo by diffusion. Now, with a typical molecular diffusivity D_a of about 10^{-9} m^2 s^{-1}, on a timescale τ of 1000 s the size of the region depleted of nutrients is of order $\sqrt{D\tau}$, about 1 mm. As time goes on, fresh nutrients have further and further to travel to get to the embryo, and waste products have to diffuse further too. But if the albumen is occasionally *sheared*, as happens in a rotation of the yolk, the picture changes. Now the depleted region is stretched out, as in Figure 21.2, and there can be access for new nutrients from the upstream side of the embryo. This is a version of what is called Taylor diffusion[4] and is another plausible explanation for why eggs are turned.

This model is a very recent and tentative one. It is an example of a complicated biological situation where the mathematical analysis of key physical mechanisms can show, on quantitative grounds, that a proposed explanation is implausible. Whether the conjectures above on the

---

[4] Yes, G. I. Taylor. This occurs when a substance (dye, say) is released into liquid flowing along a pipe at a point. The velocity profile is parabolic and so the initial dose of dye is stretched longitudinally by the flow and diffuses laterally. The net effect is an apparent longitudinal diffusion coefficient that is much larger than the molecular diffusivity. A similar effect occurs when a contaminant is advected in a groundwater flow through porous rock: the shearing that takes place in the flow through the pores greatly enhances diffusion.

effects of turning are correct needs further theoretical and experimental study.

21.3 Exercises

1 Rigid body motion in Stokes flow. The Stokes-flow equations are

$$\mu \nabla^2 \mathbf{u} = \nabla p, \qquad \nabla \cdot \mathbf{u} = 0.$$

Show that rigid body motion

$$\mathbf{u} = \mathbf{U}(t) + \boldsymbol{\Omega}(t) \wedge \mathbf{x}$$

satisfies these equations exactly and that $p = 0$ (this does not mean that there are no stresses in the fluid). This result is a consequence of the neglect of inertia in the Stokes-flow limit.

2 Rise of a buoyant yolk. Consider two-dimensional flow between two rigid circular cylinders. The inner cylinder is free to move (by both translation and rotation) and has radius R and centre $(X(t), Y(t))$; the outer cylinder, which is fixed, has radius $R(1 + \epsilon)$, $0 < \epsilon \ll 1$, and centre $(0, 0)$. The inner cylinder is acted upon by an upwards vertical force F. The gap is filled by liquid of viscosity μ.

Take polar coordinates r, θ centred at the origin. Let $\phi(t)$ denote the angle between a line fixed in the inner cylinder and the ray $\theta = 0$. Scaling all distances with R, and writing $X = Rx, Y = Ry$, show that the thickness of the gap is

$$\epsilon \left(1 - x(t) \cos \theta - y(t) \sin \theta \right) = \epsilon h(\theta, t), \quad \text{say.}$$

Show that, in suitable dimensionless variables and making the lubrication theory approximation, the pressure $p(\theta, t)$ satisfies

$$\frac{\partial}{\partial \theta} \left(\frac{h^3}{12} \frac{\partial p}{\partial \theta} \right) = \frac{\dot\phi}{2} \frac{\partial h}{\partial \theta} + \frac{\partial h}{\partial t},$$

where $\dot\phi = d\phi/dt$.

Show that, neglecting the inertia of the inner cylinder, the force balance on the inner cylinder is

As part of the lubrication analysis, you should explain why the contribution to the horizontal and vertical force balance of the drag force from the shearing of the fluid layer is neglected.

$$\int_0^{2\pi} p(\theta, t) \sin \theta \, d\theta = 1, \qquad \int_0^{2\pi} p(\theta, t) \cos \theta \, d\theta = 0,$$

where the '1' is the dimensionless version of the vertical force F. Integrate these equations by parts once and use the equation for p to show that they reduce to equations of the form

$$A_1(x, y)\dot\phi + A_2(x, y)\dot x + A_3(x, y)\dot y = 1,$$
$$B_1(x, y)\dot\phi + B_2(x, y)\dot x + B_3(x, y)\dot y = 0,$$

where the A_i and B_i are complicated integrals involving x and y as parameters.

Lastly calculate the velocity component u_θ along the layer and hence the shear force (neglected above). Assuming that the rotational inertia of the inner cylinder is negligible, so that the net torque on it vanishes, show that

$$\int_0^{2\pi} \frac{\dot{\phi}(t)}{h(\theta, t)} + \frac{h(\theta, t)}{2} \frac{\partial p}{\partial \theta} \, d\theta = 0,$$

which is the third equation and closes the system.

Suppose that $x(0) = 0$, $y(0) = y_0$, so that the cylinder rises vertically. Show that x and ϕ are constant and that

$$\dot{y} = \frac{(1 - y^2)^{3/2}}{12},$$

and solve this equation. What is the behaviour as $y \to 1$?

3 **Rotation of the yolk.** Suppose that the inner cylinder of the previous question is neutrally buoyant ($F = 0$) and centred at the origin, but that it is acted upon by a small couple equivalent to a vertical force f applied at a fixed point on its perimeter. Adapt the analysis of the previous question to find a solution in which $h(\theta, t)$ is constant and equal to 1 but $\phi(t)$ varies. Find the relevant dimensionless scales and show that, with a suitable choice of origin for ϕ, $\phi(t)$ satisfies

$$\cos\phi = 2\pi\dot{\phi}$$

(the cosine on the left comes from taking the component of the vertical force tangent to the cylinder). Solve this equation and analyse its large-time behaviour.

'The square root of a circle is a flat plane.'

22

Multiple scales and other methods for nonlinear oscillators

We end this book with two short chapters on other aspects of asymptotic expansions. This chapter deals with nonlinear oscillators and similar problems (planetary motion is an early example), and the final chapter covers the WKB or ray approach to slowly modulated linear systems, which, although related to the methods of this chapter, is important enough to deserve its own treatment.

Boundary layer techniques are designed to help when a regular expansion goes wrong in a small layer near a boundary (or elsewhere). A quite different situation arises for many nonlinear oscillator models. Here a regular expansion works well for an $O(1)$ time, but it drifts away from the true solution as small errors accumulate. A battery of methods has been devised to handle this problem, and we look at one rather limited but easy technique, the method of Poincaré and Linstedt, and one very general technique, the method of multiple scales.

22.1 The Poincaré–Linstedt method

Let us recall our discussion of small-amplitude oscillations of a pendulum in Section 13.6. We aim to solve

$$\frac{d^2\theta}{dt^2} + \sin\theta = 0$$

Use an energy argument to prove this.

with $\theta = \epsilon a_0$ and $d\theta/dt = 0$ at $t = 0$. The smallness of ϵ guarantees the smallness of the oscillations. We showed in Section 13.6 that a regular expansion leads to a solution of the form

$$\theta(t) \sim \epsilon a_0 \cos t + \epsilon^3 \theta_3(t) + \cdots,$$

292

where $\theta_3(t)$ contains a secular term proportional to $t \sin t$. When $t = O(1/\epsilon^2)$, the combination $\epsilon^3 t \sin t$ is the same size as the first term $a_0 \cos t$ and the expansion is invalid. We need to sort this out.

Let us first write $\theta(t) = \epsilon \phi(t)$ as we should have done in the first place. Then the first two terms in the expansion of $\sin \phi$ give

$$\frac{d^2\phi}{dt^2} + \phi = \tfrac{1}{6}\epsilon^2 \phi^3 + o(\epsilon^2),$$

or, writing $\delta = \tfrac{1}{6}\epsilon^2$,

$$\frac{d^2\phi}{dt^2} + \phi = \delta\phi^3 + o(\delta)$$

with

$$\phi = a_0, \qquad \frac{d\phi}{dt} = 0 \qquad \text{at} \quad t = 0.$$

The trick here is to expand the period (or frequency) in powers of δ as well as expanding ϕ. So, we seek a solution $\phi(t)$ such that

$$\phi(t + 2\pi/\omega) = \phi(t)$$

for all t, where ϕ and ω have expansions

$$\phi \sim \phi_0 + \delta\phi_1 + \cdots, \qquad \omega \sim \omega_0 + \delta\omega_1 + \cdots ;$$

obviously $\omega_0 = 1$ but we'll derive this *en route* to more useful results.

We introduce a scaled time $\tau = \omega t$, so that we have 2π-periodicity in τ: $\phi(\tau + 2\pi) = \phi(\tau)$. This gives us

$$\omega^2 \frac{d^2\phi}{d\tau^2} + \phi = \delta\phi^3.$$

Substituting the expansions for ω and ϕ and collecting terms of $O(1)$ and terms of $O(\delta)$, we get

$$\omega_0^2 \frac{d^2\phi_0}{d\tau^2} + \phi_0 = 0,$$

so the periodicity gives $\omega_0 = 1$ and then $\phi_0 = a_0 \cos \tau$: no surprises there. At $O(\delta)$, we find

$$\begin{aligned}
\frac{d^2\phi_1}{d\tau^2} + \phi_1 &= \phi_0^3 - 2\omega_1 \frac{d^2\phi_0}{d\tau^2} \\
&= a_0^3 \cos^3 \tau + 2\omega_1 \cos \tau \\
&= \tfrac{1}{4} a_0^3 \cos 3\tau + \left(\tfrac{3}{4} a_0^3 + 2\omega_1 a_0\right) \cos \tau.
\end{aligned}$$

This time, we *can* eliminate the secular terms, which arise because the factor $\cos \tau$ on the right-hand side of the equation is also a solution

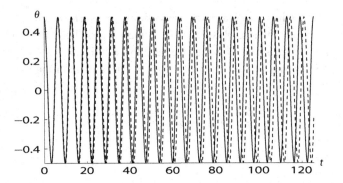

Figure 22.1 Gradual drift of a solution of the linear pendulum equation (solid curve) from the full solution (dashed curve). Here the initial amplitude is 0.5. As expected, the frequency of the full solution is slightly lower.

of the homogeneous equation, giving a resonance effect. We just set the coefficient of $\cos \tau$ equal to zero, and obtain

$$\omega_1 = -\frac{3a_0^2}{8}.$$

The phase drift is illustrated in Figure 22.1. It's then straightforward to show that

$$\phi_1 = \frac{a_0^3}{32} \left(\cos \tau - \cos 3\tau\right).$$

If you want, you can verify this result by integrating the full pendulum equation exactly and then expanding the period for small initial amplitude.

This simple example illustrates the Poincaré–Linstedt method, which consists simply in expanding the frequency of oscillation in powers of ϵ. There are examples for practice in the exercises; we move on now to greater things.

22.2 The method of multiple scales

The Poincaré–Linstedt method is rather limited in its applicability. Much more powerful is the method of multiple scales, which is built around the recognition that the solution changes in different ways over different timescales (or space scales: the method is a flexible tool). Thus, in the pendulum problem the basic oscillation is on an $O(1)$ timescale, while the phase drifts away from its linearised-theory value on a scale of $O(1/\epsilon)$. The novel idea in the method of multiple scales is to introduce new *independent* variables, one to describe each timescale, and to regard the solution as a function of this extended set of independent variables. This looks really weird until you get used to it, but it works.

Consider, then, the pendulum problem in the form

$$\frac{d^2\phi}{dt^2} + \phi = \delta\phi^3 + o(\delta),$$

with

$$\phi = a_0, \qquad \frac{d\phi}{dt} = 0 \qquad \text{at} \quad t = 0.$$

Introduce the new *slow time*

$$t_1 = \delta t,$$

which varies by only $O(1)$ if t (the *fast time*) varies by $O(1/\delta)$. Then regard ϕ as a function $\Phi(t, t_1)$ *of both time variables*. This entails replacing the time derivative d/dt by a partial derivative; using the chain rule, we get

$$\frac{d}{dt}(\Phi(t, t_1(t))) = \frac{\partial \Phi}{\partial t} + \frac{\partial \phi}{\partial t_1}\frac{\partial t_1}{\partial t} = \frac{\partial \phi}{\partial t} + \delta \frac{\partial \Phi}{\partial t_1},$$

and so

$$\frac{d}{dt} \rightarrow \frac{\partial}{\partial t} + \delta \frac{\partial}{\partial t_1}.$$

Thus, we have

$$\frac{\partial^2 \Phi}{\partial t^2} + 2\delta \frac{\partial^2 \Phi}{\partial t \partial t_1} + \delta^2 \frac{\partial^2 \Phi}{\partial t_1^2} + \Phi = \delta \Phi^3 + o(\delta),$$

with

$$\Phi = a_0, \qquad \frac{\partial \Phi}{\partial t} = 0 \qquad \text{at} \quad t = 0.$$

Now, we use a regular expansion $\Phi \sim \Phi_0 + \delta \Phi_1 + \cdots$ to find, at leading order,

$$\frac{\partial^2 \Phi_0}{\partial t^2} + \Phi_0 = 0, \qquad \Phi_0 = a_0, \qquad \frac{\partial \Phi}{\partial t} = 0 \qquad \text{at} \quad t = 0.$$

The solution of this problem is

$$\Phi_0(t, t_1) = A(t_1)\cos(t + \psi(t_1)),$$

where $A(t_1)$ and $\psi(t_1)$ are undetermined. All we know about these functions is that $A(0) = a_0$, $\psi(0) = 0$. Think about this: if $t = 0$ then certainly $t_1 = 0$, and indeed if $t = O(1)$ then t_1 is small. These conditions are effectively matching conditions joining the initial 'fast' regime onto the subsequent 'slow' regime. Put another way, while t is $O(1)$, the leading-order solution in a regular expansion is just $a_0 \cos t$, consistent with $A(0) = a_0$, $\psi(0) = 0$.

We have to go to $O(\delta)$ to find $A(t_1)$ and $\psi(t_1)$. We have

$$
\frac{\partial^2 \Phi_1}{\partial t^2} + \Phi_1 = -2\frac{\partial^2 \Phi_0}{\partial t \partial t_1} + \Phi_0^3
$$
$$
= 2\left(\frac{dA}{dt_1}\sin(t+\psi) + A\frac{d\psi}{dt_1}\cos(t+\psi)\right)
$$
$$
+ \tfrac{1}{4}A^3\left(3\cos(t+\psi) + \cos 3(t+\psi)\right).
$$

This is to be solved with $\Phi_1 = \partial \Phi_1/\partial t = 0$ at $t = 0$. Inspecting the right-hand side, we see that secular terms will appear in Φ_1 unless we choose

$$
\frac{dA}{dt_1} = 0, \qquad 2A\frac{d\psi}{dt} + \tfrac{3}{4}A^3 = 0.
$$

Thus, A is constant and equal to a_0, while $\psi = -\tfrac{3}{8}a_0^2 t_1 = -\tfrac{3}{8}a_0^2 \delta t$. This is exactly the same result as we found using Poincaré–Linstedt.

There are several things to say about this calculation. The most important is that, although the manipulation looks very similar to Poincaré–Linstedt, the method of multiple scales is completely different in spirit and *much* more general and powerful. Poincaré–Linstedt is essentially limited to analysis near periodic orbits, while multiple scales can handle many other problems (see Exercises 3 and 4 for the contrast). In particular, the amplitude and phase of the oscillation are allowed to drift by $O(1)$ as functions of the slow time t_1, rather than just being close to their leading-order values. Multiple scales can also handle problems such as heat conduction in a medium with a thermal conductivity that varies rapidly (on the fast scale), and many others.

The process of eliminating secular terms is, essentially, the Fredholm Alternative, in that it works by ensuring that the right-hand side of the equation for Φ_1 is orthogonal to the periodic solutions ('eigensolutions') of the differential operator on the left. In this way, the apparent indeterminacy that is introduced by replacing the single independent variable t by the pair t and t_1, which manifests itself in the unknown functions A and ψ, is resolved by the application of the Fredholm alternative at the next order in the expansion; this is an idea we saw in Chapter 18. If we had continued with the calculation above, we would have found that Φ_1 contained further arbitrary functions, which have to be found by going to higher order. However, if we want to go to $O(\delta^2)$ then we have to introduce a new, even slower, timescale $t_2 = \delta^2 t$ (we also have to expand $\sin\theta$ in the original problem to higher order) and the complexity increases very rapidly.

22.3 Relaxation oscillations

In all the previous examples, we have looked at systems whose behaviour is, to leading order, a linear spring system (simple harmonic motion), with a small nonlinearity in the spring, damping or both. The primary balance is thus between the acceleration and spring force terms. Our final example of oscillatory behaviour is what is called a *relaxation oscillator*, in which the inertia of the system is small and a priori the balance is between the spring and the damping, giving a first-order system with multiple states. The acceleration term only comes in when the system has a rapid transition from one state of the first-order system to another. A quick glance at an EEG trace strongly suggests that physiological systems can show such behaviour, and the boom–bust populations of various species (lemmings, for example) suggest that they occur in population-dynamics models as well. Relaxation oscillators are used to model both kinds of system (see [43]).

For an example, consider the following version of the *FitzHugh–Nagumo* system, which has been used to model electrical activity in neurons. In the form of the second-order equation

$$\epsilon \frac{d^2 u}{dt^2} + \frac{du}{dt}\left(\epsilon - f'(u)\right) + u = 0, \quad f(u) = u(1 - u^2), \quad 0 < \epsilon \ll 1,$$

it looks unpromising, but written in what is sometimes called Liénard form[1] as the first-order system

$$\epsilon \frac{du}{dt} = f(u) - v, \tag{22.1}$$

$$\frac{dv}{dt} = u - v, \tag{22.2}$$

some structure is apparent.

---

[1] Consider the differential equation

$$\frac{d^2 u}{dt^2} + g(u)\frac{du}{dt} + u = 0.$$

It can be written as a first-order system by putting $du/dt = V$, $dV/dt = -g(u)V - u$. However, if g is not well behaved, as might be the case for example in a model of stick–slip oscillations, this system may not handle well numerically. Liénard's idea was to write the original equation as

$$\frac{d}{dt}\left(\frac{du}{dt} + G(u)\right) + u = 0;$$

then $v = du/dt + G(u)$ leads to the system

$$\frac{du}{dt} = v - G(u), \qquad \frac{dv}{dt} = -u.$$

Because integration is a smoothing operator, the new system should have better properties.

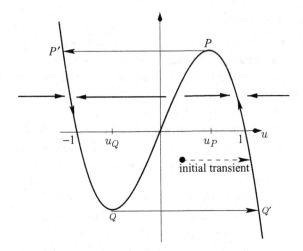

Consider Figure 22.2, which shows the graph of $f(u) = u(1 - u^2)$.
We rewrite (22.1) as

$$\frac{du}{dt} = \frac{1}{\epsilon} (f(u) - v)).$$

We then see that if we start the system off at any point in the uv-plane
away from $v = f(u)$, consideration of the sign of $f(u) - v$ on the right-
hand side of this equation shows that u immediately changes on a fast
timescale of $O(\epsilon)$ until the point $(u(t), v(t))$ lies on one of the outer parts
of the curve $v = f(u)$, from $u = -\infty$ to $u = u_Q = -1/\sqrt{3}$ or from
$u = u_P = 1/\sqrt{3}$ to $u = \infty$. The middle part of this curve is unstable.
Moreover, during this rapid transition v remains approximately constant.

Make the change of variable
$t = \epsilon\tau$ to see that $du/d\tau$ is $O(1)$
but $dv/d\tau$ is $O(\epsilon)$.

Thus, after an initial fast horizontal transient, the general idea is that
'most of the time', we can ignore the left-hand side of (22.1), which
therefore reduces to the algebraic relation $v = f(u)$. Under these *slow
dynamics*, the solution travels along the curve $v = f(u)$ according to the
equation

$$\frac{dv}{dt} = f'(u)\frac{du}{dt} = u - v = u - f(u).$$

On $Q'P$, $f(u) < u$ and so v can only increase, until point P, where
$u = u_P = 1/\sqrt{3}$ and $f'(u) = 0$. It then turns left and makes a sud-
den transition, on a timescale of $O(\epsilon)$, to the point P' on the other
stable branch. During this transition, which is essentially an interior
layer for the system, v again remains approximately constant, and so
P' is $(-2/\sqrt{3}, 2/3\sqrt{3})$. From here, the solution proceeds down the left-
hand stable branch of $v = f(u)$ and makes another transit to the right-
hand branch at Q. In this way, the oscillation proceeds by a series of
slow changes separated by rapid transitions (relaxations). The period of

The left turn can be analysed as
an inner region of size $O(\epsilon^{2/3})$,
as can the region around P'.

Solving $f(u) = f(1/\sqrt{3})$, we
know there is a double root at
$u = 1/\sqrt{3}$ so the other root is
easy to find.

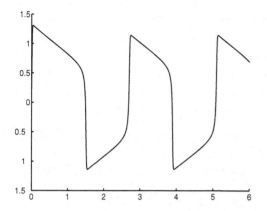

Figure 22.3 The
FitzHugh–Nagumo system:
the evolution of $u(t)$. Here
$\epsilon = 0.02$. Matlab's routine
ode45 had no trouble even
though ϵ is small enough to
warrant using a stiff solver.

oscillation is, to within $O(\epsilon)$, just twice the time taken to go from P' to Q, namely

$$2 \int_{u_{P'}}^{u_Q} \frac{f'(u)}{u - f(u)} \, du = 2 \int_{-2/\sqrt{3}}^{-1/\sqrt{3}} \frac{1 - 3u^2}{u^3} \, du$$

$$= 6 \log 2 - \frac{9}{4}$$

$$\approx 1.91.$$

If the curve $v = f(u)$ were not antisymmetric, we would have to work out two integrals, from P' to Q and Q' to P.

Figure 22.3 shows a numerical solution of the equation with $\epsilon = 0.01$ and $u(0) = 0$, $v(0) = -1$. The actual period is slightly greater than our prediction because of the turnaround near P and Q.

22.4 Exercises

1 Exact pendulum. Multiply the undamped pendulum equation

$$\frac{d^2\theta}{dt^2} + \sin \theta = 0$$

by $d\theta/dt$ and integrate, using the initial conditions $\theta = \epsilon a_0$ and $d\theta/dt = 0$. Separate the variables in this first-order equation to get an expression for half the period (if you want to look it up, it's an elliptic integral). Expand the integrand for small ϵ and integrate to confirm the Poincaré–Linstedt result.

2 Precession of the perihelion of Mercury. Recall that under Newtonian theory the planets move around the sun under the central force $-GM/r^2$ per unit mass, where M is the sun's mass and G is the universal gravitational constant (the forces due to other planets are ignored). Suppose that when it is nearest the sun (at perihelion), Mercury is at a distance a from the sun and is travelling with speed v.

Show that the equations of motion in plane polar coordinates,

$$\ddot{r} - r\dot{\theta}^2 = -\frac{GM}{r^2}, \qquad r^2\dot{\theta} = av,$$

and the substitution $u = 1/r$, lead to

$$\frac{d^2u}{d\theta^2} + u = \frac{GM}{a^2v^2}, \qquad \text{with} \quad u = \frac{1}{a}, \quad \frac{du}{d\theta} = 0 \qquad \text{at} \quad \theta = 0,$$

Why is there no $\sin\theta$ term?

and show that $u = A + B\cos\theta$ for some A and B that you should find. Note that the orbit is 2π-periodic in θ.

The theory of general relativity gives the modified equation

$$\frac{d^2u}{d\theta^2} + u = \frac{GM}{a^2v^2}\left(1 + \frac{3v^2a^2}{c^2}u^2\right),$$

where c is the speed of light. Writing $\epsilon = v^2/c^2$, find the solution up to $O(\epsilon)$ with the same initial conditions, and show that it is not 2π-periodic in θ. Show that the next perihelion (i.e. the next value of θ at which $du/d\theta = 0$) occurs at $\theta \sim 2\pi(1 + 3(GM/av^2)^2\epsilon)$. (Note that if $u(\theta;\epsilon) \sim u_0(\theta) + \epsilon u_1(\theta) + \cdots$ and $u_0'(\theta_0) = 0$, then the value of θ at which $u' = 0$ is found by writing it as $\theta \sim \theta_0 + \epsilon\theta_1 + \cdots$ and expanding the equation $u_0'(\theta_0 + \epsilon\theta_1 + \cdots) + \epsilon u_1'(\theta_0 + \cdots) = 0$ to $O(\epsilon)$. Here $\theta_0 = 2\pi$.) Confirm your analysis by carrying out the Poincaré–Linstedt expansion. Then do the problem by multiple scales.

This result has been used as a test of general relativity. If $a = 46 \times 10^6$ km, $v \approx 60$ km s^{-1} and $G = 6.6710^{-11}$ N m^2 kg^{-1}, how big is the shift per (Mercury) year?

3 Van der Pol and Rayleigh, I.   The Van der Pol equation is

$$\frac{d^2x}{dt^2} + \epsilon(x^2 - 1)\frac{dx}{dt} + x = 0, \qquad \epsilon > 0.$$

It was written down as a model for a spontaneously oscillating valve circuit: by considering the damping term explain why this is plausible. Where (for what values of x and dx/dt) is energy taken out and where is it put in?

Rayleigh's equation

$$\frac{d^2x}{dt^2} + \epsilon\left(\frac{1}{3}\left(\frac{dx}{dt}\right)^2 - 1\right)\frac{dx}{dt} + x = 0$$

You may need one or other of the expressions

$\sin^3\theta = \frac{1}{4}(3\sin\theta - \sin 3\theta),$

$\cos^3\theta = \frac{1}{4}(3\cos\theta + \cos 3\theta).$

was written down in connection with a model for a violin string. Show that it can be transformed into the Van der Pol equation by differentiation.

Take $\epsilon \ll 1$ in the Van der Pol equation, and show by the Poincaré–Linstedt method that a periodic solution of the form

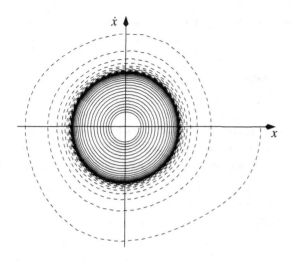

Figure 22.4 Two solutions of the Van der Pol equation, in the phase plane. Here $\epsilon = 0.05$. Both the dashed curve and the solid curve approach the limit cycle as $t \to \infty$.

$A \cos \tau + \epsilon u_1(\tau) + \cdots$, where $\tau = \omega t$ and $\omega \sim 1 + \epsilon \omega_1 + \cdots$, is only possible if $A = 2$ and $\omega_1 = 0$.

Harder: Draw the phase plane, noting the existence of this periodic solution (known as a *limit cycle*). This is shown in Figure 22.4.

4 Van der Pol and Rayleigh, II.   The Van der Pol equation

$$\frac{d^2 x}{dt^2} + \epsilon(x^2 - 1)\frac{dx}{dt} + x = 0, \qquad 0 \ll \epsilon < 1,$$

with $x = 1$ and $dx/dt = 0$ at $t = 0$ is a standard example of the application of multiple scales. Find a regular perturbation expansion correct to $O(\epsilon)$ and show that it has secular terms. Show that a multiple-scale expansion in terms of t and $t_1 = \epsilon t$ has solution

$$X(t, t_1) = A(t_1) \cos(t + \psi(t_1)), \qquad A(0) = 1, \quad \psi(0) = 0,$$

at leading order, and from the $O(\epsilon)$ terms show that

$$\frac{dA}{dt_1} = \frac{A}{8}(4 - A^2), \qquad \frac{d\psi}{dt_1} = 0.$$

Hence find $A(t_1)$. Compare with the result of Exercise 3. Use a package such as Matlab to solve the equation numerically and plot the exact and approximate solutions.

5 Relaxation oscillations for Van der Pol.   Consider the 'small-inertia' Van der Pol equation

$$\epsilon^2 \frac{d^2 x}{dt^2} + (x^2 - 1)\frac{dx}{dt} + x = 0, \qquad 0 \ll \epsilon < 1.$$

Write it in Liénard form as a first-order system and analyse the relaxation oscillations. Show that their period is approximately $3 - 2 \log 2$.

6 Two slow time scales.   Show that the solution of

$$\frac{d^2x}{dt^2} + 2\epsilon\frac{dx}{dt} + x = 0, \qquad x = 1, \qquad \frac{dx}{dt} = 0 \qquad \text{at} \quad t = 0$$

(another standard example in the subject) is

$$x(t) = e^{-\epsilon t}\cos\left(t(1 - \epsilon^2)^{1/2}\right).$$

Expand this for small ϵ, with t fixed and $O(1)$, to show that secular terms arise. Carry out a multiple-scale analysis with *two* slow times, $t_1 = \epsilon t$ and $t_2 = \epsilon^2 t$, to retrieve the approximate solution

$$x(t) \sim e^{-\epsilon t}\cos\left(t(1 - \tfrac{1}{2}\epsilon^2)\right)$$

in a systematic manner.

$$\frac{dy}{dx} = e^x \ldots \text{ looks pretty nonlinear to me.'}$$

23
Ray theory and the WKB method

23.1 Introduction

Consider the equation

$$\epsilon^2 \frac{d^2 y}{dx^2} + y = 0, \qquad 0 < \epsilon \ll 1,$$

whose solutions are

$$y(x; \epsilon) = e^{\pm ix/\epsilon}.$$

These solutions oscillate rapidly, on a fast timescale of $O(\epsilon)$. Unlike the equation

$$\epsilon^2 \frac{d^2 z}{dx^2} - z = 0,$$

whose solutions are

$$z(x; \epsilon) = e^{\pm x/\epsilon},$$

which is typical of boundary layers with their spatially limited zones of rapid change, the function $y(x; \epsilon)$ changes rapidly everywhere. Although we can find this function explicitly, we need to develop methods to solve common problems such as

$$\epsilon^2 \frac{d^2 y}{dx^2} + V(x)y = 0, \qquad 0 < \epsilon \ll 1,$$

where $V(x)$ is a smooth positive function varying on the $O(1)$ timescale. A good guess at the structure of the problem would be that the solution consists of a slowly modulated rapid oscillation, and this turns out to be correct. Although problems of this kind can be handled, with some contortions, by the multiple scale method, a specialised battery of techniques with the general name of WKB[1] (for ordinary differential equations) or ray theory (for partial differential equations) has grown up to deal with linear problems, and we now describe these.

23.2 Classical WKB theory

The equation[2]

$$\epsilon^2 \frac{d^2 y}{dx^2} + V(x)y = 0, \qquad 0 < \epsilon \ll 1, \qquad V(x) > 0,$$

occurs in many contexts, for example in quantum mechanics, where it is derived from Schrödinger's equation and ϵ is proportional to Planck's constant, or in investigations of the high-frequency eigenmodes of a vibrating system. The classical WKB approach to it is motivated by the idea that the solution should be a rapid oscillation, on a scale of $O(\epsilon)$, whose amplitude A and phase u both vary slowly, on a scale of $O(1)$. Thus, we pose an *ansatz* (an assumed form of solution)[3]

$$y(x; \epsilon) \sim A(x; \epsilon)e^{iu(x)/\epsilon},$$

where the amplitude $A(x; \epsilon)$ may be expanded as a regular series in ϵ, although one very rarely goes to the trouble of finding more than the first term. Differentiating and substituting into the differential equation, we have

$$\epsilon^2 \left(-\frac{(u')^2}{\epsilon^2} A + \frac{i}{\epsilon}(Au'' + 2A'u') + A'' \right) e^{iu/\epsilon} + V(x)e^{iu/\epsilon} = 0,$$

where a prime indicates d/dx. We can now expand in the form $A(x, \epsilon) \sim A_0(x) + \epsilon A_1(x) + \cdots$ and collect terms, to find at leading order that

$$\text{from } O(1), \qquad (u')^2 = V(x),$$
$$\text{from } O(\epsilon), \qquad A_0''u + 2A_0'u' = 0.$$

The first equation simply constitutes an integral to be evaluated, while the second can be integrated once to show that

$$A_0^2 u' = \text{constant}, \quad \text{so} \quad A_0 = \text{constant} \times V^{-1/4}.$$

---

[1] The letters stand for Wentzel, Kramers and Brillouin who were at least partially responsible for the theory, although other famous names such as Liouville and Green also contributed; see [27] Section 7.5. Both multiple scales and WKB are special cases of Kuzmak's method.

[2] The 'potential' V may also depend on ϵ but this is a mere complication as long as it has a regular expansion in ϵ.

[3] It is also possible to expand y as $\exp(iu_0/\epsilon + u_1 + \epsilon u_2 + \cdots)$, but this is more cumbersome.

Hence we have the two solutions in the approximate form

$$y(x;\epsilon) \sim (V(x))^{-1/4} \exp\left(\pm\frac{i}{\epsilon}\int^x (V(s))^{1/2}\,ds\right).$$

Note immediately that this expansion cannot be expected to be valid if $V(x)$ vanishes at any point in the interval of interest. Points of this kind are known as *turning points* and a separate boundary layer analysis, involving properties of Airy functions, is necessary near them. The details are beyond the scope of this book.

Why Airy? If, say, $V(0) = 0$ then in general $V(x) \sim ax + \cdots$ near $x = 0$ and the original equation is approximated by Airy's equation.

But, speaking of Airy's equation

$$\frac{d^2Y}{dX^2} - XY = 0,$$

let us at least find the behaviour of its solutions for large X. Write $X = x/\delta$, where δ is an artificial small parameter whose inverse measures the largeness of X, and then with $Y(X) = y(x)$ we get

$$\delta^3\frac{d^2y}{dx^2} - xy = 0.$$

Reading off from the general WKB formula above, the approximate solutions for $x < 0$ (for which $V(x) = -x > 0$) are proportional to

$$x^{-1/4}\exp\left(\pm\frac{2i}{3}\left(\frac{|x|}{\delta}\right)^{-3/2}\right),$$

or, in the original variables, two linearly independent solutions have asymptotic behaviour

$$Y(X) \sim X^{-1/4}\exp\left(\pm\frac{2i}{3}|X|^{-3/2}\right)$$

as $X \to -\infty$. The same line of argument can be used to show that the behaviour as $X \to \infty$ is

$$Y(X) \sim X^{-1/4}\exp\left(\pm\frac{2}{3}X^{-3/2}\right)$$

showing that one solution decays rapidly and the other grows. These are in fact constant multiples of Ai(X) and Bi(X) respectively (see Figure 4.6 on p. 60), and a more sophisticated analysis involving integral representations for Ai and Bi is needed to establish the so-called *connection formulae* that join up their behaviour for large positive X to the oscillatory behaviour for large negative X. These formulae are used to resolve the details of turning points, as mentioned above. They lead on to the fascinating topic of Stokes lines, which are intimately involved in the question of the optimal truncation of asymptotic expansions, as mentioned briefly in Chapter 12.

23.3 Geometric optics and ray theory: why do we say light travels in straight lines?

It is natural to generalise the WKB analysis to equations of the form $\epsilon^2 \nabla^2 \psi + V(\mathbf{x})\psi$ for a scalar function ψ of position $\mathbf{x}$. This equation arises immediately when we look for high-frequency time-periodic solutions of the wave equation

$$\frac{\partial^2 \phi}{\partial t^2} = c^2 \nabla^2 \phi.$$

Setting $\phi(\mathbf{x}, t) = e^{-i\omega t} \psi(\mathbf{x})$ gives the *Helmholtz equation*

$$\nabla^2 \psi + k^2 \psi = 0,$$

Note: Most treatments don't
bother with the scaling with L
and just use $1/k$ as the small
parameter.

where $k = \omega/c$ is the wavenumber, which we will take to be constant for now (that is, we take the medium to be homogeneous and isotropic). In many practical situations the wavelength of the field in question is much smaller than the length scale L of the solution domain, and nondimensionalisation with the latter scale leads to

$$\epsilon^2 \nabla^2 \psi + \psi = 0,$$

where $\epsilon = 1/Lk \ll 1$ is a measure of the ratio of the wavelength to the size of the domain. As an example, light waves have wavelengths between 400 nm and 700 nm so, for a domain of size 1 m, ϵ is smaller than 10^{-6}.

Let us work in two dimensions – the generalisation to three is conceptually easy but the working is more complicated. Motivated by the WKB analysis, we try an approximation

$$\psi(x, y) \sim A(x, y; \epsilon) e^{iu(x,y)/\epsilon},$$

which gives us

$$\epsilon^2 \left(-A \frac{|\nabla u|^2}{\epsilon^2} + \frac{i}{\epsilon} \left(2\nabla A \cdot \nabla u + A \nabla^2 u \right) + \nabla^2 A \right) + A = 0.$$

As before, we write $A(x, y; \epsilon) \sim A_0(x, y) + \epsilon A_1(x, y) + \cdots$, and at leading order we find

$$|\nabla u|^2 = 1, \tag{23.1}$$

$$2\nabla A_0 \cdot \nabla u + A_0 \nabla^2 u = 0. \tag{23.2}$$

Equations (23.1) and (23.2) are known as the *eikonal equation* and the *transport equation* respectively. The eikonal equation tells us how the phase $u(x, y)$ varies, and the transport equation determines the amplitude to leading order.

Let's have a quick preview of the general structure of the method of solution.

- First, we solve the eikonal equation by Charpit's method (see Chapter 7), with suitable boundary data on given boundary curves. The rays it gives are straight and are directed along ∇u, so that they are orthogonal to the wavefronts (curves of constant phase u, i.e. level curves of u). These are the light rays.
- Then we turn to the transport equation. This is a first-order linear partial differential equation and its characteristics are the rays (the $\nabla u \cdot \nabla A$ term tells us this). The term $\nabla^2 u$ in the transport equation can be calculated fairly easily, since u itself is determined using the same family of characteristics. The upshot is that we can find A on each ray in terms of the values of A and u (and the derivatives of u) on the boundary curve at the beginning of that ray.

These two points together explain why we say that light travels in straight lines: first the rays are straight and, second, the amplitude of the light on each ray is determined only by what happens where it originates and not by what happens on neighbouring rays.

The term $\nabla^2 u$ in the transport equation notwithstanding.

Now for some details. Let us suppose that we are given $u = u_0(s)$, $A_0 = a_0(s)$ on a curve $x = x_0(s)$, $y = y_0(s)$. Using t for the parameter along the rays, Charpit's equations for $|\nabla u|^2 = 1$, that is $p^2 + q^2 = 1$, are

Here, as before,

$$p = \frac{\partial u}{\partial x}, \qquad q = \frac{\partial u}{\partial y}.$$

$$\frac{dx}{dt} = 2p, \qquad \frac{dy}{dt} = 2q, \qquad \frac{dp}{dt} = \frac{dq}{dt} = 0, \qquad \frac{du}{dt} = 2.$$

The substitution $\tau = 2t$ removes all the 2's above, leaving

$$\frac{dx}{d\tau} = p, \qquad \frac{dy}{d\tau} = q, \qquad \frac{dp}{d\tau} = \frac{dq}{d\tau} = 0, \qquad \frac{du}{d\tau} = 1.$$

These equations have the solution

$$p(s, \tau) = p_0(s), \qquad q(s, \tau) = q_0(s), \qquad u(s, \tau) = u_0(s) + \tau,$$
$$x(s, \tau) = x_0(s) + \tau p_0(s), \qquad y(s, \tau) = y_0(s) + \tau q_0(s).$$

This shows immediately that the rays are straight (the second line above) and that their direction is

$$(p_0(s), q_0(s)) = (p(s, \tau), q(s, \tau)) = \nabla u.$$

(The question of finding $p_0(s)$ and $q_0(s)$ is more difficult, as we see below.)

Now we turn to the transport equation $2\nabla A_0 \cdot \nabla u + A_0 \nabla^2 u = 0$, which is

$$\frac{\partial u}{\partial x}\frac{\partial A_0}{\partial x} + \frac{\partial u}{\partial y}\frac{\partial A_0}{\partial y} = -\frac{A_0}{2}\nabla^2 u.$$

Written out like this, we see that the characteristics of the equation for A_0 are the same as the rays for u. It remains, however, to find $\nabla^2 u$ and solve this equation for A_0. The details of this calculation are given in Exercise 2 and we skip to the result,

$$A_0(s, \tau) = a_0(s)\sqrt{\frac{T(s)}{\tau + T(s)}},$$

where

$$T(s) = \frac{q_0(s)x_0'(s) - p_0(s)y_0'(s)}{q_0(s)p_0'(s) - p_0(s)q_0'(s)}.$$

This is very interesting formula, because the amplitude has the possibility of blowing up if $T(s)$ (whose meaning is not yet clear) is negative. It is really an energy conservation statement, relating A_0^2 to its initial value a_0^2 by a factor that tells us, among other things, how the spreading out or convergence of the rays dilutes or concentrates this energy. For the special case in which $u_0(s) = s$ and $a_0(s) = 1$, so that the whole initial curve oscillates with the same amplitude, it can be shown (see Exercise 2) that

$$A_0^2(s, \tau) = a_0^2(s)\frac{1}{\kappa_0(s)\tau + 1},$$

where $\kappa_0(s)$ is the curvature of the initial curve at the starting point of the ray labelled by s. The term $\kappa_0(s)\tau$ is exactly the rate at which neighbouring rays spread out or converge.

In interpreting this blow-up, we have to bear in mind that there is no requirement that $u(x, y)$ should be a single-valued function (unlike the case where the eikonal equation was used as a model for the height of a sandpile). It is perfectly possible to have several rays, from different families, through a given point, and the total wavefield is just the sum of the individual contributions. What causes blow-up in geometric optics is that the rays form an envelope, known as a *caustic*, a curve to which all the rays of a given family are tangent. It is this coincidence of rays that leads to the build-up in amplitude. Common examples of caustics are the bright lines you see at the bottom of a swimming pool in sunshine, and the nephroid curve seen when a cup of milky coffee is illuminated by a parallel beam (we look at this case below). Of course, the amplitude of the full Helmholtz problem does not blow up. The singularity in A_0 merely indicates that the approximation we have made is not valid near the caustic. The discrepancy can be resolved by an inner expansion on a smaller length scale, which also reveals that the field on the 'dark' side of the caustic is exponentially small (the two sides of the caustic are linked via an Airy function), but the details are beyond the scope of this book.

Let us now do two examples.

The parabolic reflector.   Suppose that a constant-amplitude parallel beam along the vector $(-1, 0)$ hits a perfectly conducting reflector in the shape of the parabola $y^2 = 4x$. We will demonstrate the well-known property that all the rays pass through the focus $(1, 0)$.

The incident wavefield is $\phi_i(x, y, t) = e^{-i\omega(t + x/c)}$, and we write the whole wavefield as $\phi_i + \phi_r$, the latter being the reflected field. Because the reflector is a perfect conductor, the wavefield vanishes on it. When we make the geometric optics approximation, this means that we need $e^{-ix/\epsilon} + a_0 e^{iu_0/\epsilon} = 0$ on the reflector, where the second term is the initial value of the reflected field. Thus, the initial values are

$$u_0(s) = -x_0(s), \qquad a_0(s) = -1, \qquad \text{on} \quad x = x_0(s), \quad y = y_0(s).$$

The parabola can be parametrised as $x_0(s) = s^2$, $y_0(s) = 2s$, and the next job is to find $p_0(s)$ and $q_0(s)$. Since $p_0^2 + q_0^2 = 1$, it is convenient to write $p_0(s) = \cos\theta(s)$, $q_0(s) = \sin\theta(s)$. If we differentiate the equation $u_0(s) = -x_0(s)$ with respect to s, we find

$$-x_0' = p_0 x_0' + q_0 y_0';$$

the prime denotes differentiation by s. Putting $x_0'(s) = 2s$, $y_0'(s) = 2$, we find the relation $s = -\tan\frac{1}{2}\theta$ between p_0, q_0 and s. The rays are then

$$x(s, \tau) = x_0(s) + \tau p_0(s), \qquad y(s, \tau) = y_0(s) + \tau q_0(s),$$

and they are in the direction $(p_0(s), q_0(s))$.

Before proceeding, note that the vector (x_0', y_0') is tangent to the reflector. Its dot product with the unit vector $(-1, 0)$ along the incident beam is $-x_0'$ while with the unit vector (p_0, q_0) along the ray it is $x_0' p_0 + y_0' q_0$. As these are equal, we have shown that *the angle of incidence equals the angle of reflection*, a result sometimes called the law of *specular reflection*; it holds for any reflector shape.

Let us return to the parabola. Eliminating τ between the ray equations, we find the family of curves

$$x\sin\theta(s) - y\cos\theta(s) = s^2 \sin\theta(s) - 2s\cos\theta(s) = \sin\theta(s).$$

Some trigonometric-identity bashing to be done to show this.

Thus, $y\cos\theta = (x - 1)\sin\theta$, and all the rays pass through the focus $(1, 0)$ as promised.

The nephroid.   Let us consider a semicircular reflector $x^2 + y^2 = 1$, $x < 0$, with a plane wave e^{-ikx} incident from $x = +\infty$. As above, we solve the ray equations with $u = -x$ on the circle (corresponding to a zero field on the reflector). Parametrising the circle with its arclength s, the ray passing through $(\cos s, \sin s)$, $\pi/2 < s < 3\pi/2$, has direction (p_0, q_0), where a short calculation shows that $p_0 = \cos 2s$, $q_0 = \sin 2s$,

Figure 23.1 Rays and the caustic inside a cup of coffee.

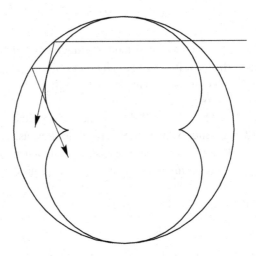

and this ray is therefore given by

$$x = \cos s + \tau \cos 2s, \qquad y = \sin s + \tau \sin 2s.$$

Eliminating τ, we have

$$x \sin 2s - y \cos 2s = \sin s,$$

and differentiating this equation with respect to s we have two parametric equations for the envelope of the rays (see Exercise 4). After tidying up, we find the envelope in the form

$$x = \sin s \sin 2s + \tfrac{1}{2} \cos s \cos 2s, \qquad y = -\sin s \cos 2s + \tfrac{1}{2} \sin 2s \cos s.$$

This curve is shown in Figure 23.1; when you put a mug of milky coffee in the sun, you see a bright caustic in this shape. It is called a *nephroid* from its resemblance to a kidney. There are four reflected rays through each point outside the nephroid, and fewer inside.

Before we move on, let us return to the question of caustics. We stated above that the apparent singularity can be resolved by an inner expansion near the caustic. There are many other situations where local expansions of this kind are necessary. One example is the shadow boundary between light and dark regions. Another occurs when an incident field hits a sharp corner or just grazes a body; from these regions spring not only shadow boundaries but also 'creeping rays', which travel round onto the shadowed side along the body (for concave bodies these rays give the whispering gallery effect), and 'evanescent rays', which leave the body, taking with them an exponentially small field (so that no shadow is truly dark). This is an active field of research (a typical application, regrettably military, is the calculation of possible interference between airborne radar and the receivers on the same aircraft), and we leave it with the thought that one can contemplate complex solutions of the eikonal equation $|\nabla u|^2 = 1$, giving growing or decaying exponential fields to complement the oscillatory ones associated with real solutions, and giving a degree of the unification that is hinted at by comparing the equations

$$\epsilon^2 \frac{d^2 y}{dx^2} + y = 0, \qquad \epsilon^2 \frac{d^2 z}{dx^2} - z = 0$$

mentioned at the beginning of the chapter. This theory of *complex rays* is in its infancy, but a good place to read about it is [8].

23.4 Kelvin's ship waves

For our final example, we turn to a famous example from hydrodynamics. If you feed the ducks, or if you watch a sailing boat from above, you will know that you see a prominent V-shaped wave pattern behind any body moving with constant velocity on the surface of the water. The pattern differs in detail from body to body, but the outline is always the same: the wake is contained in a V whose angle appears to be insensitive to the size of the body or the details of its shape, method of propulsion etc. Why? In this section we show the remarkable result, due to Lord Kelvin, that the far-field wave pattern is contained in a wedge of half-angle $\sin^{-1} \frac{1}{3} \approx 19°28'$.

Let us assume that the water is very deep and that we are far from the shore, so that the water can be taken to occupy the region $z < 0$, $-\infty < x, y < \infty$. Suppose that a ship is moving on the surface along the x-axis with speed V. We can do very well with the standard linear water-wave model (see subsection 13.5.2 with an obvious generalisation to three dimensions). The flow is inviscid and irrotational, so the velocity potential $\phi(x, y, z, t)$ satisfies

$$\frac{\partial^2 \phi}{\partial x^2} + \frac{\partial^2 \phi}{\partial y^2} + \frac{\partial^2 \phi}{\partial z^2} = 0 \qquad \text{in the water } z < 0,$$

with

$$|\nabla \phi| \to 0 \quad \text{as} \quad x^2 + y^2 + z^2 \to \infty$$

and the linearised kinematic and Bernoulli conditions

$$\frac{\partial \phi}{\partial z} = \frac{\partial h}{\partial t}, \qquad \frac{\partial \phi}{\partial t} + gh = 0 \qquad \text{on } z = 0 \text{ away from the ship,}$$

A small parameter measuring the size of the elevation has put in such a fleeting appearance that it was never written down ...

where $h(x, y, t)$ is the surface elevation. Eliminating h, these can be combined into

$$g \frac{\partial \phi}{\partial z} + \frac{\partial^2 \phi}{\partial t^2} = 0 \qquad \text{on } z = 0 \text{ away from the ship,}$$

Let us consider solutions that are steady in a coordinate system moving with the ship, so that the velocity potential can be written as $\phi(x - Vt, y, z)$. Let us also look for variations on a large horizontal scale and correspondingly large timescales, so that x, y and t are scaled by

$$x - Vt = \frac{X}{\epsilon}, \qquad y = \frac{Y}{\epsilon}, \qquad t = \frac{T}{\epsilon}$$

where $\epsilon \ll 1$. This ϵ is an artificial small parameter that measures the largeness of the distances involved in comparison with the size of the ship. In these coordinates, the problem for travelling-wave solutions of the form $\phi(x - Vt, y, z) = \Phi(X, Y, z)$ is

$$\epsilon^2 \left(\frac{\partial^2 \Phi}{\partial X^2} + \frac{\partial^2 \Phi}{\partial Y^2} \right) + \frac{\partial^2 \Phi}{\partial z^2} = 0, \qquad z < 0, \qquad (23.3)$$

Remember that $\partial/\partial T$ can be replaced by $-V\partial/\partial X$ for a travelling wave.

with

$$g \frac{\partial \Phi}{\partial z} + \epsilon^2 V^2 \frac{\partial^2 \Phi}{\partial X^2} = 0 \qquad (23.4)$$

on $z = 0$ away from the ship, which is at $X = 0$.

In the spirit of ray methods, we look for a solution in the form

$$\Phi(X, Y, z) \sim A(X, Y, z) e^{iu(X,Y,z)/\epsilon}$$

as $\epsilon \to 0$. Putting this into (23.3) gives

$$\epsilon^2 \left(-\frac{A}{\epsilon^2} \left(\frac{\partial u}{\partial X} \right)^2 + \cdots - \frac{A}{\epsilon^2} \left(\frac{\partial u}{\partial Y} \right)^2 + \cdots \right)$$

$$- \frac{A}{\epsilon^2} \left(\frac{\partial u}{\partial z} \right)^2 + \frac{i}{\epsilon} \left(2 \frac{\partial A}{\partial z} \frac{\partial u}{\partial z} + A \frac{\partial^2 u}{\partial z^2} \right) + \frac{\partial^2 A}{\partial z^2} = 0$$

(the dots indicate smaller terms that do not contribute at this order). Expanding to give $A \sim A_0 + O(\epsilon)$ as usual, we see from the $O(\epsilon^{-2})$ terms that

$$\frac{\partial u}{\partial z} = 0,$$

and then it follows at $O(1)$ that

$$\frac{\partial^2 A_0}{\partial z^2} - (p^2 + q^2)A = 0, \qquad \text{where} \qquad p = \frac{\partial u}{\partial X}, \quad q = \frac{\partial u}{\partial Y}.$$

This is an ordinary differential equation (the dependence on X and Y is parametric) whose solution vanishing at $z = -\infty$ is

$$A_0(X, Y, z) = a_0(x, y) \exp \left(z(p^2 + q^2) \right)^{1/2}.$$

Putting this into (23.4), we find that

$$V^4 p^4 = g^2(p^2 + q^2).$$

We have arrived at a nonlinear equation for $u(X, Y)$. The ray equations (Charpit's equations) for it are

$$\frac{dx}{dt} = 4V^4p^3 - 2gp = g\frac{2p^2 + q^2}{p}, \qquad \frac{dy}{dt} = -2gq,$$

$$\frac{dp}{dt} = \frac{dq}{dt} = 0, \qquad \frac{du}{dt} = 4V^4p^4 - 2gp^2 - 2gq^2 = 2V^4p^4.$$

As p and q are constant on the rays, the rays are straight and their slope is

$$\frac{dy}{dx} = -\frac{pq}{p^2 + 2q^2} = -\frac{\lambda}{\lambda^2 + 2},$$

where $\lambda = p/q$. Now, whatever values λ takes, elementary calculus shows that the slope of the rays can never leave the interval $(-1/(2\sqrt{2}), 1/(2\sqrt{2}))$. That is, they are confined to a wedge whose half-angle is $\sin^{-1}(1/3)$; remarkably, this result is independent of V (and g). More work is needed in order to solve Charpit's equations and establish the full ray pattern (the rays are curves $u = $ constant), but roughly speaking the lines $Y = \pm X/(2\sqrt{2})$ are lines of singularities at which the amplitude of the wake is large, and in addition to rays emanating from the region near the ship (which carry information about the details of its hull shape and other features), there are rays that form arcs, bowed towards the ship, across its line of travel, and these too can easily be seen. It is a major aim of designers of high-speed sea transport to minimise energy loss to waves in the ship's wake, and the calculation we have just given is a part of that endeavour. Wave drag is probably less of a problem for ducks, such as the two shown in Figure 23.2, with which we end this chapter and indeed the main text of the book.

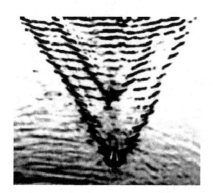

Figure 23.2 Duck waves. For an analysis of beetle waves, in which surface tension is important, see [56]. (Photographs by Tom R. Laman.)

23.5 Exercises

1 An eigenvalue problem.   Suppose that λ is an eigenvalue for the problem

By analogy with linear algebra, we say λ is an eigenvalue because there is always the trivial solution $y = 0$, and a non-trivial solution exists only for certain values of λ.

$$-(1 + x^2)\frac{d^2 y}{dx^2} = \lambda^2 y, \qquad -\infty < x < \infty, \qquad y(\pm\infty) = 0.$$

Show that the large eigenvalues are of the form $2k$, with odd associated eigenfunction, and $2k + 1$, with even associated eigenfunction. Where k is a large positive or negative integer, as follows.

Write $\lambda = 1/\epsilon \ll 1$, and show that the WKB solutions are proportional to

$$(1 + x^2)^{1/4} \exp\left(\pm\frac{i}{\epsilon} \tan^{-1} x\right),$$

which is equivalent to

$$(1 + x^2)^{1/4} \cos\left(\frac{\tan^{-1} x}{\epsilon}\right), \qquad (1 + x^2)^{1/4} \sin\left(\frac{\tan^{-1} x}{\epsilon}\right).$$

Deduce that y can vanish at $\pm\infty$ only if $\pi/(2\epsilon) = (2k + 1)\pi/2$ for the cosine solutions and $\pi/(2\epsilon) = k\pi$ for the sine solutions. Verify that the prefactor $(1 + x^2)^{1/4}$ does not prevent y from tending to zero at $\pm\infty$.

2 Solution of the transport equation.   The rays for the eikonal equation are

$$x(s, \tau) = x_0(s) + \tau p_0(s), \qquad y(s, \tau) = y_0(s) + \tau q_0(s).$$

Define

$$J = \frac{\partial(x, y)}{\partial(s, \tau)} = \begin{pmatrix} x_0' + \tau p_0' & y_0' + \tau q_0' \\ p_0 & q_0 \end{pmatrix},$$

where the prime again denotes differentiation with respect to s. Show that

$$\det J = (q_0 p_0' - p_0 q_0')(\tau + T(s)),$$

where

$$T(s) = \frac{q_0(s)x_0'(s) - p_0(s)y_0'(s)}{q_0(s)p_0'(s) - p_0(s)q_0'(s)}.$$

Show from the transport equation that

$$\frac{2}{A_0}\frac{\partial A_0}{\partial \tau} = -\nabla^2 u.$$

Invert J and use the relation

$$\nabla^2 u = \frac{\partial p}{\partial x} + \frac{\partial q}{\partial y}$$

$$= p_0'(s)\frac{\partial s}{\partial x} + q_0'(s)\frac{\partial s}{\partial y}$$

to show that $\nabla^2 u = J^{-1}\partial J/\partial \tau$. Deduce that

$$\frac{\partial}{\partial \tau}\left(A_0^2 J\right) = 0, \qquad \nabla^2 u = \frac{1}{\tau + T(s)}.$$

Finally, show that

$$A_0(s, \tau) = a_0(s)\sqrt{\frac{T(s)}{\tau + T(s)}}.$$

Suppose that $u_0(s) = s$, $a_0(s) = 1$ on the initial curve. Show that on each ray

$$A_0(s, \tau) = \sqrt{\frac{1}{\kappa_0(s)\tau + 1}},$$

where $\kappa_0(s)$ is the curvature of the initial curve at the point at which the ray leaves it. (It will help to take s to be the arclength along Γ, so that $\kappa_0(s) = y_0'' x_0' - x_0'' y_0'$.)

This last result shows explicitly how the spreading out or contracting of the rays is determined by the curvature of the initial curve.

3 Snell's law.   The *refractive index* of a medium is defined as the ratio of the speed of light in a vacuum to the speed of light in the medium. The refractive index of a vacuum is thus 1. Show that the eikonal equation for waves of frequency ω in a medium of refractive index n is $|\nabla u|^2 = n^2$.

The half-plane $x < 0$ is made of material whose refractive index is n_1, and the material of $x > 0$ has refractive index n_2. Light rays in $x < 0$ are along the vector $(\cos \theta_1, \sin \theta_1)$. Show from the eikonal equation, assuming that the phase is continuous across $x = 0$, that in $x > 0$ they are along $(\cos \theta_2, \sin \theta_2)$, where (Snell's law)

$$\frac{\sin \theta_1}{\sin \theta_2} = \frac{n_2}{n_1}.$$

If $n_2 < n_1$, deduce that rays can only propagate into $x > 0$ if θ_1 is such that $\sin \theta_1 < n_2/n_1$ (rays at larger angles are totally internally reflected).

4 Envelopes and string art.   Suppose that the curves in a family $f(x, y; s) = 0$, parametrised by s, all touch a curve $F(x, y) = 0$ called the *envelope* of the family $f(x, y; s) = 0$. (For example, the curves

$y = \left(\frac{1}{3}(x - s)\right)^3$ of Exercise 6 at the end of Chapter 3 are all tangent to the x-axis, which is the singular solution of $dV/dx = -V^{2/3}$ discussed there.)

Draw a picture and explain why the tangency condition means that the intersection of a given member of the family $f(x, y; s) = 0$ with $F(x, y) = 0$ is a double root. Hence show that $F(x, y)$ is found by eliminating s between the two expressions

$$f(x, y; s) = 0, \qquad \frac{\partial f}{\partial s}(x, y; s) = 0.$$

(Alternatively, these equations give a parametric representation of the solution with s as the parameter.)

Consider the family of lines generated by joining the points $(s, 0)$ and $(0, 1 - s)$. Find its envelope. (This is $(x - y)^2 + 1 = 2(x + y)$; why is it obviously a parabola?).

A rod of length 1 slides with one end on the y-axis and the other on the x-axis. Use the angle that it makes with one of the axes as a parameter to show that the envelope of its positions is the astroid $x^{2/3} + y^{2/3} = 1$.

(There was a short-lived phase of 'string art' made by hammering lines of nails into a board and stretching highly-coloured shiny string between them, the endpoints of successive lengths of string being related as in the examples above. The result was a complicated web with (a discrete approximation to) an envelope, often in the shape of a sailing boat or similar object. They can occasionally be seen in charity shops and holiday cottages even now and will doubtless at some point become highly collectable.)

5 Rays in an ellipse. Consider the ray equations for $u_x^2 + u_y^2 = 1$ inside the ellipse $x^2/a^2 + y^2/b^2 = 1$, with $u = 0$ on its boundary $(x_0(s), y_0(s))$. Show that the direction of a ray is $(p_0(s), q_0(s))$ and, by differentiating the condition $u = 0$ on the ellipse with respect to s, show that the rays are normal to the boundary. Use the parametric form $(a \cos s, b \sin s)$ for the boundary to show that the normals are

$$ax \sin s - by \cos s = (a^2 - b^2) \sin s \cos s$$

and deduce that there is a caustic on the envelope of these curves, namely

$$x = \frac{a^2 - b^2}{a} \cos^3 s, \qquad y = \frac{a^2 - b^2}{b} \sin^3 s.$$

Sketch this curve (it is a scaled astroid). Note that the caustic itself can have cusp singularities; they are associated with maximum and minimum values of the curvature of the boundary. Note also the difference between the geometric optics interpretation of the solution and the sand-pile interpretation, in which there is a ridge line joining the foci of the ellipse. Show that the rays intersect the ridge line before they meet the caustic.

6 A string art nephroid. Show that the rays in the nephroid caustic above have the equations $x \sin 2s - y \cos 2s = \sin s$, and hence that they meet the circle again at the point $(\cos \alpha, \sin \alpha)$ where $\alpha = 3s - \pi$. Hence explain how to make a string art nephroid.

Find the envelope obtained by replacing 3 by 2 in the definition of α (i.e. the envelope of the lines from $(\cos s, \sin s)$ to $(\cos(2s - \pi), \sin(2s - \pi))$); it is a *cardioid*, with one cusp, resembling a heart). What happens if you replace 3 by 1?

'It's because the minuses are upside down.'

References

[1] Acheson, D. J., *From Calculus to Chaos*, Oxford University Press (1997).

[2] Acheson, D. J., *Elementary Fluid Dynamics*, Oxford University Press (1990).

[3] Addison, J., The electrostatic deposition of powder coatings, D. Phil. thesis, Oxford University (1997).

[4] Barenblatt, G. I., *Scaling, Self-similarity and Intermediate Asymptotics*, Cambridge University Press (1996).

[5] Bender, C. M. & Orszag, S. A., *Advanced Mathematical Methods for Scientists and Engineers*, McGraw–hill (1978).

[6] Carrier, G. F., Krook, M. & Pearson, C. E., *Functions of a Complex Variable*, Hod Books (1983).

[7] Cartwright, D. E., *Tides: A Scientific History*, Cambridge University Press (1999).

[8] Chapman, S. J., Lawry, J. M. H., Ockendon, J. R. & Tew, R. M., On the theory of complex rays, *SIAM Review* **41**, 417–509 (1999).

[9] Cumberbatch, E. & Fitt, A. D. (eds.), *Mathematical Modeling: Case Studies from Industry*, Cambridge University Press (2001).

[10] Deeming, D. C., Principles of artificial incubation for game birds (a practical guide), in *Proc. Ratite Conference* (2000).

[11] Dewynne, J. N., Howison, S. D. & Ockendon, J. R., The numerical solution of a continuous casting problem, in an *Internat. Series Num. Math.* vol. 95, pp. 36–45, Birkhauser (1990).

[12] Dewynne, J. N., Ockendon, J. R. & Wilmott, P., On a mathematical model for fiber tapering, *SIAM J. Appl. Math.* **49**, 983–990 (1989).

[13] Driscoll, T. A. & Trefethen, L. N., *Schwarz–Christoffel Mapping*, Cambridge University Press (2002).

[14] Edwards, C. M., Ovendon, N. & Rottschäfer, V., Some cracking ideas on egg incubation, ESGI report (2003). Available from www.math-in-industry.org.

[15] Engl, H. *Mathematics and Its Applications: Regularization of Inverse Problems*, Kluwer (2003).

[16] Fauvel, J., Flood, R. & Wilson, R. (eds.), *Music and Mathematics*, Oxford University Press (2003).

[17] Fletcher, N. H. & Rossing, T. D., *The Physics of Musical Instruments*, Springer–Verlag, New York (1998).

[18] Fowkes, N. D. & Mahoney, J. J., *An Introduction to Mathematical Modelling*, Wiley UK (1994).

[19] Fowler, A. C., *Mathematical Models in the Applied Sciences*, Cambridge University Press (1997).

[20] Fowler, A. C., Frigaard, I. & Howison, S. D., Temperature surges in current-limiting circuit devices, *SIAM. J. Appl. Math.* **52**, 998–1011 (1992).

[21] Frigaard, I. & Scherzer, O., Spraying the perfect billet, *SIAM J. Appl. Math.* **57**, 649–682 (1007).

[22] Fulford, G. R. & Broadbridge, P., *Industrial Mathematics*, Cambridge University Press (2002).

[23] Gershenfeld, N., *The Nature of Mathematical Modeling*, Cambridge University Press (1999).

[24] Goldstein, A. A., Optimal temperament, in [38], pp. 242–251.

[25] Helbing, D., Traffic and related self-driven many-particle systems, *Rev. Mod. Phys.* **73**, 1067–1141 (2001).

[26] Hildebrand, F. B., *Methods of Applied Mathematics*, Dover (1992).

[27] Hinch, E. J., *Perturbation Methods*, Cambridge University Press (1991).

[28] Howell, P. D., Models for thin viscous sheets, *Europ. J. Appl. Math.* **7**, 321–343 (1996).

[29] Howison, S. D., 'If I remember rightly, $\cos \pi/2 = 1$', *Bull. Austr. Math. Soc.* **19**(5), 119–122 (1992).

[30] Isacoff, S., *Temperament*, Faber & Faber, London (2002).

[31] Jackson, J. D., *Classical Electrodynamics*, third edition, Wiley (1998).

[32] Jordan, D. & Smith, P., *Nonlinear Ordinary Differential Equations: An Introduction to Dynamical Systems*, Oxford University Press (1999).

[33] Keener, J. P., *Principles of Applied Mathematics*, second edition, Addison–Wesley (1999).

[34] Keller, J. B., Diffusion at finite speed and random walks, *Proc. Nat. Acad. Sci.* **101**, 1120–1122 (2004).

[35] Kevorkian, J., *Partial Differential Equations*, Wadsworth & Brooks/Cole (1990).

[36] Kevorkian, J. & Cole, J. D., *Perturbation Methods in Applied Mathematics*, Springer (1981).

[37] Kingman, J. F. C., *Poisson Processes*, Oxford University Press (1993).

[38] Klamkin, M. (ed.), *Mathematical Modelling: Classroom Notes in Applied Mathematics*, SIAM, Philadelphia (1987).

[39] Lighthill, M. J., *Introduction to Fourier Analysis and Generalised Functions*, Cambridge University Press (1958).

[40] Lin, C. C. & Segel, L. A., *Mathematics Applied to Deterministic Problems in the Natural Sciences*, SIAM (1988).

[41] McMahon, T. A., Rowing: a similarity analysis, *Science* **173**, 349–351 (1971).

[42] McMahon, T. A., *Muscles, Reflexes and Locomotion*, Princeton University Press (1984).

[43] Murray, J. D., *Mathematical Biology*, second edition, Springer, New York (1997).

[44] Ockendon, J. R., Howison, S. D., Lacey, A. A. & Movchan, A. B., *Applied Partial Differential Equations*, Oxford University Press (revised edition 2003).

[45] Ockendon, H. & Ockendon, J. R., *Viscous Flow*, Cambridge University Press (1995).

[46] Ockendon, H. & Ockendon, J. R., *Waves and Compressible Flow*, Springer, New York (2004).

[47] Ockendon, J. R. & Tayler, A. B., The dynamics of a current collection system for an electric locomotive, *Proc. Roy. Soc.* **A322**, 447–468 (1971).

[48] Olver, F. W. J., *Introduction to Asymptotics and Special Functions*, Academic Press (1974).

[49] O'Malley, R. E., *Thinking about Ordinary Differential Equations*, Cambridge University Press (1997).

[50] Peletier, L. A., *Spatial Patterns: Higher Order Models in Physics and Mechanics*, Birkhauser (2001).

[51] Rice, J. A., *Mathematical Statistics and Data Analysis*, Wadsworth (1988).

[52] Richards, J. I. & Youn, H. K., *Theory of Distributions*, Cambridge University Press (1990).

[53] Robinson, F. N. H., *Electromagnetism*, Oxford University Press (1973).

[54] Rodeman, R., Longcope, D. B. & Shampine, L. F., Response of a string to an accelerating mass, *J. Appl. Mech.* **98**, 675–680 (1976).

[55] Schwarz, L., *Théorie des Distributions*, vols. I & II, Hermann et Cie, Paris (1951, 1952).

[56] Segel, L. A. & Handelsman, G. H., *Mathematics Applied to Continuum Mechanics*, Dover (1987).

[57] Stakgold, I., *Green's Functions and Boundary Value Problems*, second edition, Wiley (1998).

[58] Tayler, A. B., *Mathematical Models in Applied Mechanics*, Oxford University Press (1986, reissued with corrections 2003).

[59] Taylor, G. I., The formation of a blast wave by a very intense explosion. I, Theoretical discussion; II, The atomic explosion of 1945, *Proc. Roy. Soc.* **A201**, 159–174 and 175–186 (1950).

Index

Printed in the United States
By Bookmasters